GSM System Engineering

The Artech House Mobile Communications Series

John Walker, Series Editor

For a complete listing of *The Artech House Telecommunications Library,* turn to the back of this book.

GSM System Engineering

Asha Mehrotra

Artech House, Inc.
Boston • London

Library of Congress Cataloging-in-Publication Data
Mehrotra, Asha.
  GSM System Engineering / Asha Mehrotra.
    p.   cm.
  Includes bibliographical references and index.
  ISBN 0-89006-860-7 (alk. paper)
  1.Global system for mobile communications.   I. Title.
TK5103.483.M45 1996
621.3845'6—dc21   97-4029
                                                   CIP

British Library Cataloguing in Publication Data
Mehrotra, Asha
  GSM System Engineering
  1. Cellular radio   2. Mobile communication systems.
  I. Title
  621.3'8456

  ISBN   0-89006-860-7

Cover design by Jennifer Makower.

© 1997 ARTECH HOUSE, INC.
685 Canton Street
Norwood, MA 02062

International Standard Book Number: 0-89006-860-7
Library of Congress Catalog Card Number: 97-4029

10 9 8 7 6 5 4 3 2

To the memory of my father-in-law, Bhagwati Pd. Dhawan,
and my elder sister, Madhuri Mehrotra

▼▼▼

CONTENTS

▼▼▼

PREFACE

Cellular communications is one of the fastest growing and most challenging telecommunication applications ever. Today, it represents a large and continuously increasing percentage of all new telephone subscribers around the world. In the long term, cellular digital technology may become the universal way of communication.

The mobile communications market has experienced rapid growth in European Post Offices and Telecommunication (CEPT) Europe. This has been driven by market forces, technological development, and new forms of cooperation in the areas of standardization and implementation of new systems. A major product of this standards work within CEPT Europe has been the GSM standard. The Global System for Mobile Communication was developed as the next generation digital cellular mobile communication system for CEPT Europe.

The European cellular market was, up to the 1980s, characterized by a large number of incompatible analog systems (TACS, NMT, etc.). This created a situation where service generally was limited to national territories and where the economy of scale was largely lost. This situation has made it clear that for the future a common system is required for the widespread use of mobile telephones all over Europe. GSM is the Pan-European digital mobile telephony standard specified by European Telecommunication Standards Institute (ETSI) and provides a common standard. This means that cellular subscribers can use their mobile telephones all over Europe.

The essential elements of GSM system consist of the following: Mobile users, Base Station System, Mobile Switching Center, and the Public Voice and Data networks. Some other important elements of the system include the Operations and Management Center, the Billing Center, and the various networks (SS7 and X.25) interconnecting these subsystems. Therefore, we arrange the discussion based on different aspects of the subsystem as follows.

Chapter 2 deals with the architecture of the GSM system. We discuss the functional requirements of Mobile Station; Base Station System, which includes Base Transceiver and Base Station Controller; Mobile Switching Center and its variation known as Gateway Switching Center; Operations and Network Management Centers; and Billing Center. Chapter 3 provides a detailed discussion of time and frequency axis representation of the system. We discuss the functions and characteristics of physical and logical channels and provide the reasons for the flexibility of GSM system. Chapter 4 is fully devoted to the mobility management issues and provides the timing diagrams for different types of calls, including incoming call to mobile, mobile origination of a call, and mobile-to-mobile call. Chapter 5 brings out various security aspects. In Chapter 6 we discuss technical characteristics of the system which include all the important aspects of speech coding, modulation, and error encoding applied to the GSM system. In Chapter 7 we discuss subscription management, billing and accounting, and some important aspects of network maintenance. Chapter 8 deals with GSM protocols based on the ISO and SS7 standards. Chapter 9 elaborates on Chapters 2 and 3 and discusses subsystem configuration and architecture of MSC, BSS, and OCC, and MS. Chapter 10 provides existing systems as predecessors to PCS systems. Chapter 11 points out the potential shortfalls of existing PCS systems and provides some solutions in this direction.

I believe the best way to prove that one has understood the contents of the chapters is to answer pertinent questions. With this in mind, review questions are arranged at the end of each chapter.

This book is mainly intended for engineers working in the area of cellular communication, and in particular, with GSM, or closely related TDMA-based systems. The book can also be used as a special course for TDMA systems with examples of GSM, as this is probably the most intricate mobile communication systems of the world today. For practicing engineers, a BS degree in electrical engineering and some working knowledge of mobile communications is assumed. For students taking the specialized TDMA course, it is assumed that they have taken senior-level courses in communication engineering.

I would like to thank my students at George Washington University during the summer of 1995, and my graduate students during the summer of 1996 at Virginia Tech at Washington Graduate Center, who have contributed by asking the right questions and, in general, by improving the script. I also must thank the reviewers of this book for their excellent comments, which have tremendously improved the book.

Finally, a project of this type can never be completed without the continuous support of one's family. In this regard, thanks are due to my wife, Nisha, my daughters, Anuja, Sonia, Vinita, and my son Neil. Special thanks to my daughters Anuja, who was mainly responsible for arranging the script before she got married, and Sonia, who carried the responsibility until the book was completed.

Asha Mehrotra
March 1997

CHAPTER 1

▼▼▼

INTRODUCTION TO GLOBAL SYSTEM FOR MOBILE COMMUNICATIONS

1.1 INTRODUCTION [1–5]

Cellular telecommunications is one of the fastest growing and most challenging telecommunication applications ever. Today, it represents a large and continuously increasing percentage of all new telephone subscribers around the world. In the long term, cellular digital technology may become the universal way of communication.

The mobile communications market has experienced rapid growth in *European Post Offices and Telecommunication* (CEPT). This has been driven by market forces, technological development, and new forms of cooperation in the areas of new systems standardization and implementation. A major product of this standards work within CEPT Europe has been the *Global System for Mobile Communication* (GSM) standard. The GSM was developed as the next-generation digital cellular mobile communication system for CEPT Europe. The standardization work for the first implementation in 1991 was completed in early 1990. Network operators in 17 CEPT countries signed the *memorandum of understanding* (MOU) and are committed to introduce GSM systems by 1991.

In 1987 a Group Special Mobile Conference under the auspices of CEPT took place. The goal of this conference was to define a Pan-European standard for digital cellular communications that would be implemented beginning in 1991. Thirteen countries were involved in the development of the initial recommendations. Since that time, GSM development has been characterized by remarkable progress and cooperation. Eighteen European nations decided to adopt the standard initially. Currently GSM recommendations provide country "Color Codes" for 26 European nations. Hong Kong and Australia have also adapted the GSM systems. In Europe there are several large cellular systems in operation, such as *Nordic Mobile Telephone* (NMT) in the Nordic countries and *Total Access Communication System* (TACS) in the United Kingdom. Other countries in Western Europe also offer mobile services as shown in Table 1.1. Quality, capacity, and area of coverage vary widely, but

Table 1.1
Major Analog Systems in Europe

Country	Systems	Freq. Band	Date of Launch	Operator	Subs × 1000 (Yr. 1991)
United Kingdom	ETACS	900	1985	Cellnet	1521.4
			1987	Vodafone	1521.4
Sweden	NMT	450	1981	Comvik (Millicom)	11.5
	NMT	900	1986	Telia Mobitel	694.58
Norway	NMT	450	1981	Tele-Mobil	180.5
	MT	900	1986	Teli-Mobile	279.6
Finland	NMT	450	1981	Telecom Finland	186.8
	NMT	900	1986	Telecom Finland	352.2
Denmark	NMT	450	1981	Tele Denmark Mobil A/S	38.02
		900	1986	Tele Denmark Mobil A/S	234.8
France	Radicom 2000	450,900	1985	France Telecom	291
	NMT	450	1989	Ligne SFR (SBC, Bell South)	143.9
	NMT	900	1990	France Telecom	15.0
Italy	RTMS	450	1985	SIP	17.5
	ETACS	900	1990	SIP	2191.9
Germany	CNETZ		1986	DETeMobil	733.891
	NMT	450	1992	DeTeMobile	Not Available
Switzerland	NMT	900	1987	PTT	292.3
The Netherlands	NMT	450	1985	Royal PTT	22.7
	NMT	900	1989	Royal PTT	231.3
Austria	CNETZ	450	1984	PTV	47.6
	TACS	900	1990	PTV	219.07
Spain	NMT	450	1982	Telefonica	36.5
	TACS	900	1990	Telefonica	377.45

demand in most cases has exceeded estimates almost in every country. Most systems however, are national, which makes it impossible to use the mobile telephones abroad. This situation has made it clear that for the future a common system is required for the widespread use of mobile telephones all over Europe. GSM is the Pan-European digital mobile telephony standard specified by the *European Telecommunication Standards Institute* (ETSI) and provides a common standard; thus, cellular subscribers can use their mobile telephones all over Europe. GSM growth from 1991 to 1994 is shown in Figure 1.1.

Before the 1980s, the European cellular market was characterized by a large number of incompatible analog standards (such as TACS and NMT), leading to a situation where service generally was limited to national territories and where the economy of scale was largely lost. At the same time, mobile communications expanded very rapidly and the development showed clear signs of accelerated future growth. These factors in combination were of course very unfortunate, and it was realized, in view of the increasing mobility of the subscribers all over Europe and their expectations of service being offered even in foreign countries, that unless a remedy could be found, the result would be a very difficult situation in the 1990s.

In order to solve the problem, the European telecom authorities made the following three decisions.

In 1982, two frequency bands, 890 MHz to 915 MHz and 935 MHz to 960 MHz, were reserved primarily for use by cellular systems; and a next-generation European cellular system for a newly allocated band of 2 MHz × 25 MHz was formed.

In 1985, the decision was made to implement a digital system. The next step was to choose between narrowband and wideband solutions.

In 1987, GSM concluded that digital technology working in the *Time Division Multiple Access* (TDMA) mode would provide the optimum solution for the future system. The narrowband TDMA solution was chosen (less than 10 channels per carrier frequency is generally regarded as a narrowband TDMA system), in consider-

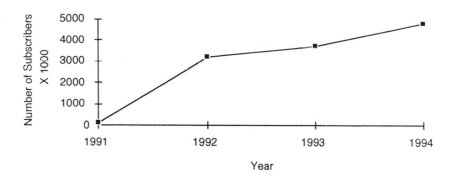

Figure 1.1 GSM growth in Europe.

ation of its several advantages. Specifically, a TDMA system has the following advantages.

- Offers a possibility of channel splitting and advanced speech coding in the future, resulting in improved spectrum efficiency;
- Offers much greater variety of service than the analog;
- Has ISDN capability;
- Is strongly favored by modern component development, which leads to lower system cost;
- Allows considerable improvements to be made with regards to the protection of information in the system.

Developing such a common system would allow a subscriber to use his own set all over Europe. From a user's point of view, the Pan-European system would appear as one system, although, in fact, it would consist of many systems run by independent operators. This standardization would be applied to certain key interfaces within the fixed parts of the system to avoid development of a large number of proprietary interfaces, resulting in loss of economy of scale. Hence, the system would be defined in terms of functional building blocks and their interfaces. This initiative was taken at the right time, with regard to the technology, the demand, and the need for standardization. There was and still is a strong interaction between the developments in the fields of telecommunications and politics, in Europe as elsewhere, that has lead to a strong drive toward standardization in many fields. Standardization was felt to be the necessity for the development of a single European market.

CCIR Study Group 8 realized in the middle of the last decade that different *Future Public Land Mobile Telecommunication Systems* (FPLMTS) were presently under study in different countries. It was obvious that CCIR had the task of coordinating the different developments and forming a list of recommendations. CCIR Study Group 8 decided to establish a special team, that is, an *Interim Working Party* (IWP), to study this important subject and determine the overall objectives of the FPLMTS systems, which include the suitable choice of frequency band or bands and defining the essential characteristics of the systems. The system should provide a wide range of services, such as voice, data, and others including compatibility for national and international roaming.

1.1.1 Objectives of FPLMTS

Some primary objectives for FPLMTS are as follows.

1. To provide a framework for continuing the extension of mobile network services and access to services and facilities of the fixed network (*public-*

switched telephone networks / integrated service digital networks (PSTN/ISDN)) subject to the constraints of radio transmission, spectrum usage, and system economics;

2. To allow mobile and fixed network users to use the services regardless of location (that is, national and international roaming);
3. To provide an open architecture that will permit the easy introduction of new technology advancements as well as different applications;
4. To allow the coexistence and interconnection with mobile systems that use direct satellite links;
5. To provide for unique user identification and PSTN/ISDN numbers in accordance with appropriate CCITT Recommendations;
6. To offer the services available in the PSTN/ISDN and other public networks, as far as possible, bearing in mind the differences in the characteristics of the fixed network and mobile radio environments;
7. To provide frequency commonality, which should allow for the desired level of operational compatibility on the systems. In principle, a complete commonality of one frequency band on a worldwide basis would be desired, but reasonable commonality could also be obtained through a common signaling band and sufficient overlap of the traffic bands to ensure compatibility.

With these reasons for developing the GSM system, we will highlight the GSM background, operational requirements, and technical requirements. Section 1.5 will describe the different services offered by GSM, followed by the contents of the book and conclusions.

1.2 GSM BACKGROUND

The Joint West-European venture of specifying the GSM system started within the CEPT Organization, where the cooperative work in the different areas of telecommunications was formally adopted in the form of recommendations. However, it was a national decision whether or not each recommendation, or set of recommendations, for a given service should be implemented by the administration of each country.

There was a need, however, for a more demanding obligation as far as the GSM was concerned. The system needed to include the concept of international roaming, and success in one country would therefore depend upon the system roll-out in other countries. The development and production of system equipment would require large industrial investments that could hardly be justified unless the different national markets evolved somewhat simultaneously. A memorandum of understanding was thus prepared during mid-1987 and signed by the first 12 countries in September 1987.

The GSM system specification work was later transferred to ETSI, an organization controlling standardization and the MOU, which served as an adequate forum for discussions on pure operational matters. The main purpose of the GSM-MOU was to provide a framework for all the necessary measures to be taken by the signatories together to ensure the opening of a commercial service in their respective countries by 1991. Also, the network operators would plan the progressive implementation of the networks in each country so that transport routes between the countries of the signatories could be brought early into the coverage of the respective systems. Priority for coverage would be given to all capital cities including the principal airport by 1991, introduction of special services by 1993, and the full European roaming market by 1995, as shown in Table 1.2.

1.2.1 Important Dates

Important dates in the development cycle of the GSM project are shown in Table 1.3. The project started in 1982 with the creation of the Group Special Mobile within CEPT. The pivotal year was 1987, when digital system was adopted and field trials were completed. The system was finally put into operation in 1991.

1.2.2 MOU and Different Working Groups

The GSM-MOU established several subgroups within GSM. The task of the individual groups was to organize the work in the following areas of concern, as shown in Table 1.4.

1.3 GSM OPERATIONAL REQUIREMENTS [6–10]

A list of operational requirements was developed that consisted of the following.

- High audio quality and link integrity;
- High spectral efficiency;
- Identical system in all countries (European harmonization and standardization);

Table 1.2
Milestones for a GSM System

Year	Occurrence
Mid-1991	Capitals;
	Voice, emergency calls, call forwarding, and barring
Mid-1993	Intercapitol roaming;
	Short message services (SMS), FAX, and call holding
Mid-1995	European roaming;
	Real-time information, call waiting, and conference calls

Table 1.3
Important Dates in the GSM Project

Year	Occurrence
1982	Group Special Mobile created within CEPT
1985	France/German agreement at Nice to support GSM, thus triggering the process to resolve European differences of opinion in favor of a unified Pan-European digital cellular service.
1986	GSM establishes a permanent nucleus in Paris.
1986	The Heads of State meeting in London in December requested agreement on standards and commitment of operators. The European Council of Ministers issued a draft directive on radio spectrum use.
1987	Field trials were completed in Paris in February following a precedent-setting decision to conduct a single set of field trials of spectrum efficiency, voice quality, and the radio interface of all the proposed systems. All systems were tested under the same conditions.
	At the Madrid GSM conference in February, it was agreed that the system would be digital, narrowband TDMA access using voice coders.
	The Ministerial meeting was held in Bonn on May 3. The United Kingdom, Germany, France, and Italy agreed at the Bonn conference on May 19 to standards, 1991 roll-out of the system, and competition concentration of industry, and asked operators for a Memorandum of Understanding.
	At Copenhagen on September 7, operators signed Memorandum of Understanding, agreeing on procedures and schedules to procure, build, and test systems.
1988	Tenders were issued by GSM member countries in March.
1989	Prototype (validation) systems were on the air. GSM became an ETSI Technical Committee.
1990	Pre-operational systems came on the air. DCS 1800 adaptation started.
1991	The United Kingdom, France, Germany, and Italy introduced digital cellular service.
1992	Motorola started the first commercial GSM system.
1993	Conference was called in Finland to explore the GSM migration toward UMTS/FPLMT. Phase 2 GSM specifications were frozen. Contracts were awarded in Asia, the Middle East, and Europe for GSM systems. The new name "Global System for Mobile Communications" was given to GSM.
1994	Phase 2 implementation began.

- Intersystem roaming (international roaming needs standardized air interface);
- High degree of flexibility (open architecture that will allow new services to be introduced at a future date);
- Economy in both sparsely and heavily populated areas;
- Integration with ISDN;
- Other security features;
- A range of additional features, such as short message service and use of facsimile system;
- Easy to introduce the system;
- Low-cost infrastructure.

The specific requirements for the system were not clearly defined by the superior committees in CEPT. On the contrary, a great deal of freedom was given to GSM

Table 1.4
Main Areas of Concern of Different Working Groups

Working Group	Areas of Concern
MOU-BARG (Billing and Accounting)	All commercial and administration principles and procedures to support European roaming including: (1) Administration of subscribers, (2) Billing harmonization, (3) Credit control, (4) Fraud prevention intersystem, (5) Accounting operation, (6) Statistics, and (7) Definitions of harmonized billing and accounting software requirements.
MOU-MP (Marketing Planning)	(1) Presentation of coverage information, (2) Identification of selling features to guide system development, (3) Coordination of awareness campaign public relations, and (4) GSM name and logo.
MOU-P (Procurement)	Harmonization of procurement policy.
MOU-EREG (European Roaming)	Coordination of all technical and operational procedures principles and plans for the support of European roaming, including: (1) Mobile numbering plans; (2) Routing of mobile terminated calls and of signaling messages; (3) Technical implications of tariff principles on international interworking; (4) Establishment of international signaling links; and (5) Interworking between PLMN utilizing different work functions, quality and availability of service.
MOU-CONIG (Conformance of Network Interfaces)	(1) List/definition of tests for conformance of interfaces "A" and "Abis" and (2) harmonization of test activities.
MOU-TAP (Type Approval Administrative Procedures)	(1) Harmonization of procedures regarding type approval, (2) Review of existing or emerging directives and identification of possible difficulties, and (3) Control and issue of IMEIs.
MOU-TADIG (Transfer Account Data Interchange)	(1) To specify the detailed file: Interchange mechanism to tape and data transfer between billing entities to facilitate the transfer account procedures as defined by relevant GSM recommendations, taking into account the necessary security and quality of service requirements; (2) To specify the format of data records to be exchanged either by tape or data transfer; and (3) To specify standard sets of protocols for such data transfer
MOU-SERG (Services)	(1) Maintenance of GSM recommendations following transfer of responsibilities from ETSI/GSM, (2) Allocation/revision of status of implementation categories of services and dates for introduction, and (3) Review of compatibility of services for roaming.
MOU-SG (Security)	(1) Administration of nondisclosure undertaking for algorithms, (2) Maintenance of algorithms and test sequences, (3) Monitoring of adequacy of system security and proposals for enhancements as required.
MOU-RIC (Radio Interface Coordination)	(1) Coordination of technical aspects of type approval and identification of problems affecting type approval as a result of validation and conflicting interpretations of recommendations, (2) Resolution of technical problems with regard to type approval in different countries, and (3) Review of the System Simulator activities.

in order for the committee to find the best compromise between the conflicting requirements of such things as high spectrum economy, low cost, and high speech quality. One reason for this flexibility was that, at the time, there was a great deal of uncertainty as to what would be the major use of the system. It seemed reasonable then to expect that while the major use in the early 1990s would be speech communication, so that the system would then mainly be a voice system, there would gradually be a need for the system to offer advanced data services. Primarily, the system had to be at least equal to the existing first-generation systems with respect to spectrum efficiency, cost of the mobile unit, cost of the network infrastructure, availability of handheld stations, quality of speech transmission, and the ease of the introduction of new services. It was also realized that in order to be able to compete with the first-generation systems, the GSM system had to be superior to those systems in at least one of the areas. Gradually, one has come to realize that in the long run the system must offer improvements in all the other areas.

One important question was how far GSM should go in its specification work; that is, to what degree the system had to be specified so as to be identical in all countries, and how much could be left to the operators and suppliers to agree upon. Clearly, without identical air interfaces in all networks, the subscribers are not going to have free roaming between networks. This was considered to be the absolute minimum degree of standardization, and these requirements were favored. Conceivably, some people might have seen it as advantageous to specify everything in the system, including the hardware of the *mobile station* (MS) and even of other parts of the system. It was agreed upon that there would be no attempt to specify the system in such detail. Basically, only the functional interfaces between the major building blocks would be specified. This approach had several advantages, perhaps the most important of which is that for each major building block, the principle of functional specification offers each operator, and thus the customer, the opportunity to purchase whatever make of equipment he wants, thus setting the stage for maximum competition between manufacturers. For instance, the fact that an operator has purchased an exchange from a certain supplier does not force him to go on buying, equipment from the same supplier. Standardized electrical interfaces as well as protocols are provided for both the fixed and subscriber equipment. These include standardized rate adaptations compatible with conventional ISDN definitions. The imposed open network architecture with defined and standardized interfaces requires interoperability among equipment from multiple vendors, which guarantees compatibility and interworking between systems and further gives the operator flexibility in selecting equipment providers at the subsystem level rather than merely the overall system level.

GSM (Group Special Mobile, or Global System for Mobile communications), the new Pan-European digital cellular telecommunications standard, will also solve present limitations of analog systems. In fact, capacity will increase two to three times due to better frequency usage and techniques that utilize smaller cells, thus greatly increasing the number of subscribers that can be served. The increased capacity of the GSM system over its predecessor analog system is illustrated in Figure 1.2 [11]. As seen from the figure the capacity increases by two to three times over the first-generation analog system. GSM is the Pan-European digital mobile telephony standard specified by ETSI and provides a common standard. This means that cellular subscribers can use their mobile telephones all over Europe. With the full introduction of GSM roaming in Europe, one can take a mobile telephone on a journey and use it in another country. The system will automatically update information in one's home system about location. One will be able to make calls and receive incoming calls. The caller does not need to know your location. A high degree of flexibility is achieved by having different base transceiver and base station controller configurations. The optimum configuration based on traffic requirements can be chosen.

As previously stated, one major force for change to fully digital cellular systems is the need for higher system capacities. A further driving force is the worldwide digitization of the telephone network and its progress to ISDN. Digital cellular systems, forming extensions of the PSTN, called PLMN, will be an extension of the ISDN, using digital radio techniques for the short trip between the cellular infrastructure and the mobile subscriber terminal equipment. The use of increasingly sophisticated services will require digital transmission capabilities throughout the entire telephone system, including cellular. This ultimately facilitates mobile integration into many different services offered by the common digital network. As an extension of the PSTN network, GSM relies heavily on *Signaling System #7* (SS7). Having its signaling extrapolated from those of the regular telephone network, GSM is considerably more complex than conventional analog cellular systems. The messaging requirements imposed by SS7 are massive.

1.4 GSM TECHNICAL REQUIREMENTS [9–12]

In this section we summarize further relevant characteristics of GSM system, details of which are covered in Chapter 6. As shown in Figure 1.3, GSM uses both TDMA and *Frequency Division Multiple Access* (FDMA) to transmit and recover information. These systems use data packets at specific times at specific frequencies. Thus, several conversations take place simultaneously and at the same frequency using different time slots. Systems are also frequency duplex so that the transmit and receive frequencies are different, and both sides of the transmission (Mobile-to-Base and Base-to-Mobile) are concurrent.

As discussed in the last section, the main reason for designing a new system in Europe is the incompatibility between different analog systems. GSM also demands compatibility with ISDN, improved user privacy, and a well-defined *Open Network Architecture* (ONA). The characteristics of GSM system are shown in Table 1.5.

The spacing between the carriers in GSM system is 200 kHz. Eight time slots carry speech and data in a GSM system. It is expected that within five years, the

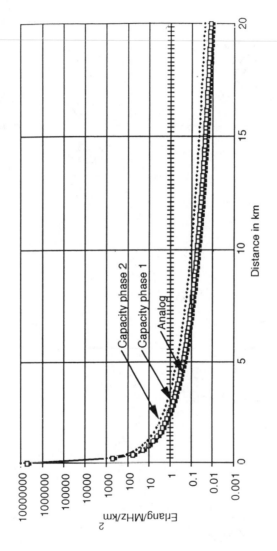

Figure 1.2 GSM system capacity.

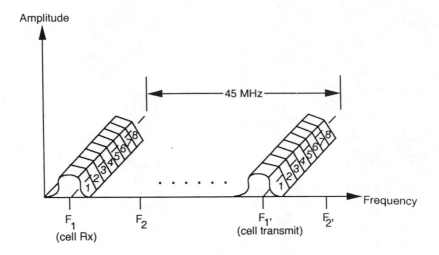

Figure 1.3 Typical TDMA/FDMA frame structure.

Table 1.5
Specifications for a GSM System

Specification	European Cellular Mobile GSM
Access method	TDMA/FDMA
Frequency bands (MHz)	
Mobile to cell	890 to 915
Cell to mobile	935 to 960
Channel bandwidth (kHz)	200
Modulation	GMSK
Bit rate (Kbps)	270.833, frequency hopped
Filter	0.3 Gaussian
Voice channel coding	RPE-LTP/Convolutional
	13 Kbps
Voice frame (ms)	4.6
Interleaving (ms)	40
Associate control channel	Extra frame
Handoff method	MAHO
Adaptive equalization	Yes
Users per channel	8
Subscriber unit power level	0.8W, 2W, 5W, 8W, 20W
Number of channels	124
Market size	15 million (conservative)

voice and channel coder rate can be compressed to 50% of its present value and will still deliver equivalent quality of voice.

The bandwidth for the GSM system is 25 MHz, which provides 125 carriers each having a bandwidth of 200 kHz. Due to interference to other systems, the very first carrier is not used, thus reducing the number of carriers to 124. Channel layout and frequency bands of operation are shown in Figure 1.4. With eight users per channel there are about 1,000 actual speech or data channels. The number of channels will double to about 2,000 as the half rate speech coder is introduced. The frequency band used for the uplink is 890 MHz to 915 MHz (from MS to base station) and for the downlink 935 MHz to 960 MHz (from base station to MS).

The modulation method in GSM is *Gaussian Minimum Shift Keying* (GMSK), which facilitates the use of narrow bandwidth and coherent detection capability. In GMSK the rectangular pulses are passed through a Gaussian filter prior to their passing through a modulator. This modulation scheme almost satisfies the adjacent channel power spectrum density requirement of −60 dBc specified by CCIR. The normalized pre-Gaussian bandwidth is kept at 0.3, which corresponds to a filter bandwidth of 81.25 kHz for an aggregate data rate of 270.8 Kbps. With 200 kHz of carrier spacing and this data rate, the spectral efficiency of the system is 1.35 b/s/Hz (270.8/200). With the bit interval of 3.7 μs, the GSM signal will encounter significant intersymbol interference in the mobile radio path due to multipath (multipath minimal delay spread 3 ms to 6 ms in urban areas). As a consequence, an adaptive equalizer is used. There are eight time slots in a frame and 26 or 51 frames in a multiframe.

With 270.8 Kbps divided among eight users in GSM, the per user data rate is 33.85 Kbps. The speech coder is a *regular pulse excitation with long-term predictor*

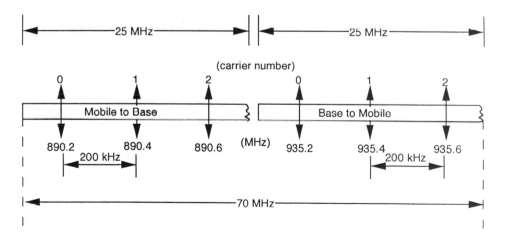

Figure 1.4 Channel layout and frequency bands of operation.

(RPE-LTP) for a full rate speech that converts speech to 13 Kbps. In the near future, a "half rate" coding scheme at a rate of roughly 7 Kbps will be used. The data transmission rates use 12 Kbps, 6 Kbps, 4.8 Kbps, and 2.4 Kbps rates, respectively, plus the control bits with each of these. Each base station is equipped with a certain number of preassigned carrier frequencies.

There are five different categories of mobile telephone units specified for the European GSM system: 0.8W, 2W, 5W, 8W, and 20W. The power level can be adjusted to vary between 3.7 mW to 20W. To optimize cochannel interference, each BS individually directs MS to use the minimum power setting that is necessary for reliable transmission. The setting is determined by BS and provided to the MS. The GSM air interface allows for frequencies to be hopped to prevent multipath problems resulting in excessive bit error rates. Both the mobile and the base station will use *Discontinuous Transmission* (DTx). This will allow the mobile to save the battery life and the base station to reduce cochannel interference.

1.5 GSM-PROVIDED SERVICES [5,8–10]

Services are defined as anything the end user explicitly sees as worth paying for. Services are classified into three groups: (1) teleservices, (2) bearer services, and (3) supplementary services.

Teleservices is a type of telecommunication service that provides complete capability, including terminal equipment functions, for communication between users according to established protocols.

Bearer services, on the other hand, is a type of service that provides the capability for the transmission of signals between user network interfaces. Some bearer services are as follows.

- Data services;
- Short message service (SMS);
- Cell broadcast;
- Local features.

Supplementary services are defined as add-ons, that is, additional features to both teleservices and bearer services. In a way, these are value-added services on top of teleservices and bearer services. Figure 1.5 provides access points for bearer- and teleservices. Details of tele-, bearer, and supplementary services are provided in Appendices 1A to 1C, respectively.

1.6 FUTURE DEVELOPMENTS [4,9–11,13]

Ten years ago mobile communications experts were frequently saying that the only frequencies of interest to them were located in the 30-MHz to 1000-MHz band.

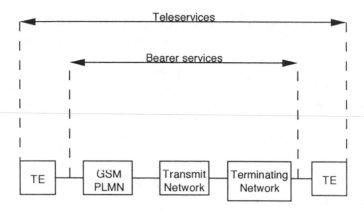

Figure 1.5 Bearer and teleservices.

Since then, semiconductor technology has progressed, making frequencies above 1 GHz much more attractive. Operators have also gained extensive experience with the 900-MHz band and realize that, in propagation also, there is no physical barrier at 1 GHz. Thus, mobile communications now claim their share of the spectrum above 1 GHz.

It is perhaps worthwhile to spend a few moments speculating on future developments after GSM has been put into operation. There is no doubt that the development in the field of mobile communications will remain very strong. We already see a great deal of interest among users, industries, and operators to go into one or another form of *Personal Communications Network* (PCN), where the emphasis will be on much smaller, very lightweight terminals and consequently much smaller cells than the systems in use today. This development will probably occur almost parallel with GSM and will in many ways not be very different from GSM. The decision has already been made to scale up GSM to a frequency band somewhere in the 1800-MHz range. The specifications for the first phase of this system were completed by the end of 1990, while a second phase was defined by the end of 1991. From a technical point of view, this system will be very strongly influenced by GSM, even in its second phase, which is natural in view of the fact that the two systems belong to the same generation.

It is well known that in the radio field, the systems tend to come in generations separated by perhaps by ten years or so. The GSM system is often referred to as a second-generation system, and the obvious question is: What will the third-generation system be like, and when will it arrive? Will it be a satellite system, or will it be land based? From what we know today, satellites could be a valuable supplement to land-based PLMN; since the medium covers large areas, it will remain a supplement that can be used to cover sparsely populated and inaccessible areas of the world.

Due to the fact that they are unable to support a very small cell structure (required for high capacity), it will not be suitable in a situation where spectrum economy is the overriding concern.

It seems that a more likely scenario for the future land mobile system is aiming for a Universal Mobile Telecommunication Service (UMTS), as presented by several bodies such as the CCIR. In such a system, the number of users and cells will far exceed what present systems can handle. Some of the salient features that are likely to be part of the future system include the following.

- Common standards for public cellular systems and cordless telephones with full interworking preferably even trunked and paging systems;
- Mass market pocket telephones;
- High-capacity infrastructure with both public and private calls connected to each other and integrated;
- Broadband Communications Network;
- Possibility of carrying a wide range of data services.

It is expected that the question of how much penetration we can attain in a future UMTS is less dependent upon communication technology than upon the shortage of radio spectrum. Demand will be highly dependent upon user satisfaction, that is, the ability of the system to meet the needs of the users. Consequently, insufficient spectrum availability could be a serious obstacle to the success of the system. Therefore, early action is a must in order to secure the necessary amount of radio spectrum.

Clearly, the creation of the UMTS will not happen without a great deal of effort from major standards bodies such as the CCIR and ETSI. Given that the willingness exists to achieve the goal, we will probably see a system, highly integrated technically with the fixed networks and capable of accepting many types of terminals with different capabilities, around the turn of the century.

By that time we will probably be able to say that a truly global system has come. Will such a system be accepted by all the major countries around the world, or will we still have regional systems as in the past? Global acceptance will depend upon the political climate of the world.

However, if the system planning work starts early enough, and if there is agreement at an early stage on what frequency band to use and on the other basic characteristics, no major technical obstacles to such a system can be seen. In view of the ever-increasing number of travelers everywhere, there is no doubt that the demand will exist. We should keep in mind that in the future the majority of MSs will probably not be installed in cars, as is the case today. The trend toward smaller, cheaper, handheld stations will continue, and the traveler will find it just as natural to take the handheld station along on a trip as it is to take along other personal effects.

We must recognize that the international standardization work (for example, UMTS and FPLMTS) has already started on the so-called third-generation wireless systems that are due to emerge by the turn of the century. The work on frequency assignment is already complete. In 1990, by request of the United Kingdom, the specification of a version of GSM adapted for the 1800-MHz frequency band was added to the scope of the standardization group, with a frequency band of 2 MHz by 75 MHz. This variant, referred to as *Digital Cellular System 1800* (DCS1800) is aimed at reaching higher capacities in urban areas for the type of mass-market approach known as PCN. Spectrum has already been allocated in the United States in the frequency band from 1850 MHz to 1990 MHz. Recently, the FCC divided the band into six areas categorized as major and minor trading areas. We shall discuss PCS systems in detail in Chapters 10 and 11.

1.7 ORGANIZATION OF THE BOOK

This book is based on the illustration of Figure 1.6. As seen from this figure, the essential elements of a GSM system consist of: mobile users, a *base station system* (BSS), a mobile switching center, and public voice and data networks. Other important elements of the system not shown in the figure are the operations and management center, billing center, and the various networks (SS7 and X.25) interconnecting these subsystems. Therefore, we arrange the discussion based on different aspects of the subsystem as follows.

Chapter 2 deals with the architecture of the GSM system, where we discuss the functional requirements of the MS; the BSS, which includes a base transceiver

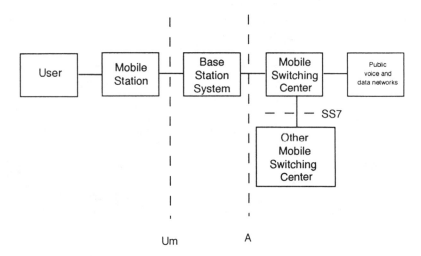

Figure 1.6 GSM interfaces.

station and base station controller; the *mobile services switching center* (MSC) and its variation known as gateway switching center; the operations and network management centers; and the billing center. Chapter 3 provides a detailed discussion of time and frequency axis representation of the system, as well as the functions and characteristics of physical and logical channels, and provides the reasons for the flexibility of this system. Chapter 4 is fully devoted to mobility management issues and provides the timing diagrams for different types of calls including incoming call to mobile, mobile origination of a call, and mobile-to-mobile call. Functions such as TMSI attach and detach, location updating, and handover are also discussed. Chapter 5 brings out the various security aspects of the GSM system. In Chapter 6 we discuss the technical characteristics of the system, including all the important aspects of modulation and coding as applied to GSM system. In Chapter 7 we discuss subscription management, billing and accounting, and the important aspects of network maintenance. Chapter 8 deals with GSM protocols based on the ISO and SS7 standards, which can be considered as additional details of Chapter 4. Chapter 9 elaborates on Chapters 2 and 3 and discusses the subsystem configuration architecture of MSC, BSS, OCC, and MS. Chapter 10 provides existing systems as predecessors to PCS systems. In conclusion, Chapter 11 points out the potential shortfalls of existing PCS systems and provides some solutions in this direction.

Problems

1.1 State reasons behind the design of GSM system.

1.2 Name major analog systems of Europe with their operating bands.

1.3 State main objectives of future PLMTS.

1.4 Justify why phased implementation has been adopted for GSM.

1.5 Using Table 1.4, list main functions of the following committees: (1) Billing and Accounting, (2) Marketing and Planning, (3) Roaming, (4) Network Interfaces, (5) Type Approval, (6) Transfer Account Data Interchange, (7) Security, and (8) Radio Interface Coordination.

1.6 List, in order of importance, the operational requirements of GSM.

1.7 Justify that the standardization of air interface can satisfy international roaming.

1.8 Give some intuitive reasoning why the capacity of GSM should be more than their analog predecessor.

1.9 Find the spectral efficiency in bps/Hz of the GSM system assuming channel bandwidth BW to be 200 kHz and the channel data rate of 270.833 Kbps. What will be the spectral efficiency if the BW is increased by 50%?

1.10 State reasons for using 0.3 Gaussian baseband filter in the GSM raw data.

1.11 Why, in actuality, are only 124 or even less channels used instead of the total available 125 channels?

1.12 Narrate advantages and disadvantages of: (1) DTx and (2) frequency hopping.

1.13 Define the following terms: (1) Teleservices, (2) Bearer services, and (3) Supplementary services.

1.14 Make a list of the potential requirements for future PCS systems.

1.15 Narrate what efforts are under way in Europe and the United States for the development of future PLMTS.

1.16 Name the following standardized interfaces: (1) air interface, (2) between BSS and MSC, (3) between MSCs, and (4) between MSC and PSTN/ISDN.

References

[1] Dechaux, C., and R. Scheller, "What Are GSM and DCS?" *Electrical Communications,* 2nd quarter, 1993, pp. 118–127.

[2] P. L., "GSM Network Architecture Issues of the Next Century," *IEEE Vehicular Technology Conf.* 1994, pp. 325–329.

[3] Temple, J., et al., "Pan-European GSM Signaling Requirements for the Abis Interface," *IEEE Vehicular Technology Conf.,* 1994, pp. 343–347.

[4] Mouly, M., and Pautet Marie-Bernadette, "Current Evolution of the GSM Systems," *IEEE Personal Communications,* October 1995, pp. 9–19.

[5] Villani, O., "CCIR Activities on Land Mobile Services," *GSM Seminar,* Budapest, October 1990, Session 1.2.

[6] Haug, T., "GSM: Targets and Achievements," *GSM Seminar,* Budapest, October 1990.

[7] Bliksrud, P., "GSM-MoU: Cooperation of CEPT's GSM Operators," *GSM Seminar,* Budapest, October 1990.

[8] Beddoes, E. W., "GSM Network Architecture," *GSM Seminar,* Budapest, October 1990, Session 2.1.

[9] Maloberti, A., et al., "Radio Subsystem and Elements," *GSM Seminar,* Budapest, October 1990, Session 3.1.

[10] Mehrotra, A., *Cellular Radio Analog and Digital Systems,* Norwood, MA: Artech House, 1994.

[11] GSM specification 2.01 Version 4.2.0, issued by ETSI, January 1993.

[12] Arnbak, J. C., "The European Revolution of Wireless Digital Networks," *IEEE Communication Magazine,* September 1993.

[13] ETSI/GSM Section 2.01, "Principles of Telecommunication Services Supported by GSM Plan," January 1993.

APPENDIX 1A TELESERVICES

Teleservices cover regular telephony, emergency calls, and voice messaging.

The most important service provided by GSM is telephony. This service enables bidirectional speech calls to be placed between GSM users and any telephone subscriber who is reachable through the general telephony network. Fixed telephone subscribers worldwide as well as mobile network subscribers or subscribers of specific networks connected to a public telephone network can be reached.

Following the GSM official terminology, emergency calling is a distinct service derived from telephony. It allows the user of a MS to reach a nearby emergency service (such as police or the fire brigade) through a simple and unified procedure, by dialing 112 (the number 112 is the standard emergency number throughout Europe and is similar to 911 in the United States).

Another service derived from telephony is voice messaging. The specifications do not identify this as a separate service, but many operators will offer it as a basic feature. It enables a voice message to be stored for later retrieval by the mobile recipient, either because he was not reachable at the time of the call or because the calling party choose to access the voice mailbox of the GSM subscriber directly.

APPENDIX 1B BEARER SERVICES

The following bearer services are defined for a GSM system.

All standard rates up to 9.6 Kbps are supported. The connection may be made to other MSs or to other data users on circuit-switched or packet-switched data networks. A suitable data terminal or computer is connected directly to the mobile phone.

In the case of circuit-switched connections to the PSTN, a modem suitable for MSC is automatically selected at the GSM for the link to the similar modem at the remote end. No modem is required at the MS.

For connection to the data packet-switched networks, asynchronous access is made via a *Packet Assembly/Disassembly* (PAD). This PAD may be located in the mobile network or a pubic PAD may be used. Packet-switched synchronous data transmission at 2.4 Kbps, 4.8 Kbps, or 9.6 Kbps is also possible.

When used in the error-correcting mode, very low data error rates are possible, even under difficult radio conditions.

A short message can be up to 160 alphanumeric characters long and can be sent to or from a mobile telephone. A service center where a calling subscriber can dictate a message may be employed. The message will then be transmitted to the destination (B-subscriber). If the B-subscriber is outside the coverage area of the system, the message is stored and forwarded when the B-subscriber is available. Short messages can be received and sent both in idle state or during a call. Short messages of up to 93 alphanumeric characters may be sent to all phones (broadcast mode) in a geographical area. This service can be used for various purposes, such as traffic information and weather forecasts.

A GSM MS is quite a complex piece of machinery and includes the capacity of a small computer. As an intelligent terminal, it can offer a number of functions locally without the help of a network. Examples include the dialing of abbreviated numbers, the storage of received short messages, the edition of short messages, the automatic repeat of failed calls, and the automatic answering of calls.

Another point worth noting is the existence of the + key, which is specified as a harmonized shortcut replacing the international prefix. For instance, when in Sweden, a GSM user can call somebody in Italy by dialing +39 followed by the national number, instead of dialing 00939. . . . Another important advantage in so doing is that the stored 11 + 39 . . . number will be recognized correctly by all GSM PLMNs (including in Italy) and, therefore, remains valid irrespective of roaming.

APPENDIX 1C SUPPLEMENTARY SERVICES

A comprehensive list of supplementary services are also available in GSM. They are listed below.

1. *Advice of charge (Symbol AoC, Date of introduction 1992):* Advice of charge is a service where the user is informed of the real-time information on progress of the cost of the call.
2. *Barring of all outgoing calls (BAOC, 1992):* This service makes it possible for a mobile subscriber to prevent all outgoing calls or those calls associated with his basic service.
3. *Barring outgoing international calls except those directed to the Home PLMN Country (BOIC-exHC, 1992):* This service makes it possible for a mobile subscriber to prevent all attempted outgoing international calls except those directed to the home PLMN country on his subscription. These calls may be associated with all services or with specific basic service.
4. *Barring of all incoming international calls except those directed to the home PLMN country (BAIC, 1992):* This service makes it possible for a mobile subscriber to prevent all incoming calls, or just those associated with a specific basic service that would otherwise be terminated at his directory number. This is similar to barring outgoing international call.
5. *Barring of all roaming calls when roaming outside the home PLMN country (BIC-Roam, 1992):* This service makes it possible for a mobile subscriber to prevent all incoming calls, or just those associated with a specific basic service, that would otherwise be terminated at his directory number. This only applies to the case when the mobile subscriber roams outside his home PLMN country.
6. *Call forwarding, unconditional (CFU, 1994):* This service permits a called mobile subscriber to have the network send all incoming calls, or just those associated with specific basic services, which are addressed to the called mobile subscriber's directory number, to another directory number.
7. *Call forwarding on mobile subscriber busy (CFB, 1994):* This service permits a called mobile subscriber to have the network send all incoming calls, or just those associated with a specific basic service, which are addressed to the called mobile subscriber's directory number and which meet mobile subscriber busy, to another directory number.
8. *Call forwarding on no reply (CFNRy, 1994):* This service permits an added mobile subscriber to have the network send all incoming calls, or just those

associated with a specific basic service, which are addressed to the called mobile subscriber's directory number and which meet no reply, to another directory number.

9. *Call forwarding on MS not reachable (CFNRc, 1994):* This service permits a called mobile subscriber to have the network send all incoming calls, or those calls associated with a specific basic service, which are addressed to the called mobile subscriber's directory number and not reachable to another directory number.

10. *Call forwarding on radio congestion:* This service permits a mobile subscriber to have the network send all incoming calls, or just those associated with a specific basic service, which are addressed to the called mobile subscriber's directory number and which meet congestion on the radio path, to another directory number.

11. *Call hold (HOLD, 1993):* The call hold service allows a served mobile subscriber to interrupt communication on an existing call and then subsequently, if desired, to reestablish communication. In this case the traffic channel remains assigned to the mobile subscriber after the communication is interrupted, which allows the origination, or possible termination, of other calls.

12. *Call waiting (CW, 1993):* Call waiting provides a mobile subscriber with the possibility to be notified of an incoming call while his MS is in the busy state. Subsequently, the subscriber can either answer, reject, or ignore the incoming call. Both the call waiting and call hold options are the same as those offered by PSTN.

13. *Call transfer (CT, 1995):* The call transfer supplementary service enables the served mobile subscriber to transfer an established incoming or outgoing call to the third party. This service differs from the call forwarding supplementary service in that call forwarding deals only with incoming calls.

14. *Completion of calls to busy subscribers (Camp-on) (CCBS, 1995):* The supplementary service allows a calling mobile subscriber who encounters a busy called subscriber to be notified by the system operator when the busy called subscribers becomes free and have the operator re-initiate the call if the calling MS so desires.

15. *Closed user group (CUG, 1995):* Closed user group provides for the possibility for a group of subscribers, connected to the PLMN and or the ISDN, to communicate only among themselves. If required, one or more subscribers may be provided with incoming/outgoing access to subscribers outside this group.

16. *Calling number identification presentation (CNIP, 1994):* Calling number identification presentation provides for the ability to indicate the ISDN number of the calling party with possible additional address information to the called party. This identity is provided to the called subscriber before answer-

ing, thus enabling him to make the decision of whether to take the call or not.

17. *Calling number identification restriction (CNIR, 1994):* Calling number identification restriction function is a supplementary service offered to the calling party to restrict presentation of the calling party's number, with possible additional address information to the called party. This is just the opposite of calling number identification presentation.

18. *Connected number identification presentation restriction (CoNP, 1994):* Connected number identification presentation provides the GSM caller with the phone number he has reached.

19. *Freephone service (FPH, 1992):* A mobile subscriber can be allocated a special (freephone) number and the charge for all calls to this number are paid by him instead of by the callers. This is similar to an 800 number in the United States by PSTN.

20. *Malicious call identification (MCI, 1994):* This supplementary scheme enables the mobile subscriber to request to register, at the discretion of the network operator, malicious, nuisance, or obscene incoming calls.

21. *Three-party service (3PTY, 1994):* The three-party service enables a mobile subscriber to establish a three-party conversation (conference call). A mobile subscriber who is active on a call is able to hold to the existing conversation, originate an additional call to a third party, and switch from one call to the other as desired.

CHAPTER 2

▼▼▼

GSM ARCHITECTURE

2.1 INTRODUCTION

A GSM system is basically designed as a combination of three major subsystems: the network subsystem, the radio subsystem, and the operation support subsystem. In order to ensure that network operators will have several sources of cellular infrastructure equipment, GSM decided to specify not only the air interface, but also the main interfaces that identify different parts. There are three dominant interfaces, namely, an interface between MSC and the *Base Station Controller* (BSC), an A-bis interface between BSC and the *Base Transceiver Station* (BTS), and an Um interface between the BTS and MS. These three interfaces are shown in Figure 1.6.

The network subsystem includes the equipment and functions related to end-to-end calls, management of subscribers, mobility, and interface with the fixed PSTN. In particular, the switching subsystem consists of MSCs, *Visitor Location Register* (VLR), *Home Location Register* (HLR), *Authentication Center* (AUC), and *Equipment Identity Register* (EIR). The MSC provides call setup, routing, and handover between BSCs in its own area and to/from other MSC; an interface to the fixed PSTN; and other functions such as billing. The HLR is a centralized database of all subscribers registered in a PLMN. There may be more than one HLR in PLMN, but the individual subscriber has entry to only one of them. The VLR is a database of all mobile, currently roaming in the MSC's area of control. As soon as an MS roams into a new MSC area, the VLR connected to that MSC will request data about the MS from the HLR. At the same time, the HLR will be informed as to which MSC area the MS resides. If, at a later time, MS wants to make a call, the VLR will have all the information needed for the call setup without having to interrogate the HLR each time. Thus, VLR in one sense is a distributed HLR. VLR also contains more exact information about the mobile location. The AUC is connected to the HLR. The function of the AUC is to provide HLR with authentication parameters and ciphering keys that are used for security purposes. The EIR is the database where the *International Mobile Equipment Identity* (IMEI) numbers for all registered mobile equipment are stored.

Some other components of the network are Echo Canceler, which reduces the annoying effect caused by the mobile network when connected to a PSTN circuit; and the network *Interworking Function* (IWF), which is the interface between MSC and other networks such as PSTN and ISDN.

The radio subsystem includes the equipment and functions related to the management of the connections on the radio path, including the management of handovers. It mainly consists of a BSC, BTS, and the MS. MS is traditionally listed as a part of the radio subsystem even though it is always one end of the conversational path and holds dialogues with the network subsystem for the management of its mobility. The MS includes both the capabilities of network termination and user termination. The GSM system is realized as a network of radio cells, which together provide complete coverage of the service area. Each cell has a BTS with several transceivers. A group of BTSs are controlled by one BSC. There are various configurations of BSC-BTS. Some configurations are best suited for high traffic, and some are meant to serve moderate-to-low traffic areas. A BSC controls such functions as handover and power control. BSS and BTS together are known as a BSS, which is viewed by the MSC through a single interface as being the entity responsible for communication with MSs in a certain area. A BSS is associated with the radio channel management, transmission functions, radio link control, and quality assessment and preparation for handover. BSS ensures the coverage of N cells, where N can be one or more.

The *Operational and Maintenance Center* (OMC) subsystem includes the operation and maintenance of GSM equipment and supports the operator network interface. It is connected to all equipment in the switching system and to the BSC. OMC performs GSM's administrative functions (for example, billing) within a country. One of the OMC's most important functions is the maintenance of the country's HLR. Depending upon the network size, each country may have more than one OMC. The global and centralized management of the network is provided by the Network Management Center, while the OMC is responsible for the regional management of the network. Two typical GSM architectures are shown in Figures 2.1(a, b).

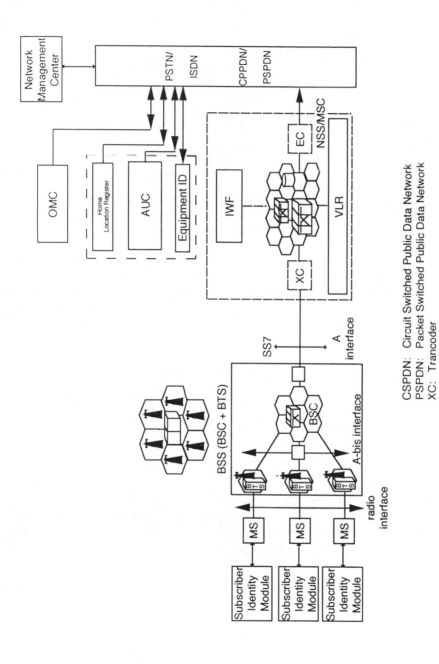

CSPDN: Circuit Switched Public Data Network
PSPDN: Packet Switched Public Data Network
XC: Trancoder

(a)

Figure 2.1 (a) Typical GSM architecture; (b) GSM system model.

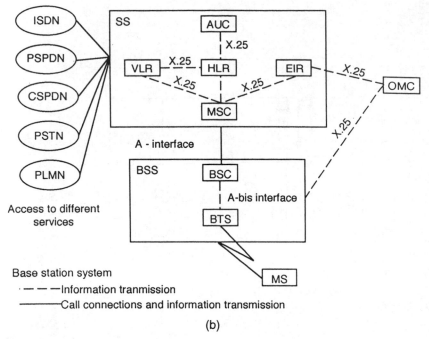

Figure 2.1 (continued).

In view of this introduction, the objective of this chapter is to discuss the functions of various network subsystems, radio subsystems, operational and maintenance subsystems, and billing subsystems. Before we take up the study of these subsystems let us first discuss the layout of the GSM network.

2.2 GSM NETWORK STRUCTURE

Every telephone network needs a well-designed structure in order to route incoming calls to the correct exchange and finally to the called subscriber. In a mobile network, this structure is of great importance because of the mobility of all its subscribers [1–4]. In the GSM system, the network is divided into the following partitioned areas.

- GSM service area;
- PLMN service area;
- MSC service area;
- *Location area* (LA);
- Cells.

The GSM service area is the total area served by the combination of all member countries where a mobile can be serviced. The next level is the PLMN service area. There can be several within a country, based on its size. The links between a GSM/PLMN network and other PSTN, ISDN, or PLMN networks will be on the level of international or national transit exchanges. All incoming calls for a GSM/PLMN network will be routed to a Gateway MSC. A Gateway MSC works as an incoming transit exchange for the GSM/PLMN. In a GSM/PLMN network, all mobile-terminated calls will be routed to a Gateway MSC. Call connections between PLMNs, or to fixed networks, must be routed through certain designated MSCs called a gateway MSC. The gateway MSC contains the interworking functions to make these connections. They also route incoming calls to the proper MSC within the network. The next level of division is the MSC/VLR service area. In one PLMN there can be several MSC/VLR service areas. MSC/VLR is a sole controller of calls within its jurisdiction. In order to route a call to a mobile subscriber, the path through the network links to the MSC in the MSC area where the subscriber is currently located. The mobile location can be uniquely identified since the MS is registered in a VLR, which is generally associated with an MSC. The PLMN and MSC/VLR service areas are shown in Figures 2.2(a,b).

The next division level is that of the LAs within a MSC/VLR combination. There are several LAs within one MSC/VLR combination. A LA is a part of the MSC/VLR service area in which a MS may move freely without updating location information to the MSC/VLR exchange that controls the LA. Within a LA a paging message is broadcast in order to find the called mobile subscriber. The LA can be identified by the system using the *Location Area Identity* (LAI). The LA is used by the GSM system to search for a subscriber in active state.

Lastly, a LA is divided into many cells. A cell is an identity served by one BTS. The MS distinguishes between cells using the *Base Station Identification Code* (BSIC) that the cell site broadcasts over the air. The division of LA into many cells is shown in Figure 2.2(c). The overall relationship between cells, LA, MSC service area, PLMN service area, and GSM service area is shown in Figure 2.2(d). Coding for LA is shown in Figure 2.2(e).

2.2.1 Cell Layout and Frequency Planning

From the system design point-of-view we consider the cell to be hexagonal, but in real life it happens to be irregular in shape and size due to uneven propagation in different radials. At the cell boundary, the power received from the BTS should be adequate for reception with an acceptable *bit error rate* (BER). In general, this can be achieved at most locations from more than one BTS, leading to an overlap of the actual cell coverage. The cell boundaries are the line of equal power from two adjacent *base stations* (BS); therefore, the cell sizes will vary as the BTS power is

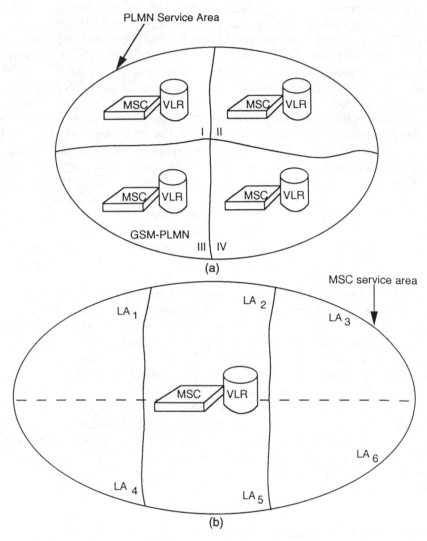

Figure 2.2 (a) PLMN service areas I to IV; the division of a MSC/VLR service area into (b) LAs and (c) cells; (d) relation between areas in GSM; (e) LA coding.

adjusted. The layout of a typical MSC/VLR LA with overlapping cell sites is shown in Figure 2.3(a).

GSM has chosen the four-cell repeat pattern for the frequency reuse cell sets. In most cases, each cell is divided into 120-deg sectors, with three base transceiver subsystems in each cell. Each base transceiver subsystem has a 120-deg antenna. In some low-traffic areas, an omnidirectional setup can be used.

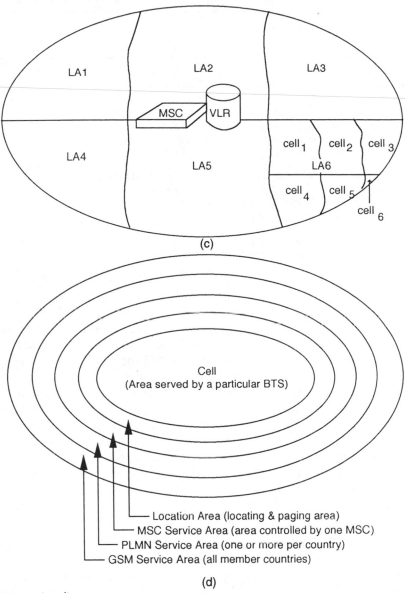

(c)

Cell
(Area served by a particular BTS)

Location Area (locating & paging area)
MSC Service Area (area controlled by one MSC)
PLMN Service Area (one or more per country)
GSM Service Area (all member countries)

(d)

Figure 2.2 (continued).

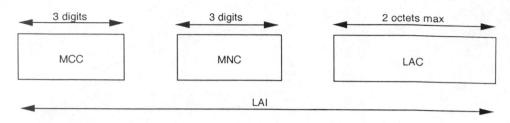

MCC - Mobile Country Code: identifies the country in which the GSM
PLMN is located. Thus the value of the MCC is the same as the 3 digit
MCC contained in the IMSI of any subscriber belonging to the country
presently visited by the mobile station.

MNC - Mobile Network Code: is a code identifying the GSM PLMN in
that country. Thus the value of the MNC is the same as the 3 digit MNC
contained in the IMSI of any subscriber belonging to the country
presently visited by the mobile station.

LAC - Location Area Code: it is flexible length code, up to 2 octets
maximum, identifying a location area within a GSM PLMN.

(e)

Figure 2.2 (continued).

With a four-cell repeat pattern and three-sectored antenna per cell site, there
are 12 sectors (called cells in a GSM system) in which the total number of 124
channels needs to be divided, assuming that the full spectrum is available from the
operator of the system. The four-cell coverage pattern is shown in Figure 2.3(b).
For optimum frequency separation, within the same sector (cell), assign to the Kth
sector (cell) K, $K + N$, $K + 2N$, $\ldots$, $K + nN$, where n is an integer such that
$(124 - nN)/N < 1$. Thus, in this case where $N = 12$ for the first sector, the assigned
channels are $1, 13, 25, 37, \ldots, 121$. Since there are 12 sectors (called cells in GSM)
per repeat pattern, each sector (cell) then gets approximately one-twelfth of the
available frequencies. The frequency assignment for this example is shown in
Table 2.1. Here the pattern of cell frequency assignments $(A, B, \ldots)$ is repeated so
that frequencies are reused with an acceptably small level of cochannel interference.

2.3 MOBILE STATION

The MS includes radio equipment and the *man machine interface* (MMI) that a
subscriber needs in order to access the services provided by the GSM PLMN. MSs
can be installed in vehicles or can be portable or handheld stations. The MS may
include provisions for data communication as well as voice. A mobile transmits and
receives messages to and from the GSM system over the air interface to establish
and continue connections through the system [1,5–8].

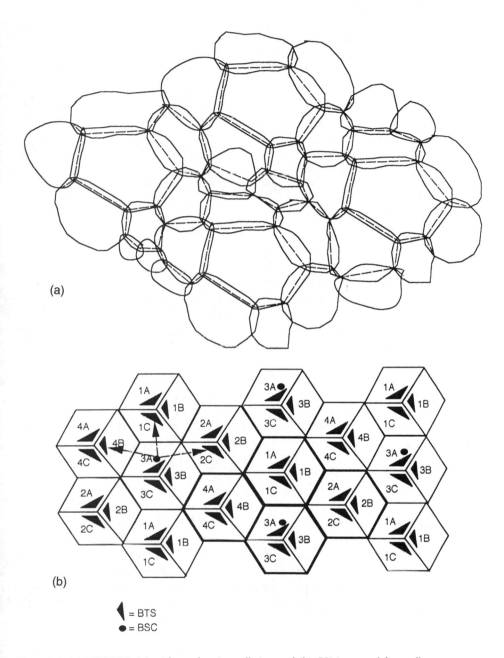

(a)

(b)

▮ = BTS
● = BSC

Figure 2.3 (a) MSC/VLR LA with overlapping cell sites and (b) GSM sectored four-cell coverage.

Table 2.1
Frequency Assignment for 12 Sectors (four-cell repeat)

1A	2A	3A	4A	1B	2B	3B	4B	1C	2C	3C	4C
1	2	3	4	5	6	7	8	9	10	11	12
13	14	15	16	17	18	19	20	21	22	23	24
25	26	27	28	29	30	31	32	33	34	35	36
37	38	39	40	41	42	43	44	45	46	47	48
49	50	51	52	53	54	55	56	57	58	59	60
61	62	63	64	65	66	67	68	69	70	71	72
73	74	75	76	77	78	79	80	81	82	83	84
85	86	87	88	89	90	91	92	93	94	95	96
97	98	99	100	101	102	103	104	105	106	107	108
109	110	111	112	113	114	115	116	117	118	119	120
121	122	123	124								

Different types of MSs can provide different types of data interfaces. To provide a common model for describing these different MS configurations, "reference configurations" for MS, similar to those defined for ISDN land stations, has been defined.

Each MS is identified by an IMEI that is permanently stored in the mobile unit. Upon request, the MS sends this number over the signaling channel to the MSC. The IMEI can be used to identify mobile units that are reported stolen or operating incorrectly. This is discussed in detail in Section 5.5.

Just as the IMEI identifies the mobile equipment, other numbers are used to identify the mobile subscriber. Different subscriber identities are used in different phases of call setup. The *Mobile Subscriber ISDN Number* (MSISDN) is the number that the calling party dials in order to reach the subscriber. It is used by the land network to route calls toward an appropriate MSC. The *International Mobile Subscriber Identity* (IMSI) is the primary function of the subscriber within the mobile network and is permanently assigned to him. The GSM system can also assign a *Temporary Mobile Subscriber Identity* (TMSI) to identify a mobile. This number can be periodically changed by the system and protects the subscriber from being identified by those attempting to monitor the radio channels.

By making a distinction between the subscriber identity and the mobile equipment identity, a GSM PLMN can route calls and perform billing based on the identity of the subscriber rather than the mobile unit being used. This can be done using a removable *Subscriber Identity Module* (SIM). A *smart card* (SC) is one possible implementation of a SIM; the other implementation can be the module mounted on the mobile equipment. The TMSI pertaining to the identity of the subscriber is stored in the SIM module itself. When the SIM is in the mobile unit, a location update procedure registers the subscriber's new location, allowing proper routing of incoming calls. Thus, in the next few sections we shall provide details of MS functions and features, SIM card usage, and various ways through which the MS is identified in the system.

2.3.1 Functions of MS

The primary functions of MS are to transmit and receive voice and data over the air interface of the GSM system. MS performs the signal processing functions of digitizing, encoding, error protecting, encrypting, and modulating the transmitted signals. It also performs the inverse functions on the received signals from the BS. Details of these are covered in Chapter 6. A list of relevant functions includes the following.

- Voice and data transmission;
- Frequency and time synchronization;
- Monitoring of power and signal quality of the surrounding cells for optimum handover;
- Provision of location updates;
- Equalization of multipath distortions;
- Display of short messages up to 160 characters long;
- Timing advance.

In order to transmit voice and data signals, the mobile must be in synchronization with the system so that the messages are transmitted and received by the mobile at the correct instant. To achieve this, the MS automatically tunes and synchronizes to the frequency and TDMA timeslot specified by the BSC. This message is received over a dedicated timeslot several times within a multiframe period of 51 frames. We shall discuss the details of this in the next chapter. The exact synchronization will also include adjusting the timing advance to compensate for varying distance of the mobile from the BTS.

The MS monitors the power level and signal quality, determined by the BER for known receiver bit sequences (synchronization sequence), from both its current BTS and up to six surrounding BTSs. This data is received on the downlink broadcast control channel. The MS determines and sends to the current BTS a list of the six best received BTS signals. The measurement results from MS on downlink quality and surrounding BTS signal levels are sent to BSC and processed within the BSC. The system then uses this list for best cell handover decisions. Unlike analog cellular systems, the measurement of adjacent signal levels by the MS rather than the BS has several benefits. Since the processing requirements are distributed to all mobile rather than being concentrated to a small number of BSs, it provides freedom and flexibility to solve the anomalous propagation problems.

MS keeps the GSM network informed of its location during both national and international roaming, even when it is inactive. As long as the system is switched on, it either provides the location update periodically or the system will force it to provide its present location. The system notes in which national PLMN, MSC zone, and BSC area the MS is currently located. This enables the system to page in its present LA.

The MS includes an equalizer that compensates for multipath distortion on the received signal. This reduces intersymbol interference that would otherwise degrade the BER. The equalizer tap setting is derived on every frame by comparing the received synchronizing sequence with the locally stored sequence within the MS. This is further elaborated in Section 3.2.4.

Finally, the MS can store and display short received alphanumeric messages on the *liquid crystal display* (LCD) that is used to show call dialing and status information. These messages are limited to 160 characters in length.

A MS must be able to adjust continuously its transmit time in order to compensate for the variation of distance between the MS and the BTS. Timing advance is determined by BTS. When the MS is on a dedicated channel (such as SDCCH or TCH), the required timing advance is sent to the MS and the actual timing advance used is reported by the MS.

2.3.2 Power Levels

There are five different categories of mobile telephone units specified by the European GSM system: 20W, 8W, 5W, 2W, and 0.8W. These correspond to 43-dBm, 39-dBm, 37-dBm, 33-dBm, and 29-dBm power levels. The 20-W and 8-W units (Peak power) are either for vehicle-mounted or portable station use. When the equipment is mounted in a vehicle, the antenna is physically mounted outside the vehicle. Vehicles may include cars, motorcycles, trains, and ships. When used as a portable station, the equipment may be hand-carried and the antenna may not be physically attached to the portion of the equipment containing the mobile equipment. The handheld unit is supposed to be carried by a person and should have a total weight less than 800 gm and volume less than 900 cm^3; the power source must last at least one hour of call duration or 10 hours of idle time. There are also intermediate-sized portable units having 8-W and 5-W power. Mobile power class, maximum power level, and its tolerance under normal and extreme operating conditions are shown in Table 2.2(a).

The MS power is adjustable in 2-dB steps from its nominal value down to 20 mW (13 dBm). This is done automatically under remote control from the BTS, which monitors the received power and adjusts the MS transmitter to the minimum power setting necessary for reliable transmission. The setting is determined by BS and provided to MS. The power level is adjusted in order to minimize cochannel interference to MSs. It also extends the MS battery lifetime between recharges. The MS power can be adjusted at one 2-dB step at a time, with one adjustment every 13 TDMA frames (60 ms). Therefore, for a sudden large required power change, a slope overload similar to that encountered in delta modulation can be created. The

Table 2.2
(a) Mobile Power Class, Maximum Power Level, and Their Tolerance

Power Class	Maximum Peak Power	Tolerance (dB) for Conditions	
		Normal	Extreme
1	20W (43 dBm)	±2	±2.5
2	8W (39 dBm)	±2	±2.5
3	5W (37 dBm)	±2	±2.5
4	2W (33 dBm)	±2	±2.5
5	0.8W (29 dBm)	±2	±2.5

(b) Mobile Power Level Control Range

Power Control Level	Peak Power (dBm)	Tolerance (dB) for Conditions		Power Control Level	Peak Power (dBm)	Tolerance (dB) for Conditions	
		Normal	Extreme			Normal	Extreme
0	43	±2	±2.5	8	27	±3	±4
1	41	±3	±4	9	25	±3	±4
2	39	±3	±4	10	23	±3	±4
3	37	±3	±4	11	21	±3	±4
4	35	±3	±4	12	19	±3	±4
5	33	±3	±4	13	17	±3	±4
6	31	±3	±4	14	15	±3	±4
7	29	±3	±4	15	13	±3	±4

MS required power level is determined by BSC, based on the reported measurements made by MS and BTS. The whole range of power control is shown in Table 2.2(b)

2.3.3 MS Configuration

An effort has been made to allow off-the-shelf terminal facsimile equipment to be connected to MSs (for instance, group here facsimile machines designed for connection to the telephone network and specific terminal adaptation functions have been specified for this purpose). This leads to the identification of three main functions listed below and shown in Figure 2.4 [2–4,8].

1. The terminal equipment carrying out functions specific to the service, without any GSM specific functions; for example, a fax machine;
2. The mobile carrying out among others all functions related to transmission on the radio interface;
3. A terminal adapter acting as a gateway between the terminal and the mobile termination, which is introduced when the external interface of the mobile termination follows the ISDN standard for a terminal installation, and the terminal equipment has a terminal-to-modem interface.

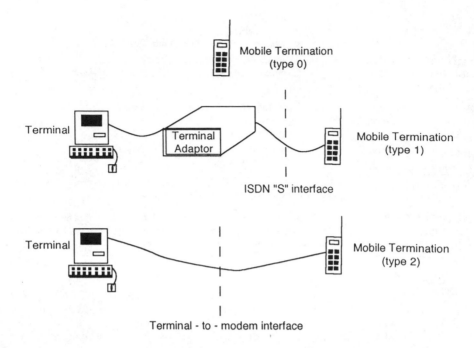

Figure 2.4 Mobile terminal types.

Based on the above functions of MS, it can be either fully integrated or include a separate terminal equipment connected to a mobile termination through a *Terminal Adaptation Function* (TAF), which can be either integrated or kept as a separate piece of equipment. Since external terminal equipment is only needed for some services (for example, fax or computer data transmission), it is a general practice not to integrate with the GSM MS. The case where the standard GSM functions, terminal equipment, and TAF are all integrated into one unit is called a Type 0 MS.

The case where the TAF contains a modem and is integrated within the MS, so that external terminal equipment can be directly connected to the MS, is called a Type 2 MS. Another possibility is when the external interface of the MS is a standard ISDN "S" Interface to which off-the-shelf ISDN terminal equipment can be connected. This connects to the MS modem port through an external TAF and is called a Type 1 MS.

2.3.4 SIM Card

As described in the first chapter, GSM subscribers are provided with a SIM card with its unique identification at the very beginning of the service. By divorcing the subscriber ID from the equipment ID, the subscriber may never own the GSM mobile equipment set. The subscriber is identified in the system when he inserts the SIM card in the mobile equipment. This provides an enormous amount of flexibility to the subscribers since they can now use any GSM-specified mobile equipment. Thus with a SIM card the idea of "Personal Communication" is already realized. A SIM card can be used to "personalize" the equipment currently in use and the respective information used by the network (location information) needs to be updated. The smart card SIM is portable between *Mobile Equipment* (ME) units. The user only needs to take his smart card on a trip. He can then rent a ME unit at the destination, even in another country, and insert his own SIM. Any calls he makes will be charged to his home GSM account. Also, the GSM system will be able to reach him at the ME unit he is currently using. It is expected that, in the future, rental cars will come equipped with GSM MEs and that many hotels will provide handheld MEs for use by their guests. When the user's ME is being serviced, the shop will give another for temporary use with one's own SIM. A sample SIM card is shown in Figure 2.5. The detailed functions and contents of the SIM card are as follows.

- Removable plastic card or the SIM module;
- Unique mobile subscriber ID through IMSI and ISDN numbers;
- PIN;
- Authentication key K_i and A3, A5, and A8 algorithms.

The SIM is a removable SC, the size of a credit card, and contains an integrated circuit chip with a microprocessor, *random access memory* (RAM), and *read-only memory* (ROM). It is inserted in the MS unit by the subscriber when he or she wants to use the MS to make or receive a call. As stated, a SIM also comes in a modular form that can be mounted in the subscriber's equipment.

The SIM contains all subscriber-specific data stored on the MS side. The SIM is basically a smart card, but it can assume the shape of an ISO-sized card or a small component, both having the same standardized electrical and functional interface with a ME. It contains few functions; they relate to security features (such as authenti-

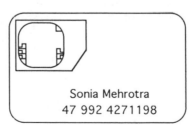

Sonia Mehrotra
47 992 4271198

Figure 2.5 SIM.

cation) and the storage of user specific data, like location information. The SIM memory contains a unique IMSI (subscription-related data) number that is necessary to identify the subscriber to the system when he attempts access to make or receive a call. It also contains a unique MSISDN, which is the unique mobile identification for PSTN/ISDN calls to the mobile. The SIM also contains various temporary subscriber related data that changes from time to time. The storage function of a SIM is very important, especially when it is removed from ME.

When a mobile subscriber wants to use the system, he or she mounts their SIM card and provide their *Personal Identification Number* (PIN), which is compared with a PIN stored within the SIM. If the user enters three incorrect PIN codes, the SIM is disabled. The PIN can also be permanently bypassed by the service provider if requested by the subscriber. Disabling the PIN code simplifies the call setup but reduces the protection of the user's account in the event of a stolen SIM.

User authentication to the network is provided by the MS/SIM sending a unique response to a random number challenge from the GSM system. The SIM contains an authentication parameter K_i uniquely assigned to the subscriber at the very beginning of the service that is also kept at the AUC and Algorithm A3 to calculate the response to the network challenge. Algorithm A8 generates the encryption key and Algorithm A5 the actual encryption. Further details of usage of these algorithms are provided in Section 5.4.

2.3.5 Mobile Identification Numbers

GSM uses a number of descriptors to identify subscribers, equipment, and fixed stations/areas. Many are temporary and used to maintain the confidentiality of fixed identities [2,3]. An understanding of these descriptors is essential when considering GSM exploitation. There are basically three numbers that identify the mobile subscriber, namely, the IMSI, MS-ISDN, and the TMSI, which is a temporary identification number that is assigned by the serving MSC/VLR combination. The TMSI is mainly used for security reasons to avoid broadcasting the IMSI over the RF air interface, thereby making it harder for eavesdroppers. The TMSI is supposed to be changed on a per-call basis as recommended by GSM specific actions. The following subsections describe the structure of these identifications.

2.3.5.1 International Mobile Subscriber Identity

An IMSI is assigned to each authorized GSM user. It consists of a *mobile country code* (MCC), *mobile network code* (MNC), and a PLMN unique *mobile subscriber identification number* (MSIN). The IMSI is not hardware-specific. Instead, it is maintained on a SC by an authorized subscriber and is the only absolute identity that a subscriber has within the GSM system. The IMSI consists of the MCC followed by the NMSI and shall not exceed 15 digits. The structural composition of IMSI along with the description of different fields is shown in Figure 2.6(a).

2.3.5.2 Temporary Mobile Subscriber Identity

A TMSI is a MSC-VLR specific alias that is designed to maintain user confidentiality. It is assigned only after successful subscriber authentication. The correlation of a TMSI to an IMSI only occurs during a mobile subscriber's initial transaction with an MSC (for example, location updating). Under certain conditions (such as traffic system disruption and malfunctioning of the system), the MSC can direct individual TMSIs to provide the MSC with their IMSI. In all cases, a subscriber will always respond to IMSI even if a TMSI has been assigned. The VLR must be capable of correlating the IMSI of a MS with its current TMSI. Since the TMSI has only local significance (that is, within the VLR and the area controlled by the VLR), the structure of this can be chosen by each administration in order to meet local needs.

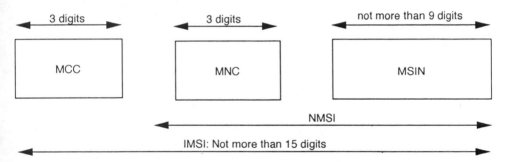

MCC - Mobile Country Code: The part of the mobile station identification uniquely identifying the country of domicile of the mobile subscriber. It consists of 3 digits. The allocation of MCCs is administered by the CCITT and is given in ANNEX A to CCITT Blue Book Recommendation E.212.

MNC - Mobile Network Code: It uniquely identifies the home GSM PLMN of the mobile subscriber. It consists of 3 digits.

MSIN - Mobile Station Identification Number: It uniquely identifies the mobile subscriber within a GSM PLMN. The first 3 digits identify the logical HLR-id of the mobile subscriber.

NMSI - National Mobile Station Identity: The NMSI consists of the MNC followed by the MSIN and is to be assigned by individual administration.

(a)

Figure 2.6 Coding for (a) IMSI, (b) MSISDN, and (c) MSRN.

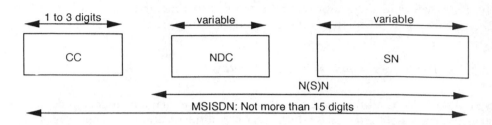

CC - Country Code: The country in which the VLR mobile station is registered.

NDC - National Destination Code: for GSM applications a NDC is allocated to each GSM PLMN. In some countries more than one NDC may be required for each GSM PLMN. The CC and the NDC fields must provide the routing information to reach the HLR of the mobile station. If further routing information is required, it should be contained in the first few digits of the SN.

(b)

SN - Subscriber Number

N(S)N - National (Significant) Number: consists of NDC and SN.

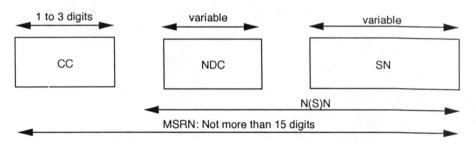

CC - Country Code: the country of the country in which the VLR is located.

NDC - National Destination Code: the NDC of the visitor GSM PLMN or numbering area.

(c)

SN - Subscriber Number: it has to have an appropriate structure for the visited area numbering plan.

N(S)N - National (Significant) Number: consists of NDC and SN

Figure 2.6 (continued).

In order to avoid a double allocation of TMSIs after a restart of a VLR, some part of the TMSI may be related to the time.

2.3.5.3 Mobile Station ISDN Number

The MS international number must be dialed after the international prefix in order to obtain a mobile subscriber in another country. The MSISDN number is composed of the *country code* (CC) followed by the *National Significant Number* (N(S)N), which shall not exceed 15 digits. Figure 2.6(b) shows the composition structure along with the explanation of various fields of the MSISDN number.

2.3.5.4 The Mobile Station Roaming Number (MSRN)

The MSRN is allocated on a temporary basis when the MS roams into another numbering area. The MSRN number is used by the HLR for rerouting calls to the MS. It is assigned upon demand by the HLR on a per-call basis. The MSRN for PSTN/ISDN routing shall have the same structure as international ISDN numbers in the area in which the MSRN is allocated. The HLR knows in what MSC/VLR service area the subscriber is located. At the reception of the MSRN, HLR sends it to the GMSC, which can now route the call to the MSC/VLR exchange where the called subscriber is currently registered. Figure 2.6(c) shows the composition structure of the MSRN and gives an explanation of various associated fields.

2.3.5.5 Intentional Mobile Equipment Identity

The IMEI is the unique identity of the equipment used by a subscriber by each PLMN and is used to determine authorized (white), unauthorized (black), and malfunctioning (gray) GSM hardware. In conjunction with the IMSI, it is used to ensure that only authorized users are granted access to the system. An IMEI is never sent in cipher mode by a MS.

2.4 BASE STATION SYSTEM

The BSS is a set of BS equipment (such as transceivers and controllers) that is in view by the MSC through a single A interface as being the entity responsible for communicating with MSs in a certain area. The radio equipment of a BSS may be composed of one or more cells [4,7–9]. A BSS may consist of one or more BSs. The interface between BSC and BTS is designed as an A-bis interface. The BSS includes two types of machines: the BTS in contact with the MSs through the radio interface and the BSC, the latter being in contact with the MSC. The function split is basically between a transmission equipment, the BTS, and a managing equipment at the BSC. A BTS comprises radio transmission and reception devices, up to and including the antennas, and also all the signal processing specific to the radio interface. A single transceiver within BTS supports eight basic radio channels of the same TDMA frame. A BSC is a network component in the PLMN that functions for control of one or more BTS. It is a functional entity that handles common control functions within a BTS.

A BTS is a network component that serves one cell and is controlled by a BSC. BTS is typically able to handle three to five radio carriers, carrying between 24 and 40 simultaneous communications. Reducing the BTS volume is important to keeping down the cost of the cell sites.

An important component of the BSS that is considered in the GSM architecture as a part of the BTS is the *Transcoder/Rate Adapter Unit* (TRAU). The TRAU is the equipment in which coding and decoding is carried out as well as the rate adaptation in case of data. Although the specifications consider the TRAU as a subpart of the BTS, it can be sited away from the BTS (at MSC), and even between the BSC and the MSC.

The interface between the MSC and the BSS is a standardized SS7 interface (A-interface) that, as stated before, is fully defined in the GSM recommendations. This allows the system operator to purchase switching equipment from one supplier and radio equipment and the controller from another. The interface between the BSC and a remote BTS likewise is a standard interface termed the A-bis. In splitting the BSS functions between BTS and BSC, the main principle was that only such functions that had to reside close to the radio transmitters/receivers should be placed In BTS. This will also help reduce the complexity of the BTS.

Based on the aforementioned argument, some of the functions that were placed in the BTS areas are as follows.

- *Broadcast Control and Common Control Channels* (BCCH/CCCH) Message Scheduling has to be made by BTS as it has the exact knowledge of BCCH/CCCH timing (not known by BSC). This includes the scheduling of paging messages on paging subchannels. Details of these signaling channels are provided in Section 3.2.
- Random access detection has to be made by BTS, which in turn sends a message to BSC. Subsequent channel assignment is made by BSC.
- Timing advance is determined by BTS. It is also signaled to MS by the BTS except after random access when it is reported to BSC for inclusion in the assignment message.
- Uplink radio channel measurements has to be made by BTS.
- The BTS has to detect the handover access burst sent by MS.
- Error protection channel coding/decoding and encryption of the radio channel has to be done in BTS. Details of coding are addressed in Chapter 6.

- Layer 2 of the radio interface (LAPDm) has to be terminated in BTS. Details of $LAPD_m$ protocol are covered in Chapter 8.

Based on this discussion, we shall first provide functions of BTS; BTS-BSC configurations; the use of a transcoder, which for all practical purposes can be considered to be the part of BTS; and lastly, functions of BSC.

2.4.1 Functions of BTS

As stated, the primary responsibility of the BTS is to transmit and receive radio signals from a mobile unit over an air interface. To perform this function completely, the signals are encoded, encrypted, multiplexed, modulated, and then fed to the antenna system at the cell site. Transcoding to bring 13-Kbps speech to a standard data rate of 16 Kbps and then combining four of these signals to 64 Kbps is essentially a part of BTS, though, it can be done at BSC or at MSC [7,8]. The voice communication can be either at a full or half rate over logical speech channel. In order to keep the mobile synchronized, BTS transmits frequency and time synchronization signals over *Frequency Correction Channel* (FCCH) and BCCH logical channels. The received signal from the mobile is decoded, decrypted, and equalized for channel impairments. Since the GSM signals are supposed to be frequency hopped, the control within the cell is actually exercised such that no two subscribers hop to the same frequency. Thus it is the responsibility of the BTS to make sure that hopping signals are kept orthogonal within the BSC serving area. A list of functions performed by BTS is as follows.

- Encodes, encrypts, multiplexes, modulates and feeds the RF signals to the antenna;
- Transcoding and rate adaptation;
- Time and frequency synchronization signals transmitted from BTS;
- Each BTS serves a single cell;
- Voice communication through full rate or half rate (future date) speech channel;
- Received signal from mobile is decoded, decrypted and equalized before demodulation;
- Frequency hopping controlled such that no two MSs in the same BSC area are hopped together;
- Random access detection;
- Timing advance;
- Uplink radio channel measurements.

Random access detection is made by BTS, which then sends the message to BSC. The channel subsequent assignment is made by BSC. Timing advance is determined by BTS. BTS signals the mobile for proper timing adjustment. Uplink radio channel measurements corresponding to the downlink measurements made by MS has to be made by BTS.

2.4.1.1 BTS-BSC Configurations

There are several BTS-BSC configurations: single site, single cell; single site, multicell; and multisite, multicell. These configurations are chosen based on the rural or urban applications. These configurations make the GSM system economical since the operation has options to adapt the best layout based on the traffic requirements. Thus, in some sense, system optimization is possible by the proper choice of the configurations. These include omnidirectional rural configurations where the BSC and BTS are on the same site; chain and multidrop loop configurations in which several BTSs are controlled by a single remote BSC with a chain or ring connection topology; rural star configurations in which several BTSs are connected by individual lines to the same BSC; and sectorized urban configurations in which three BTSs share the same site and are controlled by either a collocated or remote BSC.

In rural areas, most BSs are installed to provide maximum coverage rather than maximum capacity. High levels of traffic are not problems in those areas. The largest coverage area from the radio point-of-view (30 Km to 70 Km) can only provide a fraction of traffic for a single transceiver. Therefore, it was necessary to come up with a low-cost and low-complexity configuration for the rural areas. The MSC in this case would still be connected via the A-interface not to the BSC but to the BSS consisting of a single or multiple cells. If the cells are not colocated, the BSS will be split between BSC and BTS where BSC will then be connected to several BTSs. Between BSC and BTS is the A-bis interface. For high-traffic surroundings in urban areas, MSC can be connected to a number of BSSs via A-interfaces. Some of the BSSs are multicell (sectored) sites. Several groups of omnidirectional as well as sectorized BTSs may be tied into a common remote BSC via combinations of star, chain, and multidrop connections as shown in Figure 2.7(a). Figure 2.7(b) provides a typical MSC-BSS configuration.

2.4.1.2 Transcoder [1,5–8]

Depending on the relative costs of a transmission plant for a particular cellular operator, there may be some benefit, for larger cells and certain network topologies, in having the transcoders either at the BTS, BSC, or MSC location. If the transcoder is located at MSC, they are still considered functionally a part of the BSS. This approach allows for the maximum of flexibility and innovation in optimizing the transmission between MSC and BTS.

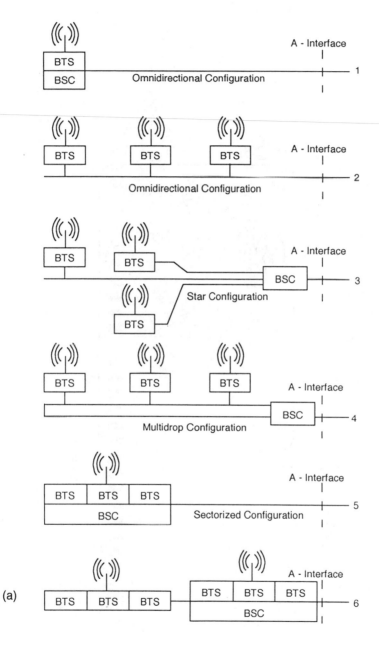

Figure 2.7 (a) Sectorized configuration with remote BSC and (b) MSC-BSS configuration.

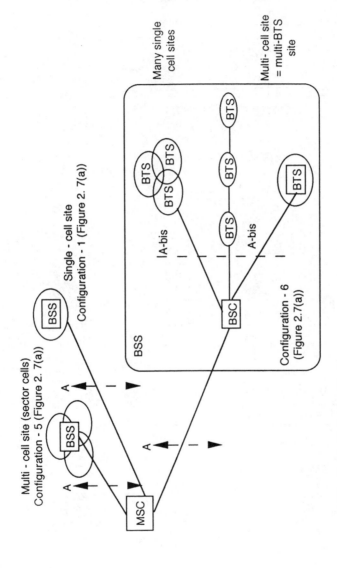

MSC: Mobile Switching Centre
BSS: Base Station System
BSC: Base Station Controller
BTS: Base Transceiver Station

Figure 2.7 (continued).

The transcoder is the device that takes 13-Kbps speech or 3.6/6/12-Kbps data multiplexes and four of them to convert into standard 64-Kbps data. First, the 13 Kbps or the data at 3.6/6/12 Kbps are brought up to the level of 16 Kbps by inserting additional synchronizing data to make up the difference between a 13-Kbps speech or lower rate data, and then four of them are combined in the transcoder to provide 64 Kbps. If the transcoder/rate adapter is placed outside the BTS (part of BSC or MSC), the A-bis interface can only operate on a 16-Kbps channel within the BSS. Four traffic channels can then be multiplexed on one 64-Kbps circuit. Thus, the TRAU output data rate is 64 Kbps. Then, up to 30 such 64-Kbps channels are multiplexed onto a 2.048 Mbps if a CEPT1 channel is provided on the A-bis interface. This channel can carry up to 120 (16 × 120) traffic and control signals. Since the data rate to the PSTN is normally at 2 Mbps, which is the result of combining 30-Kbps by 64-Kbps channels, or 120-Kbps by 16-Kbps channels. The various configurations where TRAU can be inserted in the BTS, BSC, and MSC chain are shown in Figure 2.8.

2.4.2 BSC

The BSC, as discussed, is connected to the MSC on one side and to the BTSs on the other. The BSC performs the *Radio Resource* (RR) management for the cells under

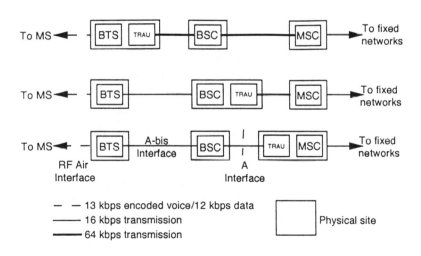

Figure 2.8 Transcoder (XC) configurations.

its control. It assigns and releases frequencies and timeslots for all MSs in its own area. The BSC performs the intercell handover for MSs moving between BTSs in its control. It also reallocates frequencies to the BTSs in its area to meet locally heavy demands during peak hours or on special events. The BSC controls the power transmission of both BSSs and MSs in its area. The minimum power level for a mobile unit is broadcast over the BCCH. The BSC provides the time and frequency synchronization reference signals broadcast by its BTSs. The BSC also measures the time delay of received MS signals relative to the BTS clock. If the received MS signal is not centered in its assigned timeslot at the BTS, the BSC can direct the BTS to notify the MS to advance the timing such that proper synchronization takes place [1,5,6,10]. The functions of BSC are as follows.

- RR management for BTSs under its control;
- Intercell handover;
- Reallocation of frequencies among BTSs;
- Power management of BTSs;
- Time and frequency synchronization signals to BTSs;
- Time delay measurement of the received signals from MSs with respect to BTS clock;
- Controls frequency hopping;
- Performs traffic concentration to reduce the number of lines from BSC to MSC and BTSs;
- Provides interface to the Operations and Management for BSS.

The BSC controls the frequency hopping of all the BTSs and MSs in its area. It establishes the hopping sequence for each BTS and directs the BTS to inform the MSs under its control of the assigned sequence.

The BSC may also perform traffic concentration to reduce the number of transmission lines from the BSC to its BTSs, as discussed in the last section.

2.5 SWITCHING SUBSYSTEMS: MOBILE SWITCHING CENTER AND GATEWAY SWITCHING CENTER

The network and the switching subsystem together include the main switching functions of GSM as well as the databases needed for subscriber data and mobility management (VLR). The main role of the MSC is to manage the communications between the GSM users and other telecommunications network users. The basic switching function is performed by the MSC, whose main function is to coordinate setting up calls to and from GSM users. The MSC has interfaces with the BSS on one side (through which MSC VLR is in contact with GSM users) and the external networks on the other (ISDN/PSTN/PSPDN) [6–9]. The main difference between a

MSC and an exchange in a fixed network is that the MSC has to take into account the impact of the allocation of RRs and the mobile nature of the subscribers and has to perform, in addition, at least, activities required for the location registration and handover.

The MSC provides the interface between the fixed and mobile networks. The MSC is the telephone switching office for mobile-originated or terminated traffic. Each MSC provides service to mobiles located within a certain geographic coverage area, and the network typically contains more than one MSC. The MSC controls the call setup and routing procedures in a manner similar to the functions of a land network end office. The MSC is the specific exchange for the GSM system. An MSC is generally connected to several BSSs, which provide radio coverage to the MSC area. The set of subscribers managed by the MSC are those located in its area (MSC area); and the connection to those subscribers is provided, when required, through one BSS. The MSC is also connected to other GSM PLMN entities such as other MSCs and HLR through a fixed network. Thus, each MSC provides service to mobiles within a certain geographic area, and the network typically has more than one MSC.

The MSC is a telephony switch that performs all the switching functions for MSs located in a geographical area designated as the MSC area. The MSC must also handle different types of numbers and identities related to the same MS and contained in different registers: IMSI, TMSI, ISDN number, and MSRN. In general, identities are used in the interfaces between the MSC and the MS, while numbers are used in the fixed part of the network, such as, for routing.

2.5.1 Functions of MSC

As stated, the main function of the MSC is to coordinate the set up of calls between GSM mobile and PSTN users. Specifically, it performs functions such as paging, resource allocation, location registration, and encryption. A list of relevant functions performed by MSC includes the following.

- Paging;
- Coordination of call set up from all MSs in its jurisdiction;
- Dynamic allocation of resources;
- Location registration;
- Interworking function with different networks;
- Handover management;
- Billing for all subscribers based in its area;
- Reallocation of frequencies to BTSs in its area to meet heavy demands;
- Encryption;
- *Echo canceler* (EC) operation control;

- Signaling exchange between different interfaces;
- Synchronization with BSSs;
- One MSC may interface several BSSs;
- Gateway to SMS.

Specifically, the call-handling function of paging is controlled by MSC. MSC coordinates the set up of calls to and from all GSM subscribers operating in its area. The dynamic allocation of access resources is done in coordination with the BSS. More specifically, the MSC decides when and which types of channels should be assigned to which MS. The channel identity and related radio parameters are the responsibility of the BSS. The MSC provides the control of interworking with different networks. It is transparent for the subscriber authentication procedure. The MSC supervises the connection transfer between different BSSs for MSs, with an active call, moving from one cell to another. This is ensured if the two BSSs are connected to the same MSC but also when they are not. In this latter case the procedure is more complex, since more than one MSC is involved. The MSC performs billing on calls for all subscribers based in its area. When the subscriber is roaming elsewhere, the MSC obtains data for the call billing from the visited MSC. Encryption parameters transfer from VLR to BSS to facilitate ciphering on the radio interface are done by MSC. The exchange of signaling information on the various interfaces toward the other network elements and the management of the interfaces themselves are all controlled by the MSC. Finally, the MSC serves as a SMS gateway to forward SMS messages from *Short Message Service Centers* (SMSC) to the subscribers and from the subscribers to the SMSCs. It thus acts as a message mailbox and delivery system.

2.5.2 VLR

The VLR collocated with an MSC is shown in Figure 2.1(a). A MS roaming in an MSC area is controlled by the VLR responsible for that area. When a MS appears in a LA, it starts a registration procedure. The MSC for that area notices this registration and transfers to the VLR the identity of the LA where the MS is situated. A VLR may be in charge of one or several MSC LAs. The VLR constitutes the database that supports the MSC in the storage and retrieval of the data of subscribers present in its area. When an MS enters the MSC area borders, it signals its arrival to the MSC that stores its identity in the VLR. The information necessary to manage the MS is contained in the HLR and is transferred to the VLR so that they can be easily retrieved if so required.

The location registration procedure allows the subscriber data to follow the movements of the MS. For such reasons the data contained in the VLR and in the HLR are more or less the same. Nevertheless, the data are present in the VLR only as long as the MS is registered in the area related to that VLR. The terms permanent and temporary, in this case, are meaningful only during that time interval when the mobile is in the area of local MSC/VLR combination. The data contained in the VLR can be compared with the subscriber-related data contained in a normal fixed exchange; the location information can be compared with the line equipment reference attached to each fixed subscriber connected to that exchange. The VLR is responsible for assigning a new TMSI number to the subscriber. It also relays the ciphering key from HLR to BSS.

Cells in the PLMN are grouped into geographic areas, and each is assigned a LAI, as shown in Figure 2.2(c). Each VLR controls a certain set of LAs. When a mobile subscriber roams from one LA to another, their current location is automatically updated in their VLR. If the old and new LAs are under the control of two different VLRs, the entry on the old VLR is deleted and an entry is created in the new VLR by copying the basic data from the HLR. The subscriber's current VLR address, stored at the HLR, is also updated. This provides the information necessary to complete calls to roaming mobiles. The VLR supports a mobile paging and tracking subsystem in the local area where the mobile is presently roaming. The detailed functions of VLR are as follows.

- Works with the HLR and AUC on authentication;
- Relays cipher key from HLR to BSS for encryption/decryption;
- Controls allocation of new TMSI numbers; a subscriber's TMSI number can be periodically changed to secure a subscriber's identity;
- Supports paging;
- Tracks state of all MSs in its area.

2.5.2.1 Data Stored in VLR

The VLR constitutes the database that supports the MSC in the storage and retrieval of the data of subscribers present in its area. When an MS enters the MSC area borders, it signals its arrival to the MSC that stores its identity in the VLR. The information necessary to manage the MS in whichever type of call it may attempt is contained in the HLR and is transferred to the VLR so that they can be easily retrieved if so required (location registration). This procedure allows the subscriber data to follow the movements of the MS [2,3,7,8]. For such reasons the data contained in the VLR and in the HLR are more or less the same. Nevertheless the data are present in the VLR only as long as the MS is registered in the area related to that VLR. Data associated with the movement of mobile are IMSI, MSISDN, MSRN, and TMSI. The terms permanent and temporary, in this case, are meaningful only during that time interval. Some data are mandatory, others are optional. Data stored in VLR are as follows.

- The IMSI;
- The MSISDN;
- The MSRN, which is allocated to the MS either when the station is registered in an MSC area or on a per-call basis and is used to route the incoming calls to that station;
- The TMSI;
- The LA where the MS has been registered, which will be used to call the station;
- Supplementary service parameters;
- MS category;
- Authentication key, query and response obtained from AUC;
- ID of the current MSC.

2.6 HOME LOCATION REGISTER

The HLR is a data base that permanently stores data related to a given set of subscribers. The HLR is the reference database for subscriber parameters. Various identification numbers and addresses as well as authentication parameters, services subscribed, and special routing information are stored. Current subscriber status, including a subscriber's temporary roaming number and associated VLR if the mobile is roaming, are maintained [2,7,8].

The HLR provides data needed to route calls to all MS-SIMs home based in its MSC area, even when they are roaming out of area or in other GSM networks. The HLR provides the current location data needed to support searching for and paging the MS-SIM for incoming calls, wherever the MS-SIM may be. The HLR is responsible for storage and provision of SIM authentication and encryption parameters needed by the MSC where the MS-SIM is operating. It obtains these parameters from the AUC.

The HLR maintains records of which supplementary services each user has subscribed to and provides permission control in granting access to these services. The HLR stores the identifications of SMS gateways that have messages for the subscriber under the SMS until they can be transmitted to the subscriber and receipt is acknowledged. The HLR provides receipt and forwarding to the billing center of charging information for its home subscribers, even when that information comes from other PLMNs while the home subscribers are roaming. Based on the above functions, different types of data are stored in HLR. Some data are permanent; that is, they are modified only for administrative reasons, while others are temporary and modified automatically by other network entities depending on the movements and actions performed by the subscriber.

Some data are mandatory, other data are optional. Both the HLR and the VLR can be implemented in the same equipment in an MSC (collocated). A PLMN may contain one or several HLRs. The permanent data stored in an HLR includes the following.

- IMSI: It identifies unambiguously the MS in the whole GSM system;
- International MS ISDN number: It is the directory number of the mobile station;
- MS category specifies whether a MS is a pay phone or not;
- Roaming restriction (allowed or not);
- *Closed user group* (CUG) membership data;
- Supplementary services related parameters: Forwarded-to number, registration status, no reply condition timer, call barring password, activation status, supplementary services check flag;
- Authentication key, which is used in the security procedure and especially to authenticate the declared identity of a MS.

The temporary data consists of the following.

- LMSI (Local MS identity);
- RAND/SRES and K_c; data related to authentication and ciphering;
- MSRN;
- VLR address, which identifies the VLR currently handling the MS;
- MSC address, which identifies the MSC area where the MS is registered;
- Roaming restriction;
- Messages waiting data (used for SMS);
- RAND/SRES and ciphering key, that is, data related to authentication and ciphering;
- MSRN.

The permanent data associated with the mobile are those that do not change as it moves from one area to another. On the other hand, temporary data changes from call to call. The HLR interacts with MSCs mainly for the procedures of interrogation for routing calls to a MS and to transfer charging information after call termination.

Location registration is performed by HLR. When the subscriber changes the VLR area, the HLR is informed about the address of the actual VLR. The HLR updates the new VLR with all relevant subscriber data. Similarly, location canceling is done by HLR. After the subscriber roams to a different VLR area, the HLR updates the new VLR with all the relevant subscriber data.

Supplementary services, as explained in Chapter 1, are add-ons to the basic service. These parameters need not all be stored in the HLR. However, it is safer to store all subscription parameters in the HLR even when some are stored in a subscriber card. The data stored in the HLR is changed only by MMI action when new subscribers are added, old subscribers are deleted, or the specific services to which they subscribe are changed and not dynamically updated by the system.

2.7 AUTHENTICATION CENTER

The AUC stores information that is necessary to protect communications through the air interface against intrusions, to which the mobile is vulnerable. The legitimacy of the subscriber is established through authentication and ciphering, which protects the user information against unwanted disclosure. Authentication information and ciphering keys are stored in a database within the AUC, which protects the user information against unwanted disclosure and access.

In the authentication procedure, the key K_i is never transmitted to the mobile over the air path, only a random number is sent. In order to gain access to the system, the mobile must provide the correct *Signed Response* (SRES) in answer to a *random number* (RAND) generated by AUC.

Also, K_i and the cipher key K_c are never transmitted across the air interface between the BTS and the MS. Only the random challenge and the calculated response are transmitted. Thus, the value of K_i and K_c are kept secure. The cipher key, on the other hand, is transmitted on the SS7 link between the home HLR/AUC and the visited MSC, which is a point of potential vulnerability. On the other hand, the random number and cipher key is supposed to change with each phone call, so finding them on one call will not benefit using them on the next call.

The HLR is also responsible for the "authentication" of the subscriber each time he makes or receives a call. The AUC, which actually performs this function, is a separate GSM entity that will often be physically included with the HLR. Being separate, it will use separate processing equipment for the AUC database functions.

2.8 EQUIPMENT IDENTITY REGISTER

EIR is a database that stores the IMEI numbers for all registered ME units. The IMEI uniquely identifies all registered ME. There is generally one EIR per PLMN. It interfaces to the various HLRs in the PLMN. The EIR keeps track of all ME units in the PLMN. It maintains various lists of messages. The database stores the ME identification and has nothing to do with the subscriber who is receiving or originating a call. There are three classes of ME that are stored in the database, and each group has different characteristics.

- *White List:* contains those IMEIs that are known to have been assigned to valid MSs. This is the category of genuine equipment.
- *Black List:* contains IMEIs of mobiles that have been reported stolen.
- *Gray List:* contains IMEIs of mobiles that have problems (for example, faulty software, wrong make of the equipment). This list contains all MEs with faults not important enough for barring.

2.9 INTERWORKING FUNCTION

GSM provides a wide range of data services to its subscribers. The GSM system interfaces with the various forms of public and private data networks currently available [4]. It is the job of the IWF to provide this interfacing capability. Networks to which IWF presently provides interface are as follows.

- PSTN;
- ISDN;
- *Circuit-switched public data networks* (CSPDN);
- *Packet-switched public data networks* (PSPDN).

The IWF, which in essence is a part of MSC, provides the subscriber with access to data rate and protocol conversion facilities so that data can be transmitted between GSM *Data Terminal Equipment* (DTE) and a land-line DTE. It will also allocate a modem from its "modem bank" when required. This will be the case when the GSM DE exchanges data with a land DE connected via an analog modem, through a narrowband PSTN circuit. The IWF also provides direct-connect interfaces for customer-provided equipment such as X.25 Pads. In order to achieve this interconnection, the IWF has to convert protocols between the GSM PLMN and the connecting network. Different protocol conversions may be required for signaling and traffic messages. This may include data rate adaptation and the addition of signaling bits reformatting. Additionally, as previously stated, this may also require an audio modem for transmission through a narrowband telephone link. Figure 2.9 provides the setup for IWF and its interface configuration with PSTN and ISDN users. A GSM interface with modems of different rates for PSTN connection is shown in Table 2.3.

2.10 ECHO CANCELER

As shown in Figure 2.10 the EC is used on the PSTN side of the MSC for all voice circuits. The EC is required at the MSC PSTN interface to reduce the effect of GSM delay when the mobile is connected to the PSTN circuit. The total round-trip delay introduced by the GSM system, which is the result of speech encoding, decoding, and signal processing, is of the order of 180 ms. Normally this delay would not be an annoying factor to the mobile, except when communicating to PSTN as it requires a two-wire to four-wire hybrid transformer in the circuit. This hybrid is required at the local switching office because the standard local loop is a two-wire circuit. Due to the presence of this hybrid, some of the energy at its four-wire receive side from the mobile is coupled to the four-wire transmit side and thus retransmitted to the mobile. This causes the echo, which does not affect the land subscriber but is an annoying factor to the mobile. The standard EC cancels about 70 ms of delay.

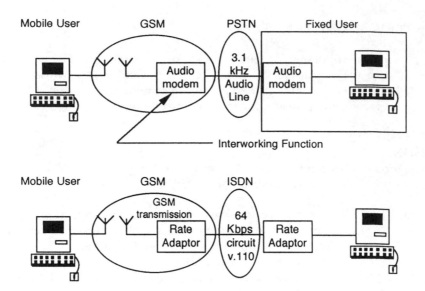

Figure 2.9 GSM to PSTN/ISDN interface.

Table 2.3
Modem Types for a PSTN Interface

Modem Type	Data Rate (bps)	Transmission Mode
V.21	300	Asynchronous
V.22	1200	Asynchronous/Synchronous
V.22bis	2400	Synchronous
V.23	1200/75	Asynchronous
V.26ter	2400	Synchronous
V.32	4800/9600	Synchronous

During a normal PSTN (land-to-land call), no echo is apparent because the delay is too short and the land user is unable to distinguish between the echo and the normal telephone "side tones." However, with the GSM round-trip delay added and without the EC, the effect would be irritating to the MS subscriber.

2.11 OPERATIONS AND MAINTENANCE CENTER

The objective of the OMC is to offer the customer cost-effective support for the centralized regional and local operational and maintenance activities required for a cellular network. The main purpose of the OMC is to perform all operations and

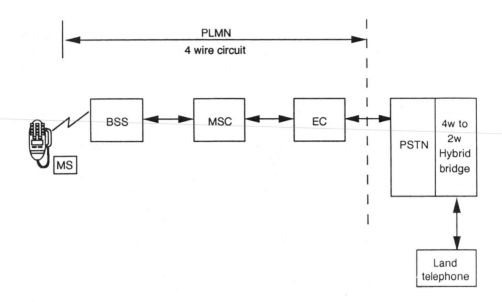

Figure 2.10 Use of EC on PSTN MSC interface.

maintenance functions on elements of the GSM PLMN system. In that sense it provides a central network overview and supports the maintenance functions of different O&M organizations. The OMC uses a separate *Telecommunications Management Network* (TMN) to communicate with the various components of the GSM system. In general, it is done through leased lines on the PSTN or other fixed networks. The OMC message and data transfers can either be carried by SS7 or X.25 protocols. SS7 protocols are mostly used within the GSM PLMN for short and medium length control message transactions through the PSTN or ISDN circuits. On the other hand, X.25 may be used for large external data transfers such as software downloading or subscriber database transfers from subscription centers.

In summary, the following network functions are performed.

- Supports for maintenance;
- X.25 interface;
- Alarm handling;
- Fault management;
- Performance management;
- Software version and configuration control;
- Network status;
- Traffic collection from network.

The OMC provides alarm handling functions to report and log alarms generated by the other network entities. The maintenance personnel at the OMC can define the criticality of the alarm. Maintenance functions cover both technical and administrative actions to maintain and correct the system operation, or to restore normal operations after a breakdown, in the shortest possible time.

The fault management functions of the OMC allow network devices to be manually or automatically removed from or restored to service. The status of network devices can be checked, and tests and diagnostics on various devices can be invoked. For example, diagnostics may be initiated remotely by the OMC. A mobile call trace facility can also be invoked. The performance management functions include collecting traffic statistics from the GSM network entities and archiving them in disk files or displaying them for analysis. Because a potential to collect large amounts of data exists, maintenance personnel can select which of the detailed statistics to be collected based on personal interests and past experience. As a result of performance analysis, if necessary, an alarm can be set remotely.

The OMC provides system change control for the software revisions and configuration data bases in the network entities. Software loads can be downloaded from the OMC to other network entities or uploaded to the OMC. The OMC also keeps track of the different software versions running on different subsystems of the GSM. Figure 2.1(b) shows the OMC interface with other GSM subsystems along with the list of its functions.

2.12 THE NETWORK MANAGEMENT CENTER

The Network Management Center provides global and centralized management for operations and maintenance of the networks supported by OMCs that are responsible for regional network management. It provides both administrative and commercial management, security management of the facilities, system change control, and physical maintenance [2,4,8]. The NMC is generally connected to the PLMN subsystems through leased lines via PSTN. The salient characteristics and features of the NMC are as follows.

- Single NMC per network;
- Provides traffic management for the whole network;
- Monitors high-level alarms such as failed or overloaded nodes;
- Performs responsibilities of an OMC when it is not staffed;
- Provides network planners with essential data for network performance.

As stated in the last section, the OMC is a regionalized management center while NMC is the global management center. OMC is used for monitoring and controlling the daily activities of the system operations, while NMC is for the long-term planning. OMC is used by network operators, while the NMC is used by network managers and planners.

2.13 BILLING CENTER

The billing center is a system provided by the PLMN administration that collects the billing data from the GSM network entities and applies that data to subscribers' accounts. The details of the billing center's operation are not addressed by GSM specifications. Billing data is made of two parts: call records produced by MSCs and events records produced by HLRs and VLRs. These event records are the results of the number of times these databases are accessed due to mobility of the user, such as location updates and the forwarding of the mobile-terminated calls. Whereas GSM recommends that each MS involved in a call produce a separate call record, the initial MSC (anchor MSC) involved with a call is assigned the role of collecting call record data from any other MSCs that were involved in the call. When the call disconnects, the initial MSC assembles the call data into a single, composite call record that is subsequently sent to the billing center. In this way, calls that enter the PLMN through a gateway MSC can be ticketed with only one call record. Mobile-originated calls switched from one MSC to another, within the PLMN, before entering the PSTN, are also ticketed with a single call record. Also, inter-MSC handover data can be included by the single record format.

The records are stored in a disk file as they are being generated. The file has a fixed size; and when the file becomes full, it is automatically closed and a new one opened. The complete file can be automatically transferred to the billing center using X.25 communication links. We shall discuss details of the billing center in Chapter 7.

2.14 NETWORKING

Signaling is an essential part of telephony. Signaling between switches is essential to seizing and releasing trunk circuits. For most of the network communications in GSM, internationally recognized standards SS7 and X.25 have been adopted. In general, the open systems interface recommendations of the seven-layer ISO protocol layers have been accepted. The GSM recommendations include detailed specifications for the radio channel ("air") interface between the MS and the BS. By standardizing the air interface between the mobile unit and the base transceiver, universal compatibility for the mobile is preserved. This specification owes heavily to analog cellular standards and X.25 concepts. However, much of the air interface is unique to GSM, which has pioneered the digital cellular system. The use of these standards throughout the mobile network provides compatibility between network elements from different manufactures [2,3,9]. The GSM procedures for completing calls to roaming mobiles and for inter-MSC handovers are also specified to allow for greater flexibility. These procedures require trunk circuits to be set up between the two MSCs. GSM specifies these procedures in such a way that the existing PSTN or ISDN networks can be used to provide the connection between two MSCs. This ability, along with the use of standard interfaces, allows these procedures to take place even when the MSCs are supplied by two different manufacturers.

The use of X.25 networks between OMC and the other subsystem is shown in Figure 2.11(b). A X.25 network communicates between OMC, VLR, HLR, AUC, NMC, MSC, and BSS. One example of the X.25 network, as shown in Figure 2.11(a) below, is built using two basic building blocks: signaling points and packet switches.

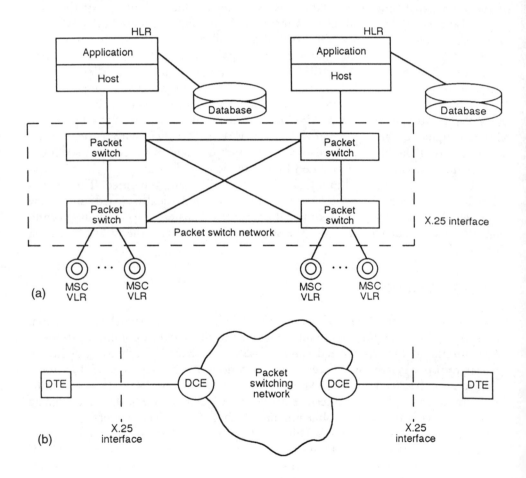

Figure 2.11 (a) X.25 network between OMC and HLR; (b) general X.25 interface between DTE and DCE; (c) a typical SS7 network interconnecting MSC and HLR [9].

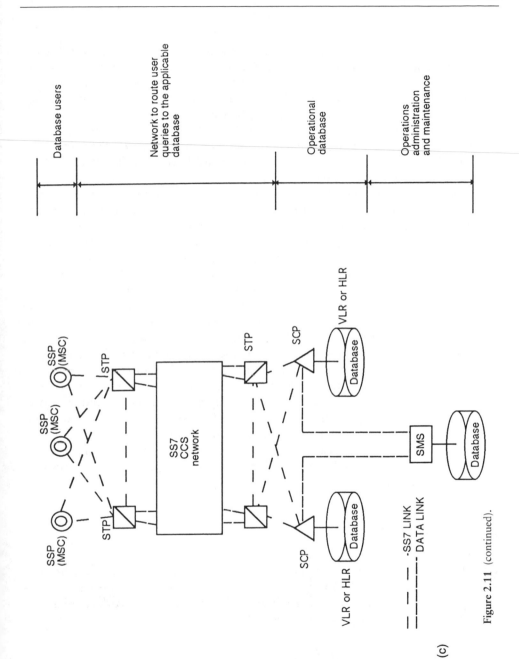

Figure 2.11 (continued).

The signaling points, in this case, are the host computers and the OMC. Host computers contain the database of the HLR and the OMC interface with the HLR. The general X.25 network interface is between the DTE and the DCE, as shown in Figure 2.11(b). Obviously, on one side of the network the host HLR plays the role of DTE, and on the other side OMC plays the role of DTE.

An example of interconnection between MSC and HLR is shown in Figure 2.11(c). The main components of the network are the *Service Signaling Point* (SSP), the *Signaling Transfer Point* (STP), the *Service Control Point* (SCP), and the *Service Management System* (SMSm).

Two types of messages flow over an SS7 network: circuit related and database related. Circuit-related messages are used to set up and tear down a call between two switches or signaling points; STP transfers these messages from the originating signaling point to the terminating signaling point. The information in these messages includes the identity of the circuit between two signaling points, the called number, an answer indication, and a circuit release indication. Data base access messages are used to access information stored in the SCP. A query message is sent from the SSP, via the STP, to the SCP requesting that required data be sent to the requesting SSP. The SCP returns the data in a response message over a SS7 network back to the originating signaling point. An SSP is a telephone switching office that is able to launch SS7 queries and react to their responses. The STP is a high-speed packet switch with powerful routing facilities, including global title translation for routing packets to the proper destination. Its routing is based on link availability, node availability, and the supplied address. The SCP contains applications and databases that provide the basic intelligence for the network services. The database contains subscriber records in two different functional entities: the HLR and the VLR. The HLR is the centralized database and may be duplicated within one PLMN for high reliability. The VLR is included with each MSC. Also, both of these databases may be included in the same MSC.

2.15 CONCLUSIONS

With the standardization of A- and A-bis interfaces, the road toward a flexible, efficient, and cost-effective BS has been created. As a result, it has formed a competitive equipment market and a future evolution of the individual network components. Thus, the following points can be made.

- Mobiles can now take their equipment anywhere in Europe and get the GSM service in the same way as if they were in their home PLMN.
- Optimum choice of BSC and BTS can now be adopted for both rural and urban areas.

- Equipment manufacturers can independently develop MSC, BSC, and BTS subsystems.
- Equipment developed by different manufacturers can now be easily interfaced.

Problems

2.1 Why is VLR regarded as a distributed HLR? What particulars about mobile will you like to keep in the VLR data base?

2.2 State the main functions of: (1) BTS and BSC, (2) MSC/GMSC, (3) OMC, and (4) NMC.

2.3 Narrate the distinction between the following service areas: GSM, PLMN, MSC, LA, and cells.

2.4 Justify that cell boundaries are lines of equal power from two adjacent BSs. Will the cell size change as the transmit power changes?

2.5 Draw tables of frequency assignments and system layouts for three- and two-cell repeat patterns.

2.6 Define the following terms: IMEI, IMSI, TMSI, and SIM.

2.7 Provide reasons behind the separation of transmission equipment and subscriber in a GSM system. Can you think of some reasons against this separation?

2.8 Add at least three more functions of MS to the list in Section 2.3.1.

2.9 What do you think are the reasons for stating both the normal and extreme tolerances for the mobile power class in a GSM system?

2.10 Provide distinctions between Type 0, 1, and 2 MSs.

2.11 What are different elements of information you would like to include in the SIM card? Do you see any disadvantage of a SIM card?

2.12 State two different types of SIM cards in use.

2.13 Justify why IMSI is standardized while TMSI is not. Why is TMSI only important in the local area?

2.14 Why do mobile units require two IDs: MSISDN and IMSI?

2.15 Why is a mobile roaming number important? Why do you need both TMSI and MSRN?

2.16 Explain White, Black, and Gray identities with respect to IMEI.

2.17 Estimate the cell capacity in terms of the number of channels, assuming five radio carriers and full-rate users. How will the capacity change using half-rate coders?

2.18 Clearly describe the functions of transcoders and IWF. What is meant by rate adaptation?

2.19 Suggest some configurations of a BTS-BSC combination with reasoning for rural, urban, and suburban areas of the United States. State some advantages and disadvantages of colocating BTS with BSC.

2.20 State some salient functions of MSC.

2.21 State at least three additional functions of BSC not covered in the list in Section 2.4.2.

2.22 State at least three additional functions of MSC not covered in the list in Section 2.5.1.

2.23 State at least three additional functions of VLR not covered in the list in Section 2.5.2.

2.24 State the main purpose of a SRES and random number generation at the AUC.

2.25 Clearly state reasons for providing EC at MSC.

2.26 State the main functions of NMC and its relationship with respect to OMC.

2.27 Why is the billing data provided in two parts: (1) event record and (2) call record?

References

[1] Haug, T., "Developing GSM Standards," *Pan-European Digital Cellular Radio Conf.*, Nice, France, 1991.

[2] Mehrotra, A., *GSM Seminars at Bell South*, Atlanta, March 27–28 and April 24–25, 1995.

[3] Mehrotra, A., *GSM Seminar at Lockheed Martin*, Philadelphia, PA, October 30 and 31, 1995.

[4] Mouly, M., and Pautet Marie-Bernadette, "Current Evolution of the GSM Systems," *IEEE Personal Communications*, October 1995, pp. 9–19.

[5] Lycksell, E., "The A and A-bis Interfaces of the GSM System," *GSM Seminar*, Budapest, October 1990, Session 3.3.

[6] Eynard Carlo, "The Switching Subsystem—Functions and Elements," *GSM Seminar*, Budapest, October 1990, Session 4.1.

[7] Maloberti, A., et al., "Radio Subsystem and Elements," *GSM Seminar*, Budapest, October 1990, Session 3.1.

[8] Beddoes, E. W., "GSM Network Architecture," *GSM Seminar*, Budapest, October 1990, Session 2.1.

[9] Mehrotra, A., *Cellular Radio Analog and Digital Systems*, Norwood, MA: Artech House, 1994.

[10] Siegmund, M. R., et al., *An Introduction to GSM*, Norwood, MA: Artech House, 1995.

CHAPTER 3
▼▼▼

TIME AND FREQUENCY AXIS REPRESENTATION

3.1 INTRODUCTION

GSM was first devised as a cellular system specific to the 900-MHz band, called "the primary band." This primary band includes two subbands of 25 MHz each, 890 MHz to 915 MHz and 935 MHz to 960 MHz. However, the whole primary band does not have to be used for GSM in a given country, especially at the start of the system when the traffic may not justify the allocation of the whole spectrum. In general, since most countries have several operators, an operator is rarely given more than a portion of this band. On the other hand, every MS must be able to use the full band so as not to impose constraints upon roaming users. Thus, the GSM PLMN has allocated 124 duplex carrier frequencies over the following bands of operation.

- Uplink: 890 MHz to 915 MHz (MS transmit, BTS receive);
- Downlink: 935 MHz to 960 MHz (BTS transmit, MS receive);
- Carrier spacing 200 kHz.

The time and frequency domain layout of a GSM system is shown in Figure 3.1. Since there are a total of eight channels per carrier, the total number of channels in

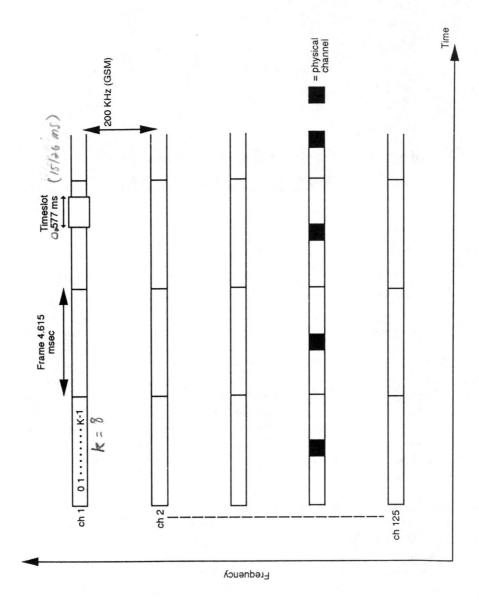

Figure 3.1 Time and frequency domain access for cellular radio [1].

a GSM system are, therefore, 125 × 8 = 1,000 channels. In actuality, only 124 carriers are used. There is a total of eight bursts during a frame cycle, which is defined as the information format during one timeslot on the TDMA channel; that is, with regular time intervals (every eighth timeslot on a TDMA channel, the user transmits or receives his information), we send a burst of information (either speech or data). Thus, a burst is considered as a period of RF carrier that is modulated by a data stream. A burst therefore represents the physical content of a timeslot. This burst period (15/26 ms) is shown in Figure 3.2.

In 1990, at the request of the United Kingdom, a second frequency band was specified. This band includes two frequency ranges: 1710 MHz to 1785 MHz and 1805 MHz to 1880 MHz; that is, twice 75 MHz and three times as much as the primary 900 MHz. MSs using this band will have different constraints from those using the primary band. Using the same ME for roaming between the two variants of the system, GSM 900 and DCS 1800, although not ruled out, is difficult to envision. Another extension of the primary band is also foreseen. It will consist of a band that is directly "below" the primary band. For instance, an 8-MHz extension would raise the 900-MHz bands from 882 MHz to 915 MHz and 927 MHz to 960 MHz, that is, twice 33 MHz.

The central frequencies of the channel slots are spread evenly at every 200-kHz interval within these bands, starting 200 kHz away from the band borders. Therefore, 124 different frequency slots are defined in 25 MHz, and 374 in 75 MHz. The modulation spectrum is somewhat wider than 200 kHz, resulting in some level of interference on adjacent frequency slots. This is a nuisance at least near the band borders, since this interference could disturb non-GSM applications in adjacent bands. The border frequencies are usually avoided. The normal practice is not to use the frequency slots at the border (those numbered 0 to 4) except when a special agreement has been reached with the users of the adjacent band. As a consequence,

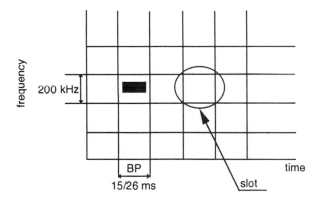

Figure 3.2 A slot in the time and frequency domain.

the number of frequency slots that can be used in 25 MHz can be limited to 122, though in most systems 124 channels have been used.

In view of this discussion, we shall examine the following concepts in this chapter. Section 3.2 provides the time domain representation of the GSM system, including details of logical channels, the physical layer description of different types of bursts, and how the speech and control channels are organized within the system. In Section 3.3 we shall discuss the frequency domain description of the system followed by some operational concepts in Section 3.4.

3.2   TIME DOMAIN REPRESENTATION

As mentioned in Section 3.1, the GSM system has a total available bandwidth of 25 MHz divided among 124 channels of 200 kHz each (as shown in Figure 3.1). It should be noted that there are a total of 125 channels, out of which the lowest channel is not used as it may interfere with other nearby non-GSM systems. Each of these 124 channels operates at different carrier frequencies. An individual channel is time division multiple accessed by users at different locations within a cell site. The frame duration is 4.615 ms divided among eight timeslots. Each of these timeslots is a physical channel occupied by an individual user. Each timeslot, or physical channel, carries control and traffic data in a burst form. The time duration of an individual channel is 3/5200 sec or 0.577 ms. At the BTS, TDMA frames on all radio frequency channels, in the downlink as well as on the uplink, are aligned. However, the start of an uplink TDMA frame is delayed with respect to downlink by a fixed period of three timeslots. Staggering TDMA frames allows the same *timeslot number* (TN) to be used in both the down and uplinks while avoiding the requirement for mobile to transmit and receive simultaneously. The TN within a frame is numbered from 0 to 7, and each TN can be referenced by a unique TN. Higher order frames, called multiframe, consist of 26 frames and have a duration of 120 ms (26 × 4.615 ms). This multiframe consists of 26 TDMA frames and carries a *traffic channel* (TCH), *slow associated control channel* (SACCH), and *fast associated control channel* (FACCH). Similarly, a 51-frame multiframe has a duration of 235.365 ms (51 × 4.615 ms). One superframe consists of 51 traffic multiframes or 26 control multiframes and consists of 51 × 26 TDMA frames with a total duration of 6.12 sec (51 × 120 ms). A 26 TDMA frame multiframe supports traffic and associated control channels, and a 51 TDMA frame multiframe supports *Broadcast Common Control* (BCC) and Stand Alone Dedicated Control Channels. The highest order frame is called a hyperframe and consists of 2,048 superframes, or 2,715,648 frames (2048 × 51 × 26). The frame number is cyclic and has a range from 0 to FN_MAX, where FN_MAX = (2048 × 51 × 26) − 1 = 2,715,647. The frame number is incremented at the end of each frame. The time duration of the hyperframe is 3 hrs, 28 min, and 52.76 sec (2,715,648 × 4.615 ms). This long period of a hyperframe is required to support encryption with high security. To organize the information transmitted on each carrier, GSM defines several time intervals ranging from 0.9 μs (quarter of one bit in the guard interval of the basic frame to be discussed in Section 3.2.4) to a hyperframe interval of over three hours. Details of frame, multiframe, superframe, and hyperframe are provided in Figure 3.3. As seen from the figure, the cycle of a multiframe and superframe is different for speech and control channels. It is not by accident that the control channel multiframe is not a direct multiple of the TCH multiframe. It can be seen that any frame number will only occur simultaneously in both multiframes every 1,326 frames (26 $\times$ 51). This arrangement means that the timing of the TCH multiframe is always moving in relation to that of the control channel multiframe, and this enables a receiver to receive and decode all the control channels along with TCH. If two multiframes were exact multiples of each other, then the control channel timeslot would be permanently "masked" by the TCH timeslot activity. This changing relationship is particularly important, for example, to a mobile unit that needs to be able to monitor and report RSSIs of up to six target cells (it needs to be able to see all the BCCHs of surrounding cells in order to do this).

3.2.1 Logical Channels

A great variety of information must be transmitted between the BTS and the MS, specifically, user data and control signaling. Depending on the kind of information transmitted, we refer to different logical channels. These logical channels are mapped onto the physical channels (slots). Digital speech is sent on a logical channel, named TCH, which during the transmission can be allocated a certain physical channel, say channel *Timeslot 7* (TS 7). In a GSM system no RF carrier and no slot is dedicated a priori to the exclusive use of anything. In other words, just about any slot of any RF carrier can be used for a number of different uses. Channelization is accomplished by the data communications notion of virtual circuits or logical channels. Logical channels are divided into the two categories, TCHs and control channels. We shall now provide details of these logical channels.

3.2.2 Logical TCHs

TCHs are intended to carry either encoded speech or user data both in the up and downlink directions in a point-to-point communication. There are two types of TCHs that are differentiated by their traffic rates and are defined as follows.

A *full rate TCH* (TCH/F), B_m , carries information (encoded speech or data) at a gross rate of 22.8 Kbps. The raw data rate for each TCH is 13 Kbps for speech and 12 Kbps, 6 Kbps, and 3.6 Kbps for data. The actual user data rates are

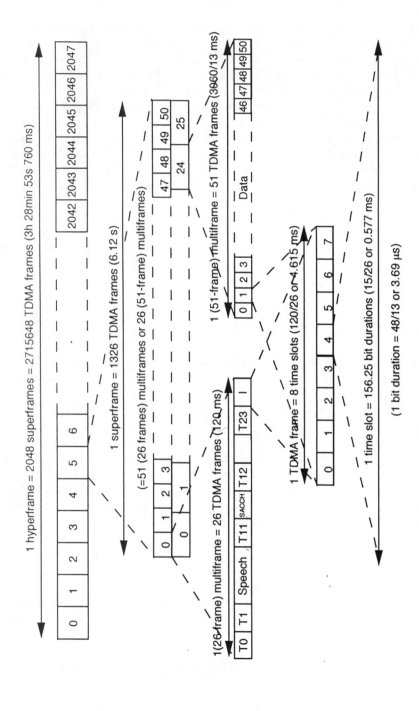

Figure 3.3   Frame, multiframe, superframe, and hyperframe for a GSM network.

9.6 Kbps, 4.8 Kbps, and 2.4 Kbps, which are padded to bring up the rates to 12 Kbps, 6 Kbps, and 3.6 Kbps.

A half rate TCH (TCH/H), L_m , carries information (encoded speech or data) at half of the full rate channel or at the gross rate of 11.4 Kbps. The user data rate associated with the half rate TCH are as follows. 4.8 Kbps and 2.4 Kbps. The allowed combinations of user data rate with full and half rate speech are as follows.

- Full rate speech (TCH/F);
- Half rate speech (TCH/H);
- 9.6 Kbps full rate data (TCH/F9.6);
- 4.8 Kbps full rate data (TCH/F4.8);
- 2.4 Kbps full rate data (TCH/F2.4);
- 4.8 Kbps half rate data (TCH/H4.8);
- 2.4 Kbps half rate data (TCH/H2.4).

3.2.3 Logical Control Channels

Most of the user services offered by GSM rely on the four transmission modes (single speech and three data modes). Control channels are intended to carry signaling or synchronization data. Three kinds have been defined below.

- *Broadcast Control Channel* (BCCH);
- *Common Control Channel* (CCCH);
- *Dedicated Control Channel* (DCCH).

Figure 3.4 shows details of different logical channels. The BCCH is a point-to-multipoint unidirectional control channel from the fixed subsystem to the MS that is intended to broadcast a variety of information to MSs, including information necessary for the MS to register in the system. The BCCH includes a *Frequency Correction Channel* (FCCH), which is used to allow an MS to accurately tune to a BS, and the *Synchronization Channel* (SCH), which is used to provide TDMA frame-oriented synchronization data to a MS. When mobile recovers both FCCH and SCH signals, we consider the synchronization to be complete.

A CCCH is a point-to-multipoint (bidirectional control channel) channel that is primarily intended to carry signaling information necessary for access management functions (e.g., allocation of dedicated control channels). The CCCH can include the following: (1) *paging channel* (PCH), which is a down channel used to page MSs; (2) *random access channel* (RACH), which is an uplink channel used for request assignment of a DCCH; and (3) *access grant channel* (AGCH), which is a downlink channel used to assign a MS to a specific DCCH.

A DCCH is a point-to-point, directional control channel. Two types of DCCHs are used: (1) *stand alone dedicated control channel* and (2) *associated control channel*

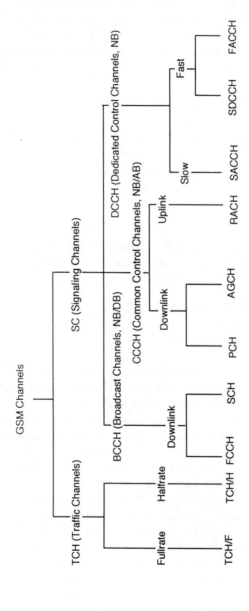

Figure 3.4 Logical channels in GSM.

(ACCH). Stand alone SDCCH is a DCCH whose allocation is not linked to the allocation of a TCH. This channel is used before the MS is assigned a TCH. The SDCCH is used to provide authentication to MS and for location updates and assignment to TCHs. A DCCH whose allocation is linked to the allocation of a TCH has ACCHs (that is, FACCH and SACCH). A FACCH or burst stealing is a DCCH obtained by preemptive dynamic multiplexing on a TCH. A SAACH, also known as a continuous data stream, is allocated together with a TCH or a SDCCH. We shall now provide details of each of these three groups of control channels.

3.2.3.1 Broadcast Control Channel

The BCCH provides general information on a per-BTS basis (cell-specific information) including information necessary for the MS to register in the system. After initially accessing the mobile, the BS calculates the required MS power level and sends a set of power commands on this channel. Other information sent over this channel includes country code, network code, local area code, PLMN code, RF channels used within the cell where the mobile is located, surrounding cells, hopping sequence number, mobile RF channel number for allocation, cell selection parameters, and RACH description. One of the important messages on a BCCH channel is CCCH_CONF, which indicates the organization of the CCCHs. From this parameter, the number of CCCHs (BS_CC_CHANS), indicating whether or not CCCHs are combined with four SDCCHs and four SACCHs on to the same physical channel called BS_CCCH_SDCCH_COMB = true or false, are derived as shown in Table 3.1.

Similarly, the 3-bit message BS_AG_BLKS_RES represents the number of blocks on each CCCH reserved for access grant messages. This channel is used to downlink point-to-multipoint communication and is unidirectional; there is no corresponding uplink. The signal strength of this channel is continuously measured by all mobiles, which may seek a handover from its present cell. Thus, it is always transmitted on a designated RF carrier (the "BCCH carrier") using timeslot 0, denoted as C0T0. This channel is never kept idle—either the relevant messages are sent or a dummy burst is sent. Other channels that belong to this group are the FCCH and SCH.

Table 3.1
CCCH Configuration Message Structure

CCCH_CONF	BS_CC_CHANS	BS_CCCH_SDCCH_COMB
000	1	false
001	1	true
010	2	false
100	3	false
110	4	false

The FCCH carries information for frequency correction of the MS downlink. It is required for the correct operation of a radio subsystem. Similar, to the BCCH, the channel is also for point-to-multipoint communication. This will allow an MS to accurately tune to a BS. FCCH sends all zeros in its burst to represent an unmodulated carrier, which we shall discuss in the next section.

The SCH carries information for the frame synchronization (TDMA-frame number) of the MS and the identification of a BTS. Similar to FCCH, this is also required for the correct operation of a mobile. The SCH channel carries a 64-bit binary sequence that is a priori known to the mobile. By correlating these bits with the internally stored 64 bits, MS achieves the exact timing (timing synchronization) with respect to a GSM frame. Specifically, the synchronization channel contains two encoded parameters: (1) *BTS identification code* (BSIC) and (2) *reduced TDMA frame number* (RFN).

The BSIC consists of six bits, out of which three bits represent the PLMN color code with a range from 0 to 7, and the other three represent BS color code with a range from 0 to 7.

The RFN is 19-bits long and consists of

T1(11 bits)	$= FN \text{ div}(26 \times 51)$	Range from 0 to 2,047
T2(5 bits)	$= FN \text{ mod}(26)$	Range 0 to 25
T3(3 bits)	$= (FN \text{ mod}(51) - 1) \text{ div}(10)$	Range 0 to 4

where FN represents the TDMA frame number. The layout of BSIC and RFN within the message part of the synchronization sequence is shown in Figure 3.5. SCH, FACCH, and BCCH channels cannot be frequency hopped, as these channels carry synchronization and system-related information whose exact location must be know to mobile.

In the BSIC, the channel is once again in the downlink direction only and is used for point-to-multipoint communication.

3.2.3.2 Common Control Channels

A CCCH is a point-to-multipoint bidirectional control channel and is primarily intended to carry signaling information necessary for access management functions (e.g., allocation of dedicated control channel). CCCH includes: (1) a PCH, which

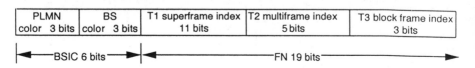

Figure 3.5 SCH message format.

is used to page (search) the MS in the downlink direction; and (2) a RACH, which is used by MS to request allocation of an SDCCH either as a page response from MS or call origination/registration from the MS. This is an uplink channel and operates in point-to-point mode (from MS to BTS). The channel operates on slotted Aloha Protocol and thus the contention possibility exists. If the mobile's request through this channel is not answered within a specified time, the MS assumes that a collision has occurred and repeats the request. Mobile must allow a random delay before reinitiating the request to avoid repeated collision.

The AGCH is used to allocate an SDCCH or a TCH directly to an MS. This channel is in the downlink direction and operates in point-to-point mode. A combined paging and access grant channel is designated as PAGCH, although the GSM specifications refer to PAGCH as two separate channels.

3.2.3.3 Dedicated Control Channels

DCCHs consist of a SDCCH and the ACCH. The SDCCH is used for system signaling during idle periods and call setup before allocating a TCH. For example, MS registration, authentication, and location update takes place through this channel. When a TCH is assigned to the MS, this channel is released. It uses 1/8 rate TCH (TCH/8); that is, its data rate is one-eighth of the full rate speech channel. This is achieved by transmitting data over this channel once every eighth frame. The channel is used for both up/downlink and is meant for point-to-point usage.

As stated, the SACCH can be linked to a TCH or an SDCCH. It is a continuous data channel carrying information, such as measurement reports, from the mobile of received signal strength for a serving cell as well as adjacent cells. This is a necessary channel for the mobile-assisted handover function. The channel is also used for power regulation of the MS and time alignment and is meant for both the up and downlink. It is used for point-to-point communication between mobile and BS.

A FACCH is associated to a TCH. FACCH works in a stealing mode. This means that if suddenly during a speech transmission it is necessary to exchange signaling information with the system at a rate much higher than the SACCH can handle, then 20-ms speech (data) bursts are stolen for signaling purposes. This is the case at handover. The interruption of speech will not be heard by the user since it lasts only for 20 ms and cannot be sensed by human ears. We shall now take up details of how the data is actually arranged within a physical channel for the logical speech and control channels.

3.2.4 Structure of a TDMA Slot With a Frame

There are five different kinds of bursts in the GSM system.

- Normal burst;
- Synchronization burst;

- Frequency correction burst;
- Access burst;
- Dummy burst.

3.2.4.1 Normal Burst

This burst is used to carry information on the TCH and on control channels, except for RACH, SCH, and FCCH, and is shown in Figure 3.6(a). Burst is divided into 156.25 bits. The first bit in the slot is numbered 0, and the last (1/4) bit period is numbered 156. The lowest bit number is transmitted first. The encrypted bits are 57 bits of data or speech plus one bit "stealing flag" indicating whether the burst was stolen for FACCH signaling or not. The training sequence is a known 26-bit pattern used by the equalizer to create a channel model. The reason why the training sequence is placed in the middle is that the channel is constantly changing. By having it there, the chances are better that the channel is not too different when it affects the training sequence compared to when the information bits were affected. If we put the training sequence at the beginning of a burst, the channel model we create might not be valid for the bits at the end of a burst. There are eight training sequences as shown in Table 3.2. The 26-bit equalization patterns are determined at the time of call setup.

The *tail bits* (TB) always equal (0,0,0), which has bit location from 0 to 2 and also from 145 to 147. They help the equalizer know the start and stop bit pattern; that is, the algorithm used in the equalizer needs a certain start/stop point. The *guard period* (bits) (GP) is empty space. Since we have a maximum of eight users per carrier, using different time slots, we must make sure that they do not overlap each other during the transmission. To synchronize the bursts with exact accuracy (with no GP) is not easy to do in practice since the mobiles are moving during the call and causing the bursts from different mobiles to "slip" a little (this can be caused by different delays, which in turn are caused by different locations of users within a cell) compared to each other. Thus, the 8.25 bits that correspond to about 30.46 μs (8.25 × 3.693) is allowed to protect the overlapping of messages at the BS. The GP also allows the transmitter to ramp-up and ramp-down within limits specified by the GSM recommendations.

3.2.4.2 Synchronization Burst

This burst is used for time synchronization of the mobile. It contains a long synchronization sequence (64-bits long). The training sequence states are: (BN42, BN43, . . ., BN105) = (1, 0, 1, 1, 1, 0, 0, 1, 0, 1, 1, 0, 0, 0, 1, 0, 0, 0, 0, 0, 0, 1, 0, 0, 0, 0, 0, 0, 1, 1, 1, 1, 0, 0, 1, 0, 1, 1, 0, 1, 0, 1, 0, 0, 0, 1, 0, 1, 0, 1, 1, 1, 0, 1, 1, 0, 0, 0, 0, 0, 1, 1, 0, 1, 1). The encrypted 78 bits carry information of the TDMA frame number

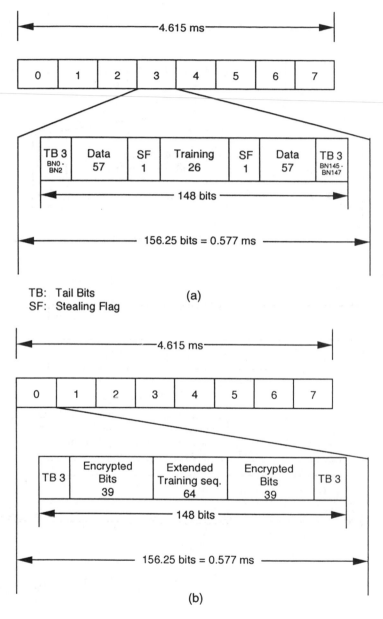

TB: Tail Bits
SF: Stealing Flag

(a)

(b)

Figure 3.6 (a) Normal burst, (b) synchronization burst, (c) frequency correction burst, (d) access burst, and (e) dummy burst.

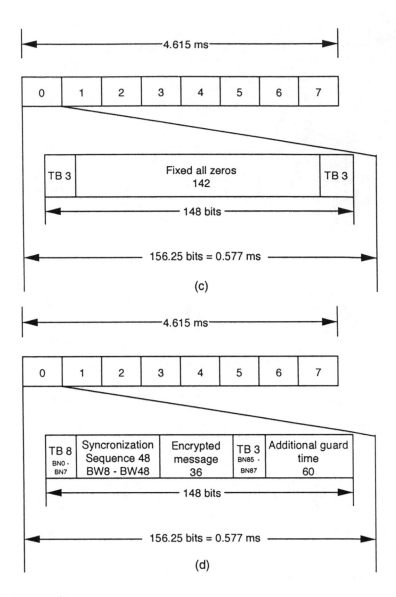

Figure 3.6 (continued).

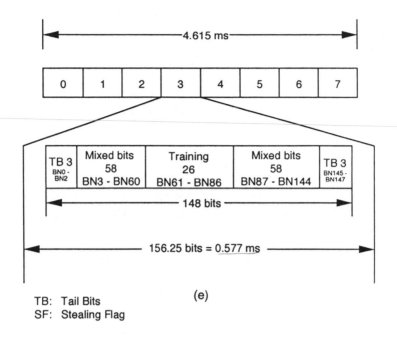

TB: Tail Bits
SF: Stealing Flag

(e)

Figure 3.6 (continued).

Table 3.2
Normal Burst

Code Number	Training Sequence Code (TSC) Sequence of 26 bits
0	00100101110000100010010111
1	00101101110111100010110111
2	01000011101110100100001110
3	01000111101101000100011110
4	00011010111001000001101011
5	01001110101100000100111010
6	10100111110110001010011111
7	11101111000100101110111100

along with the BSIC, as discussed in the previous section. It is broadcast together with the frequency correction burst. These bursts are named as SCH and shown in Figure 3.6(b). The TDMA frame is broadcast over a SCH, as discussed in Section 3.3, in order to protect user information against eavesdropping, which is accomplished by ciphering the information before transmitting it. The algorithm that calculates the ciphering key uses a TDMA frame number as one of the input parameters, and therefore, every frame must have a frame number. The numbering scheme is cyclic with a period of 3.5 hr (2715648 TDMA frames). An FN is encoded in 19 bits before channel encoding and is shown in Figure 3.5. Also, by knowing the TDMA frame number, the mobile will know what kind of logical channel is being transmitted on the control channel TS0. BSIC is also used by the mobile to check the identity of the BTS when making signal strength measurements (to prevent measurements on cochannel cells).

3.2.4.3 Frequency Correction Burst

This burst is used for frequency synchronization of the mobile. It is equivalent to an unmodulated carrier with a specific frequency offset. The repetitions of these bursts are called FCCH. The fixed input bits are all zeros, causing the modulator to deliver an unmodulated carrier with an offset of 1,625/24 kHz above the nominal carrier frequency. Here the TB and guard bits are the same as in normal burst. A frequency correction burst is shown in Figure 3.6(c).

3.2.4.4 Access Burst

This burst is used for random access and has a longer GP (68.25 bits or 252 μs) to protect for burst transmission from a mobile that does not know the timing advance when it first accesses the system. This allows for a distance of 35 Km from base to mobile. In case the mobile is far away from the BTS, the initial burst will arrive late since there is no timing advance on the first burst. The delay must be shorter to prevent it from overlapping the burst in the adjacent timeslot following this. Figure 3.6(d) shows the layout of an access burst. Here the extended tail bits are defined as modulating bits with the following states: (BN0, BN1, . . . , BN7) = (0, 0, 1, 1, 1, 0 ,1, 0). The TB are (0, 0, 0), and they occupy the position (BN85, BN86, BN87). Similarly, synch bits occupy the position (BN8 through BN48).

3.2.4.5 Dummy Burst

This burst is sent from BTS on some occasions. It carries no information. The format is the same as for a normal burst. Figure 3.6(e) shows the layout of a dummy burst. Here the TB occupy positions (BN0, BN1, BN2) and (BN145, BN146, BN147) and are (0, 0, 0) in both instances. The training sequence bits occupy positions BN61 through BN86. The mixed bits are defined as modulating bits states as

$$(BN3, BN4, . . ., BN60) = (1, 1, 1, 1, 1, 0, 1, 1, 0, 1, 1, 1, 0, 1, 1, 0, 0, 0, 0, 0,$$
$$1, 0, 1, 0, 0, 1, 0, 0, 1, 1, 1, 0, 0, 0, 0, 0, 1, 0, 0, 1,$$
$$0, 0, 0, 1, 0, 0, 0, 0, 0, 0, 0, 1, 1, 1, 1, 1, 0, 0)$$

$$(BN87, BN88, \ldots, BN144) = (0, 1, 1, 1, 0, 1, 0, 0, 1, 0, 1, 0, 0, 0, 1, 1, 0, 0, 1,$$
$$1, 0, 0, 1, 1, 1, 0, 0, 1, 1, 1, 1, 0, 1, 0, 0, 1, 1, 1,$$
$$1, 1, 0, 0, 0, 1, 0, 0, 1, 0, 1, 1, 1, 1, 1, 0, 1, 0, 1,$$
$$0)$$

In the next section we shall build upon this information by discussing different organizations of signaling channels.

3.2.5 Time Organization of Signaling Channels

In this section we shall discuss how different traffic and control channels are organized. For establishing end-to-end communication between mobile and PSTN, or between two mobiles, a large amount of data has to be exchanged between the mobile and the system before a speech channel can be assigned to the mobile. Data are exchanged between mobile and the system over different logical control channels. In this section the layout of different logical control channels will be covered. To accomplish this we shall consider the layout of the following channels [2,3].

- Full rate and half rate TCHs;
- Signaling channels (downlink only): (1) FCCH and SCH, where the mobile synchronizes its frequency and reads the TDMA frame number and the BSIC; (2) BCCH, through which the mobile reads the necessary general information about the system; and (3) PAGCH;
- Signaling channel (uplink only): RACH, through which the mobile requests service either in response to a paging or as an initial request to originate a call;
- Signaling channel (bidirectional): SACCH and FACH;
- Common channel allocation in a typical system: (1) basic case, and (2) small-capacity system.

3.2.5.1 Time Organization of Traffic Channel (Full and Half Rate)

As stated, the TCH/F carries information (encoded speech or data) at the gross rate of 22.8 Kbps. The time organization of TCH is shown in Figures 3.7(a). A TACH/F contains TCH/F and SACCH. Eight different types of TACH/F can be defined based on their phase modulo 8 (TN$_0$ to TN$_7$). SCH channel occupies the 13th frame. The 26th frame is kept idle for future use. *SACCH*

A half rate TCH carries information (encoded speech or data) at a gross rate of 11.4 Kbps. The raw data rate is 7 Kbps for speech or data rates of 6 Kbps or

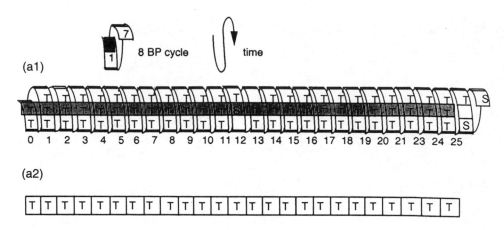

Figure 3.7 (a1) Time organization of a TACH/F channel and (a2) equivalent representation; (b) time organization of TACH/H; (c1) time organization of the FCCH and SCH and (c2) equivalent representation; (d1) FCCH, SCH, BCCH, and PAGCH/F channel pattern (uplink) and (d2) (downlink); (e1) RACH/F and (e2) RACH/H channel patterns; (f1) SDCCH and SACCH link from BTS to MS and (f2) MS to BTS; (g1) basic uplink and (g2) downlink common channel pattern; and common (h1) downlink and (h2) uplink channel pattern for small-capacity cells [4].

3.6 Kbps. The time organization of TCH/H is shown in Figure 3.7(b). Since full rate speech uses one slot every 8 slots or a frame, the TACH/H uses one slot every 16 slots and thereby reduces the effective data rate to half of the full rate channel [3, 4].

Example

Figure 3.3 shows the speech frame in which there are 24 frames in one multiframe lasting 120 ms carrying data. The number of data bits per frame, as shown in Figure 3.6(a), are 114. Thus, the data rate for full and half rate channels are

$$\frac{(114 \text{ Bits/Traf·frm}) * (24 \text{ Traffic·frms/Multiframe})}{120 \text{ ms/Multiframe}} = 22.8 \text{ Kbps}$$

The data rate for the half rate channel will be 22.8/2 = 11.4 Kbps.

Example

Figure 3.3 shows one frame of a SACH in one multiframe. The number of data bits, as shown in Figure 3.6(a), are the same as speech data. Thus, the data rate for a SACH is

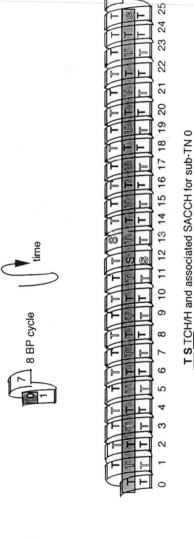

T S TCH/H and associated SACCH for sub-TN 0

TCH/H and associated SACCH for sub-TN1

(b)

Figure 3.7 (continued).

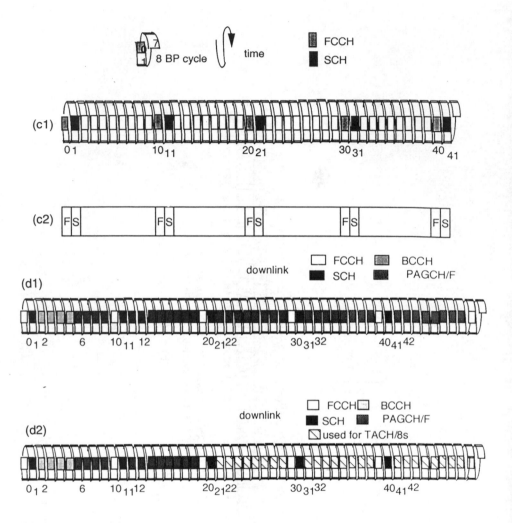

Figure 3.7 (continued).

$$\frac{(114 \text{ Bits/Traf·frm}) * (1 \text{ Traffic·frms/Multiframe})}{120 \text{ ms/Multiframe}} = 950 \text{ bps}$$

3.2.5.2 Time Organization of Control Channels

All CCCHs have been defined with the intention of grouping them together in a few combinations. Their time definitions are all based on the same cycle, that is,

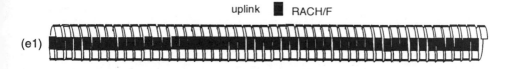

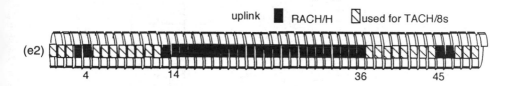

Figure 3.7 (continued).

$BP's$

51×8 ~~bps~~. This cycle and the cycle of TCHs were deliberately chosen with different values in order to allow MSs in dedicated mode to listen to the SCH and FCCH of surrounding BSs, both of which carry the information needed for MSs to become and stay synchronized with a cell. MSs are then able to receive, at least from time-to-time, a burst belonging to the FCCH or the SCH. By so doing, they acquire the synchronization information needed for surrounding cells. This functionality is referred to as presynchronization. Each BTS has n (duplex) carriers, each having 8 TSs, as shown in Figure 3.1. The carriers are named C0, C1, C2, . . ., Cn. We start with TS0, downlink, on C0. TS0 on C0 is used to map control channels only. There are 51 TSs in all. Note that the length in time is 51 TDMA frames, although we use only TS0 from each frame. This sequence repeats itself over and over again; that is, after the IDLE gap, it starts with F, S, . . . again. We shall now elaborate on how these channels are used within the system.

Signaling Channel (Downlink Only): Frequency and Time Synchronizing Channels

Both the FCCH and SCH have the same time structure: one SCH slot follows each FCCH slot eight burst slots later. Each of those two channels occupy five slots each in 51×8 bps cycle, as shown in Figures 3.7(c). A single set (FCCH + SCH) is broadcast in any given cell. In all the cells, the slots of these channels have the same position within the 8-BP cycle, that is to say, the same TN. This position is by definition called TN0.

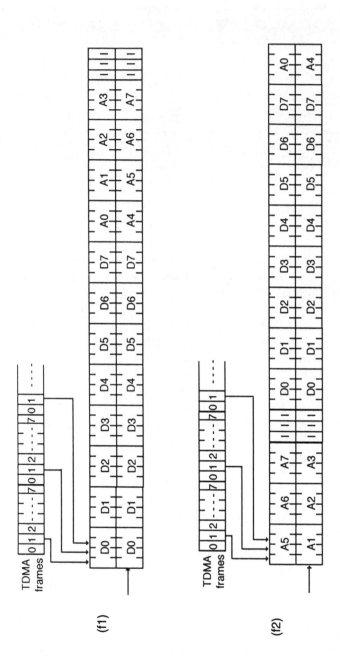

Figure 3.7 (continued).

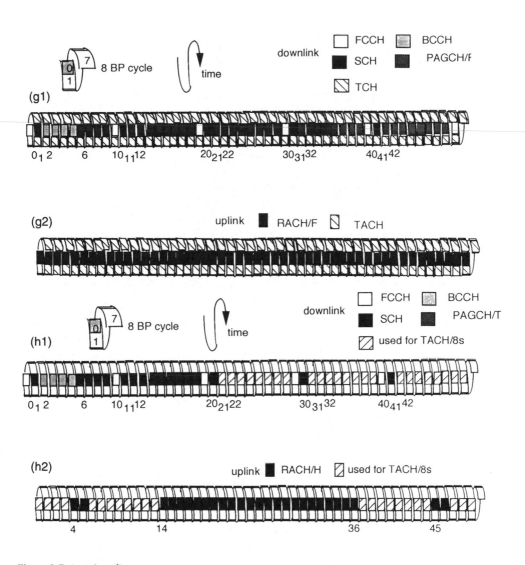

Figure 3.7 (continued).

Signaling Channel (Downlink Only): Broadcast Control Channel and Paging and Access Channels

Apart from the SCH and FCCH, the other downlink common channels include BCCH and PAGCH. The difference between these two channels lies more in their usage than in their transmission characteristics. It is indeed possible that their respective size may be allowed to vary in order for the BCCH to have some additional capacity at the expense of the PAGCH capacity. A BCCH together with a PAGCH/F uses 40 slots per 51 × 8 bp cycle, all with the same TN. Figure 3.7(d1) shows how these slots are spread. These 40 slots are built as 10 groups of four. A BCCH together with a PAGCH/T uses 16 slots per 51 × 8 bp cycle, all with the same TN. Figure 3.7(d2) shows how these slots are spread. These 16 slots are arranged as four groups of four PAGCH/T is 1/3 of PAGCH/F, as shown in Figure 3.7(d1).

Signaling Channel (Uplink Only): Random Access Channel

There are two types of RACHs: RACH/F and RACH/H. RACH/F uses one slot every 8 bps, and the RACH/H uses 27 slots in 51 × 8 cycle as shown in Figure 3.7(e). Unlike BCCH, FCCH, SCH, PCH, and AGCH, which are only downlink channels, RACH is an uplink channel. We have covered the usage of all the control channels individually. We now describe those channels that work with speech or user data transmission.

Signaling in connection with a call: slow and fast associated control channel. In order to transport signaling data parallel with the transmission of user data flow, GSM offers two possibilities. Each TCH comes with an associated low rate channel, used for the transport of signaling: the SACCH. This bidirectional channel may carry about two messages per second in each direction. It is used for nonurgent procedures, mainly the transmission of the radio measurement data needed for decisions concerning handover. The other needs for associated signaling are with messages to indicate the call establishment progress, to authenticate the subscriber, and to command a handover. Handover is commanded through the FACCH, which is used in place of TCH. The receiver is able to distinguish both these uses (TCH and FACCH) by reading binary information transmitted on the TCH, called the stealing flag. During the initialization and release phases, no user data is transmitted, and therefore signaling may use the channel without any conflict with other types of data. However, during the call, the transmission of fast associated signaling reduces the effective data rate of the user and is thus known as "stealing." The other assignment of SACCH is with a SDCCH channel as shown in Figures 3.7(f1,f2) and for downlink and uplink transmissions. In the next section we shall examine how common channels are assigned to a typical system.

Common Channel Allocation in a Typical System

We shall discuss two cases: first the channel allocation in normal surroundings, known as the basic case, and the second case for low-capacity systems.

The configuration for the basic case consists of: Downlink: FCCH, SCH, BCCH, and PAGCH/F all on TN0. The uplink on RACH/H and TCHs TACH/Fs (Traffic and associated Control Channels) on TNs 1 through 7. Channel allocation is shown in Figures 3.7(g1,g2). It should be noted that all control channels occupy TN0 slots. The configuration for the small-capacity system consists of Downlink: PAGCH/T, FCCH, SCH, BCCH plus TACH/8s, and the Uplink: RACH/H with TACH/8s. The channel configuration for this is shown in Figures 3.7(h1,h2).

Signaling Outside a Call (TCH/8)

In some cases, there is a need to establish a connection between an MS and the network just for signaling matters. It can be by the user's demand (for example, call forwarding management and transmission of short messages) or for other management needs, such as location updating. A TCH/F or a TCH/H may be used for that purpose. However, this will waste a lot of spectrum. In order to increase system efficiency, an additional type of channel has been introduced. Its rate is very low and only has specified usage for signaling and short message transmission. This channel is referred as TCH/8. If a TCH/H is considered as half a TCH/F, then this is one-eighth of a TCH/F. The other important aspect of a GSM system is its representation in frequency domain, which is the topic of our discussion in the next section.

3.3 FREQUENCY AXIS REPRESENTATION

Physical resources available to the radio subsystem in GSM are two bands of 25-MHz bandwidth as described in Section 3.1. Within this band, frequency is partitioned by *radio frequency channels* (RFCHs). A number of channels is assigned to each cell, which is known as *cell allocation* (CA). Each cell has one RFCH to carry out synchronization, which is known as a BCCH carrier [2–4]. During conversation a subset of the channel is allocated to a mobile known as *mobile allocation* (MA), which consists of a downlink and an uplink—the downlink being from BS to mobile and the uplink from mobile to BS. There is a total of 124 carrier frequencies that are related to channel numbers by the following equations.

Base to Mobile: $F_u(n) = 935.2 + 0.2*(n - 1)$ MHz $(1 \leq n \leq 124)$
Mobile to Base: $F_d(n) = 890.2 + 0.2*(n - 1)$ MHz $(1 \leq n \leq 124)$

Here *n* is called the *Absolute Radio Frequency Channel Number* (ARFCN). A GSM system is also allowed to operate in the following extended bands having 10-MHz bandwidths: 880 to 890 uplink and 925 to 935 downlink.

3.3.1 Frequency Hopping

In this section we discuss various reasons for choosing frequency hopping in a GSM system along with the algorithm used for hopping, reasons why some control channels cannot be a part of hopping, and various implementation schemes employed. In essence, the discussion is arranged as follows.

- FH improves performance in multipath fading;
- Decreases the required C/I;
- Mandatory for MS when requested by BS;
- Algorithm: cyclic and pseudorandom;
- FCCH, SCH, and BCCH cannot hop;
- Two implementations: baseband hopping and synthesizer hopping;
- Hopping offset 45 MHz between uplink and downlink.

The mobile radio channel is a frequency-selective fading channel, which means the propagation conditions are different for each individual radio frequency. Whereas channel number 1, for example, experiences problems when a mobile passes a large building, channel number 10 may not suffer any degradation in quality. To average these conditions over all the BS's available frequencies in a cell, *slow frequency hopping* (SFH) is introduced, which improves the signal quality. It is so called (to distinguish it from fast frequency hopping) because the operating frequency is changed only with every TDMA frame. The hopping rate is 216.7 hops/s, which corresponds to 1/(frame duration) = $1/(4.1615 \times 10^{-3})$. Therefore, a MS transmits at one frequency during a time slot and must hop to a different frequency before the next time slot. *Fast frequency hopping* (FFH) is used in spread spectrum systems and changes the operating frequency of a link many times per symbol rather than only one time in each frame. To perform well, a frequency synthesizer must be able to change its frequency and settle quietly on a new one within a fraction of one time slot (577 μs). Frequency hopping provides frequency diversity to overcome Rayleigh fading due to the multipath propagation environment. This environment can give 40-dB to 50-dB fades on the received signals. Since the fading will appear and disappear over a distance of one-quarter wavelength (around 3 ft), a rapidly moving MS will completely uncorrelate the fading from one burst to the next. Since interleaving spreads each encoded speech block over eight bursts, the effect of fading on any burst is greatly reduced. However, slowly moving or stationary MSs may remain in a fade for many bursts. The effect of fading is also uncorrelated over frequency differences of the order of 1 MHz, burst-to-burst frequency hopping over frequencies that are far apart will greatly reduce the effect of fading on these slow and stationary MSs.

Frequency hopping also provides interference diversity. At any time, the amount of interference on the various channels in a given cell will vary from channel to channel. A receiver set on a channel with strong interference will suffer excessive errors over long strings of bursts. By frequency hopping, the receiver will not spend successive bursts on the same high interference channel. The following paragraphs describing the operations of frequency hopping are taken from [6].

Frequency hopping reduces the *signal-to-noise ratio* (SNR) required for good communications. For a nonhopping link, the minimum required ratio is about 11 dB, whereas frequency hopping reduces the requirement to only 9 dB and is an additional increase of 2 dB of margin in the channel. The hopping also adds frequency diversity to the channel. Frequency hopping is one of the options for each individual cell. The BSS is not required to support this feature. A mobile, however, has no choice but to switch to a frequency-hopping mode when the BS tells it to do so. The mobile needs to add frequency diversity to its transmissions as it moves toward an area of high interference or at any other time the channel becomes marginal. When the BSC observes the failing channel from the mobile, it decides to tell the mobile to turn on frequency hopping by simply assigning the mobile a full set of RF channels rather than a single RF channel. The mobile performs a "dance" on the assigned set of frequencies to satisfy its SFH obligations.

Different hopping algorithms can be assigned to the MS with the channel set. One is cyclic hopping, in which hopping is performed through the assigned frequency list from the first frequency, to the second frequency, to the third, and so on until the frequency list is repeated. The other general algorithm is (pseudo) random hopping, in which hopping is performed in a random way through the frequency list. There are 63 different random dances that can be assigned to the mobile. A BTS with n frequencies available can theoretically employ up to factorial n different nonrepeating hopping sequences. For a set of n frequencies, GSM allows $64 \times n$ different hopping sequences to be built. If $n! > 64 \times n$ (that is, $n > 5$), then not all potential sequences can be used. When the MS is asked to use SFH, it is advised of the channel assignment (a set of channels) and which one of the hopping algorithms it should use with an appropriate frequency-*hopping sequence number* (HSN). The details of the hopping algorithm are provided at the end of this section.

The base channel is not allowed to hop. The base channel, confined to TN 0 and carrying the FCCH, the SCH, and the BCCH, is the beacon upon which mobiles perform their periodic signal strength measurements on neighboring cells. These channels also signal the MS to synchronize with a system as it initially seeks service or gets ready to move to another cell. The FCCH, SCH, and BCCH exist only on the base channel on timeslot 0, and all the other timeslots on the base channel's frequency are filled with some kind of data to bring the base channel's power above all the other channels in the cell.

There are also two different implementations for frequency hopping in BSs. One of the implementations is baseband hopping, which is used if a BS has several transceivers available. The data flow is simply routed in the baseband to various transceivers, each of which operates on a fixed frequency in accordance with the assigned hopping sequence. The different transceivers receive a specific individual timeslot in each TDMA frame, which transfers information destined for different MSs. The other implementation is synthesizer hopping. Figure 3.8(a) shows the implementation of baseband hopping. On the other hand, one can find BSs in remote areas fitted with only one or two transceivers but still want to use SFH. In this unusual case, the hopping is performed on the RF transceiver, which requires the transceiver to hop on the different frequencies itself. Figure 3.8(b) shows the timing conditions for a MS during frequency hopping. It is the case where SFH is applied on three different frequencies. Since the MS first receives a message from the BS and then responds to the BS three timeslots later, there are four timeslots of time during which the MS can accomplish different tasks, such as monitoring adjacent cells, before it has to hop to the next frequency. Since the number of bursts in the uplink direction derives conventionally from the one in the downlink direction by a delay of 3 TS, the hopping sequence (that is, the function that associates a frequency to each TN modulo 8) in the uplink direction is simply shifted by 45 MHz.

For a set of n given frequencies, GSM allows $64 \times n$ different hopping sequences to be built. They are described by two parameters, the *Mobile Allocation Index Offset* (MAIO), which may take as many values as the number of frequencies in the set, and the HSN, which may take 64 different values. Two channels bearing the same HSN but different MAIOs never use the same frequency on the same burst. All mobiles in a given cell or cell group sharing the same frequency set will have the same HSN, while each will have a different MAIO. The frequency assigned to each MS is also a function of the frame number. The different MAIOs assure that no two MSs in a cell are ever assigned the same frequency at the same time. Thus, the frequency hopping sequence is orthogonal within one cell and independent to the cochannel cells.

All channels in one cell use the same frequency set and HSN. Adjacent cells in the same group may use the same HSN since these cells use different frequency

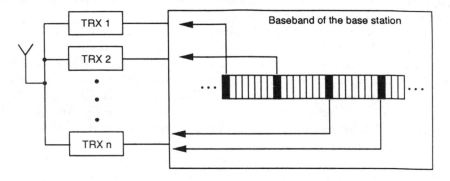

Figure 3.8   (a) Implementation of baseband frequency hopping and (b) frequency hopping within a cell [6].

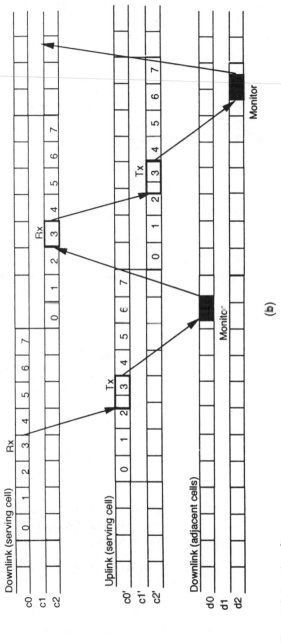

Figure 3.8 (continued).

sets. Distant cells using the same frequency set generally use different HSNs to improve the interference diversity. These channels in distant cells will then interfere on only 1/nth of the frames. We shall now describe the frequency hopping algorithm. The actual parameters for frequency hopping are as follows.

MA: the actual allocation of frequencies $1 \leq N \leq 64$ to mobile.

MAIO: values lie between 0 to $N - 1$ (represented by six bits) and determines the next frequency where the mobile will hop.

HSN: $0 \leq HSN \leq 63$, which results in cyclic hopping (increases by each frame) when HSN = 0; otherwise the hopping is random.

FN is derived out of T1, T2, and T3, where $0 \leq T1 \leq 2,047$, $0 \leq T2 \leq 25$, and $0 \leq T3 \leq 50$.

NBIN = number of bits required to represent Integer($\log_2 (N) + 1$).

RNTABLE is a look-up table of 28 integer values whose values lie between 0 and 127.

The algorithm is shown in Figure 3.9. The input parameters to the algorithm are MA, MAIO, HSN, FN(T1), FN(T2), and FN(T3). The FNs are specified by the three indices FN(T1), FN(T2), and FN(T3); and the values of theses indices are received over the SCH. If HSN = 0, the algorithm chooses the cyclic hopping path and the *mobile allocation index* (MAI) is computed as

$$MAI = (FN + MAIO) \bmod N$$

If HSN $\neq$ 0, parameters M, M', and T' are computed. If $M' \leq N$, the intermediate parameter S is set to M', otherwise $S = (M' + T') \bmod(N)$. Finally the MAI is computed by specifying the next RF channel for hopping used by the MS.

The hopping sequence is the same on the uplink (from the MS) as on the downlink (to the MS). The uplink frequencies are offset by 45 MHz relative to the corresponding received downlink frequencies.

3.4 OPERATIONAL CONCEPTS

In this section we shall discuss different operational modes of mobile, access support, and sequence of operations that the mobile has to take within the system. The basic concepts of operation are explained in the following subsections.

3.4.1 Mobile Modes

The rarity of the radio spectrum does not allow each user of the system to have his own TCH at all times. TCHs are therefore allocated to the users only when the need arises. This leads to the basic distinction of dedicated mode and idle mode, which

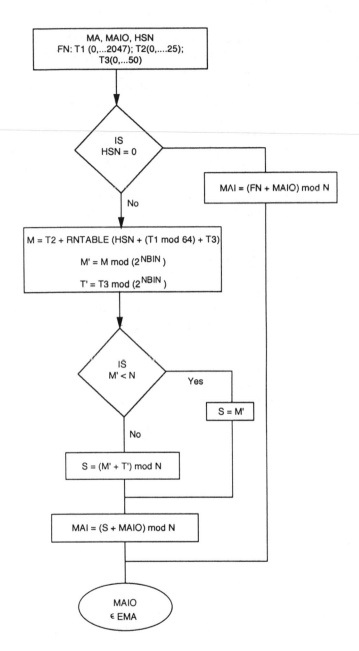

Figure 3.9 Hopping algorithm in GSM system.

are essential concepts in radiotelephony. Formally, a MS shall be considered in dedicated mode when a TCH is at its disposal. This corresponds to phases when full bidirectional point-to-point transmission is possible between the MS and the infrastructure—for instance, during an established call or during a period of location updating by mobile. TCHs and SACCHs are therefore referred to as dedicated channels in the GSM specifications.

When a MS is active (powered on) without being in dedicated mode, it is deemed to be in idle mode. However, the MS is far from being idle. It must continuously stay in contact with the BS, listen to what this base transmits in order to intercept paging messages (to know if its user is being called), and monitor the radio environment in order to evaluate the channel quality and choose the most suitable BS. In addition, there is one telecommunication service that is provided to the MSs in idle mode: the cell broadcast short message service, where the mobile is supposed to receive and interpret.

The transition between idle mode and dedicated mode requires some information exchanges between the MS and the BS (the "access" procedure). The MS indicates to the network that it needs a connection, and the network indicates in return which dedicated channel it may use. All of these uses require specific transmission means that are grouped under the terminology "common channels."

3.4.2 Access Support to Mobile

In order to get access to the GSM system, mobile makes use of the following logical channels.

- FCCH, SCH, BCCH;
- PCH, AGCH;
- RACH.

In order to communicate with a BS, an MS must first become (and stay) synchronized with it. Cell broadcasts on the FCCH and the SCH. The functions and formatting of these channels have already been discussed in Section 3.2.

MSs in idle mode require a fair amount of information to act efficiently. Most often a MS can receive and potentially be received by several cells possibly in different networks or even in different countries. It then has to choose one of them, and some information is required for the choice—for instance, the network to which each cell belongs. This information, as well as some other, is broadcast regularly in each cell, to be listened to by all the MSs in idle mode. The channel used for this purpose is the BCCH.

Paging messages have to be broadcast to indicate to some MSs that a call toward its user is being set up. The access procedure itself includes a request from the MS and an answer from the BS allocating a channel. Paging messages and messages indicating the allocated channel upon prime access are transmitted on the PAGCH.

All the common channels listed (FCCH, SCH, BCCH, PAGCH) are "downlink" unidirectional channels; that is, they convey information from the network to the MSs. The last type of common channel that allows the MSs to transmit their access requests to the network is an "uplink" unidirectional channel. It is called RACH. Its name indicates that MSs choose their emission time on this channel in a random manner, thus resulting in potential collision. A discussion with respect to how these channels are arranged within a frame has already been presented in Section 3.2. Details of this will be covered in the next chapter on mobility management.

3.4.3 Mobile Sequence of Operations

The following sequence of steps takes place from the instant mobile turns on its power.

- Mobile frequency and time synchronization;
- Access request by mobile for registration;
- Mobile detects the incoming call;
- Mobile originating a call.

A mobile receiver will try to lock to those carriers whose power is above a certain threshold (adequate C/I at the receiver input). If the carrier has FCCH channel (frequency synchronization signal, a string of 142 zeros) and if the C/I is adequate, the mobile will detect a string of zeros. The mobile receiver will measure the carrier frequency and will be offset by 1,625/24 MHz. This is called coarse synchronization (Figure 3.10[a]). The receiver will declare this to be the C0T0 location for the detected carrier. Now the receiver will look at the same location (C0T0) in the next frame and should detect a known pattern of 64 bits. If detection does take place, the receiver has achieved fine synchronization (Figure 3.10[b]). The encoded 78 bits of data provide a hyperframe TDMA frame number and BSIC (in the BSIC channel). The general information for the system can now be read over the next four frames (BCCH Figure 3.10[c]). It includes mobile country code, network code, frequencies used in the present cell (the cell where the FCCH and SCH belong), cells barred, and a description of neighboring cells. At the end of this process the receiver is synchronized and has decoded the relevant information. After initial synchronization with the system, mobile can now register with the system. Then it is ready to either receive an incoming call or originate a call. We shall discuss the case of registration followed by receiving or originating calls by mobile.

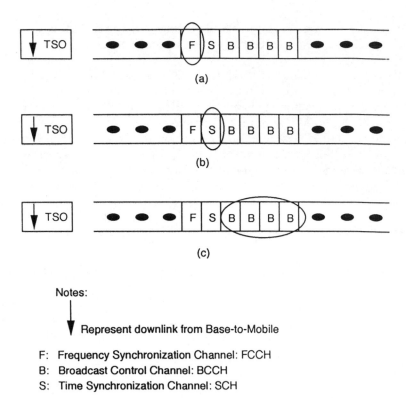

Notes:

↓ Represent downlink from Base-to-Mobile

F: Frequency Synchronization Channel: FCCH
B: Broadcast Control Channel: BCCH
S: Time Synchronization Channel: SCH

Figure 3.10 Frequency and time synchronization by mobile.

3.4.3.1 *Access Request by Mobile for Registration*

Mobile sends an access request on the RACH (Figure 3.11[a]). The system will allocate SDCCH through AGCH (Figure 3.11[b]). Registration is performed by exchanging information through SDCCH/SACCH (Figure 3.11[c]). Mobile is in the idle mode at the end of the registration and starts monitoring paging or it can also originate a call. We shall discuss both of these cases in the following subsections.

3.4.3.2 *Mobile Detects the Incoming Call*

After the fine synchronization the system can page the mobile through PCH. Since mobile is monitoring the PAGCH, it will know when it is paged (Figure 3.12[a]). Mobile responds by sending a page response on the reverse access channel (Figure 3.12[b]). As a result of mobile request through the reverse access channel,

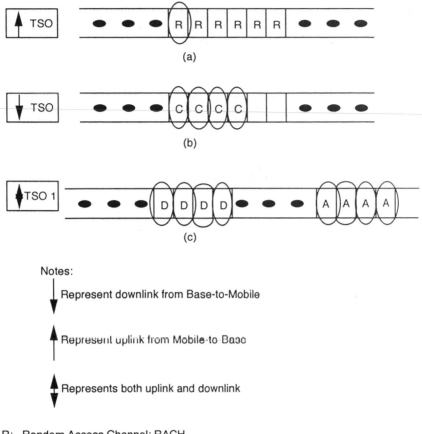

Notes:

↓ Represent downlink from Base-to-Mobile

↑ Represent uplink from Mobile-to Base

↕ Represents both uplink and downlink

R: Random Access Channel: RACH
C: Common Control Channel: BCCH: AGCH (b), PCH (b)
D: Standalone Dedicated Control Channel: SDCCH

Figure 3.11 Mobile access request for registration.

the GSM system will allocate SDCCH to the mobile via the AGCH (Figure 3.12[c]). It should be noted that since the reverse link can have a contention, the mobile may have to make a second try or even more tries before being successful. For subsequent requests, if necessary, the mobile has to wait for a random amount of time. The system and the mobile will also exchange necessary information to set up the call; for example, authentication, set-up information, measurement reports, and power control on SDCCH/SACCH (Figure 3.12[d]). Lastly, the mobile will be allocated a TCH by SDCCH. The mobile now starts conversation. Details of this are shown in Figure 3.12(e).

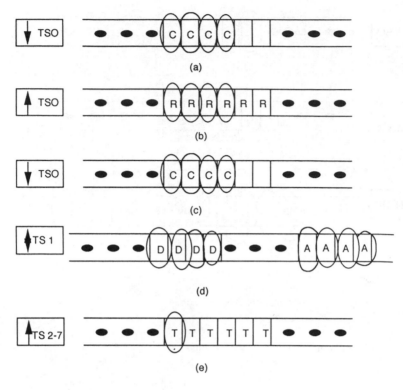

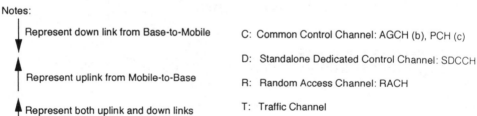

Notes:

| Represent down link from Base-to-Mobile | C: Common Control Channel: AGCH (b), PCH (c) |

Represent down link from Base-to-Mobile   C: Common Control Channel: AGCH (b), PCH (c)

Represent uplink from Mobile-to-Base   D: Standalone Dedicated Control Channel: SDCCH

Represent uplink from Mobile-to-Base   R: Random Access Channel: RACH

Represent both uplink and down links   T: Traffic Channel

Figure 3.12 Paging the mobile.

3.4.3.3 Mobile Originating a Call

Mobile sends an access request over the RACH (Figure 3.13[a]). The system allocates SDCCH to mobile via AGCH (Figure 3.13[b]). Set-up information is exchanged on SDCCH, which will include items such as authentication, measurement reports, and power control on the SDCCH/SACCH (Figure 3.13[c]). Lastly, a TCH is assigned through the SDCH where a conversation can start (Figure 3.13[d]).

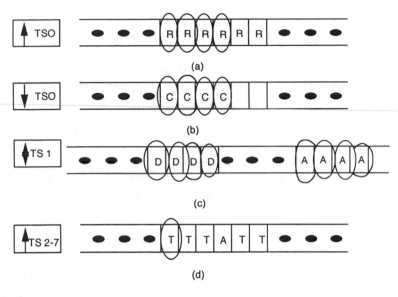

Notes:

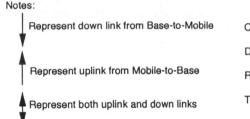

Represent down link from Base-to-Mobile

Represent uplink from Mobile-to-Base

Represent both uplink and down links

C: Common Control Channel: AGCH (b), PCH (c)

D: Standalone Dedicated Control Channel: SDCCH

R: Random Access Channel: RACH

T: Traffic Channel

Figure 3.13 Mobile organization of a call.

3.5 CONCLUSIONS

In this chapter we discussed the time and frequency domain representations of a GSM system. The flexibility of a GSM system is enormous in view of the fact that almost any channel can be assigned for any logical function. No channel is assigned a priori for any function except synchronization (frequency and time) and BCCH channels. To a large extent, the use of frequency hopping mitigates the effect of channel fading. We have also discussed some operational aspects of the system, which will be the main topics of discussion for the next chapter.

Problems

3.1 Why do you think that the whole frequency band of operation may not be allocated at the beginning of GSM system operation in a given area of a country?

3.2 State reasons to justify 75-MHz band allocation for a PCS system.

3.3 State reasons for not using the first channel in a GSM system.

3.4 Find the frame, multiframe, and the superframe rates.

3.5 Find the efficiency of normal, synchronization, frequency correction, access, and dummy bursts.

3.6 State reasons for choosing two different multiframe timings for speech and signaling channels.

3.7 List different logical channels and their associated functions.

3.8 If 912 bits of encoded data (with both block and convolutional coding) are generated every 40 ms, then find the gross data rate. If you assume this to be full rate speech data, what will be the gross rate for half rate speech?

3.9 List the messages (information) transmitted over BCCH, FCCH, and SCH. Justify why hopping cannot be used for these channels.

3.10 Actual user data rates of 9.6 Kbps, 4.8 Kbps, and 2.4 Kbps are padded to bring up the rates to 12 Kbps, 6 Kbps, and 3.6 Kbps. Find the data rate for the overhead.

3.11 Narrate the basic functions of the following channels: PCH, RACH, DCCH, AGCH, TCH, FACCH, and SACCH.

3.12 Why it is necessary to wait for a random amount of time before retransmission over RACH?

3.13 Can SACCH be linked to both TCH and SDCCH simultaneously or one after another?

3.14 Explain clearly the stealing mode of operation for FACCH.

3.15 Name different types of burst signals used in GSM. Justify the utility of including the training sequence in the middle of the normal burst. What is the reason for using TB?

3.16 State reasons for using a dummy burst over the air.

3.17 Find the burst transmission rate for FCCH and SCH based on the layout of Figure 3.7(c1).

3.18 Find the burst transmission rates of FCCH, SCH, BCCH, SACCH, and PAGCH/F based on Figure 3.7(d1).

3.19 Find the burst transmission rates of RACH/F and RACH/H based on the layout of Figure 3.7(e).

3.20 Compute the actual carrier frequency for channel numbers 30, 60, 90, 100, and 124.

3.21 Assuming the hopping is at the frame rate in a GSM system, compute the hopping rate. What happens to signal quality as the hopping rate is reduced to half or increased by a factor of two? Can we indefinitely increase the hopping rate in a GSM system?

3.22 What are idle and dedicated modes of mobile operation?

3.23 What protection does a GSM system provide against overlap of data on adjacent channels since users of these channels are widely separated from the

BS (one user close to BS and the other at the boundary of the BS coverage area)?

3.24 Why it is necessary to have a longer GP for access burst?

References

[1] Mehrotra, A., *Cellular Radio Analog and Digital Systems*, Norwood, MA: Artech House, 1994, p. 290.

[2] Haung, T., "Developing GSM Standards," *Pan-European Digital Cellular Radio Conf.*, Nice, France, 1991.

[3] Lycksell, E., "The A and A-bis Interfaces of the GSM System," Budapest, October 1990, Session 3.3.

[4] Mouly, M., et al., *The GSM System for Mobile Communications*, Michel Mouly and Marie-Bernadette Pautet, 1992.

[5] Eynard Carlo, "The Switching Subsystem—Functions and Elements," *GSM Seminar*, Budapest, October 1990, Session 4.1.

[6] Siegmund, M. R., et al., *An Introduction to GSM*, Norwood, MA: Artech House, 1995.

CHAPTER 4
▼▼▼

MOBILITY MANAGEMENT

4.1 INTRODUCTION

Mobility management entails the GSM system keeping track of the mobile while on the move. Basically, there are two different situations: mobile idle and mobile busy. These two cases together lead to all the relevant cases we need to consider here, namely, (1) MS is turned off, (2) MS is turned on but is in the idle state, and (3) MS is in the conversational mode. Let us elaborate on these three cases.

MS turned off is the case when the mobile cannot be reached by the network because it does not answer a paging message. It does not inform the system about possible changes of LA as it is simply inoperative. In this case, the MS is simply considered to be detached from the system (IMSI detached).

When the MS is turned on but in idle state, the system can page the MS successfully. This is the situation when MS is considered attached (IMSI attached). While on the move, the MS has to check that it is always connected to the best received BCCH. This is called roaming. While on the move, the mobile must also inform the system about changes of LA, which is called location updating. In summary, this is the state when the mobile is switched on, synchronized to the system, and ready to receive or place a call.

In the case that the MS is busy, the radio network has traffic channels allocated for the data flow to/from the MS. While moving, the MS must also be able to change to a new traffic channel as the signal on the traffic channel drops below an unacceptable level. This is called handover. In order to decide whether to handover, the MSC (BSC in some cases) interprets information received from the MS and BTS, which is called locating.

In view of this information, we shall discuss the complete mobility aspect of the system, which revolves around these three states. To achieve this objective we will briefly introduce the necessary aspects of signaling protocols directly applicable to this chapter. (The details of GSM protocols will be discussed in Chapter 8.) Following this we shall study the primary requirement of MS frequency and time synchronization. This is essentially the first step for successful operation of the mobile. In Section 4.4 we discuss these basic operations and form the foundation for more complex operations of MS within the GSM system. These include connection request, paging, identification of the MS within the system, authentication, ciphering, call clearing procedure, and IMSI attach and detach. More complex operations are covered in Sections 4.5 to 4.9 and include location update, mobile-to-land line call in both directions, mobile-to-mobile call, and the call handover process. For a firm understanding of the GSM system, it is essential that the reader first understand the basic steps covered in Section 4.4 before going into the more complex operations of the system covered in Sections 4.5 to 4.9. A good understanding of this chapter will enable the reader to understand the vital operations of this complex system.

4.2 SIGNALING PROTOCOLS [1,2]

GSM uses the CCITT SS7 protocols for its signaling. Figure 4.1 shows key elements of a GSM system as vertical bars, with spaces between them representing interfaces between different subsystems. The protocol is consistent with *Open System Interconnection* (OSI) layers, which are shown at the left.

At the physical layer, the air interface uses RF radio transmission. The A-bis interface between BTS and BSC uses 64-Kbps channels on wire cable, optical fiber, or microwave links based on the best availability and ease of installation. All other interfaces use the SS7 *Message Transfer Part* (MTP) level 1 with channel rates of 64 Kbps, and the medium can be coaxial cable, balanced wire pairs, or fiber optic cable. The data link layer uses ISDN *Link Access Protocol D* (LAP-D) on the A-bis interface, and a version called LAP-Dm on radio links to MSs. LAP-Dm uses the GSM RF interface framing and synchronization scheme rather than that of ISDN. All other interfaces use the SS7 MTP level 2 protocol for the data link layer.

The air and A-bis interfaces have no Network layer. All other interfaces within GSM use both MTP level 3 and the SS7 *Signaling Connection Control Part* (SCCP). MTP 3 provides routing of general connectionless messages to the various network nodes, while SCCP routes connection-oriented messages, specific to particular user

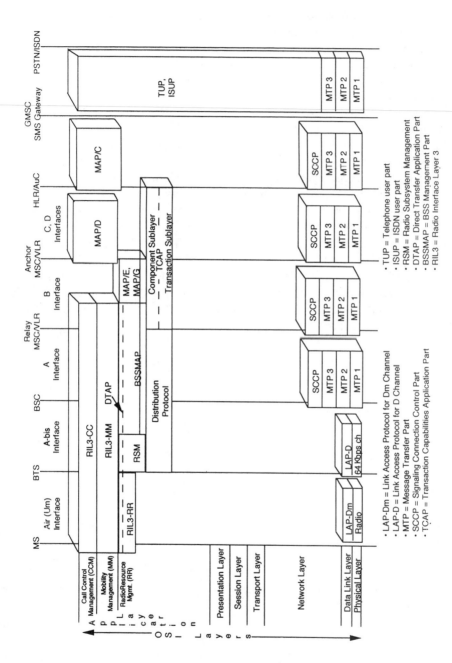

Figure 4.1 GSM signaling protocols.

transactions, all the way to the destination terminals. No GSM interfaces require the SS7 transport, session, and presentation layers. The rest of the interfaces are lumped at the application layer 3. Interfaces to the fixed networks (PSTN, ISDN) use their own standard *Telephone User Part* (TUP) and *ISDN User Part* (ISUP) protocols. At lower layers, MTP1 through MTP3 protocols are used. Once a *Radio Resource Management* (RR) connection has been established, there exists a physical point-to-point bidirectional connection between two RR entities. At one instant of time there will not be more than one RR connection available between a MS and the MSC. One RR connection may be used to support several *Mobility Management* (MM) connections.

The SS7 application layer has several sublayers specific to the GSM system. The *Radio Interface Layer 3* (RIL3) RR protocols, for example, establish and release radio connections between an MS and various BSCs for the duration of a call despite user movements, and provide system information broadcasting, inter- and intracell change of channels, and the ciphering mode setting. The *Radio Subsystem Management* (RSM) protocol provide RR functions between the BTS and BSC. The *Direct Transfer Application Part* (DTAP) protocols provide RR messages between the MS and MSC. The *BSS Management Application Part* (BSSMAP) protocols provide RR messages between the BSC and MSC. The distinction between DTAP and BSSMAP is provided by a Distribution protocol below them.

The RIL3 MM protocols deal with MS location management and security aspects of the system. Functions performed under location management include location updating, IMSI attach, and periodic updating. Some security aspect functions performed are authentication, TMSI reallocation, and identification. The RIL3 *Call Control Management* (CCM) protocols deal with mobile originating and terminating call establishment, supplementary services, and short messages. All MM and *call control* (CC) functions reside in the MSC. Thus, all CC and MM messages are interpreted by neither the BSC nor the BTS. They are transparently conveyed to its corresponding entity within the MSC. Before exchanging messages between MM entities of the MS and the network (MSC), a RR connection needs to be established.

The *Transaction Capabilities Application Part* (TCAP) provides a correlation between individual operations and structured exchanges in building up a complete transaction. The transaction sublayer manages transactions on an end-to-end basis. The component sublayer correlates commands and responses within a dialog. Non-call-related signaling between different network parts is handled by the various *Mobile Application Part* (MAP) protocols, details of which are shown in Figure 4.2. These protocols are designated as MAP/B through MAP/H. For example, signaling between MSC and the EIR is through the MAP/F. Though these messages are not directly related in completing the actual voice or data transmission path, they are essential for the proper establishment of the traffic channel.

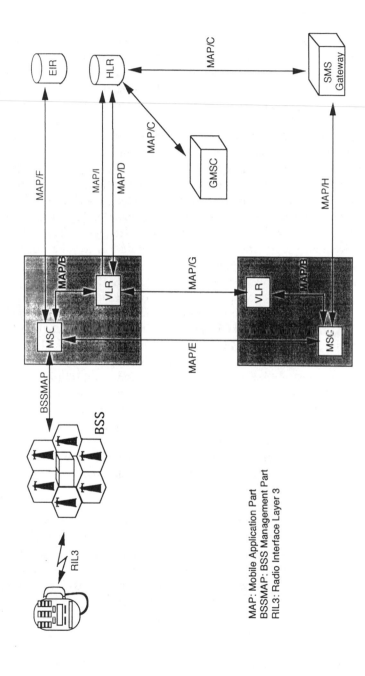

MAP: Mobile Application Part
BSSMAP: BSS Management Part
RIL3: Radio Interface Layer 3

Figure 4.2 Map protocol connections.

4.3 MOBILE INITIALIZATION

Prior to establishing any communication links to other parties, the MS must first acquire synchronization with the GSM system. This begins after the MS is turned on in a PLMN. The first step is for the MS to search for and acquire a FCCH burst on some common control frequency channel. Mobile will scan all the 124 RF channels and obtain the average signal strength of each channel. During the scanning process, several readings of the RF level have to be taken so that the mobile gets an accurate estimate of the channel power. Thus, the scanning may take several seconds.

For each of the 124 channels, starting with the one of highest signal strength level, the mobile searches for the frequency correction burst (FCCH). This is the first step of the process known as frequency synchronization. This information is present in the slot T0 (see Section 3.2.5 for a detailed description of T0). The frequency correction burst is unique and easily recognizable. The FCCH burst is a long sine wave that is offset by 67.7 kHz from the carrier frequency. The cell transmits all zeros for the frequency correction burst. The mobile has to take out this offset before an estimate of the carrier frequency can be made. This process of frequency synchronization is shown as the first step in Figure 4.3. If no frequency burst is detected, then mobile can go to a channel with the next highest signal strength level.

After the frequency correction burst is detected, the MS will try to synchronize with the time synchronization burst (SCH). The SCH always occurs in the next frame in the same timeslot as the FCCH as explained in Section 3.2.5. The occurrence of this is eight *burst periods* (BPs) later than the FCCH. The SCH contains precise timing information on the timeslot boundaries in order to permit refining the received slot timing. The SCH message also contains the current frame number, to which the MS is also synchronized, as shown in Figure 3.5. This time synchronization is generally carried out in two steps: coarse and fine. Here the internally stored synchronizing pattern is correlated, and at the peak of correlation the channel is considered to be synchronized. If synchronization does not occur, the process of frequency synchronization with the next highest power channel in the list may start. If the synchronization is successful, the mobile will read the TDMA frame number and the BSIC as discussed in Section 3.2.4, which is the part of SCH transmission.

Assuming that the mobile is synchronized, it decodes the information on a BCCH. The BCCH information contains such items like adjacent cell list, BCCH location, minimum received signal strength, and LAI and provides beacon frequencies of surrounding BTS cells. All BCCH transmissions are at a standard power level, which permits the MS to determine received power from that BTS as well as from adjoining BTSs also. Therefore, when the BCCH information is correctly decoded, the mobile will follow one of the following two paths.

If the BCCH information includes the present BCCH channel number, then the mobile will simply stay on the channel. If the current channel is not included in the BCCH information list or the received signal strength level (BCCH level) is below

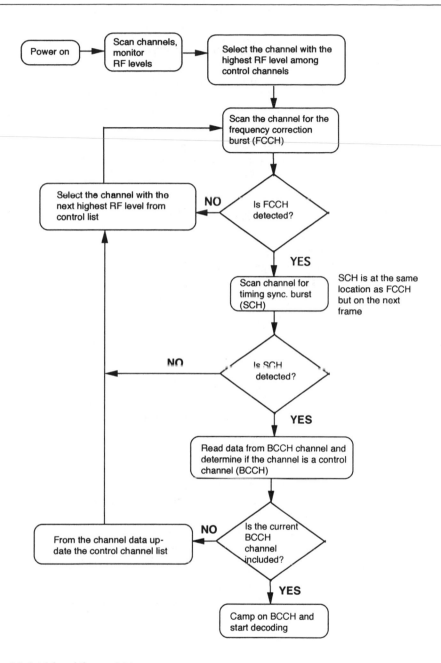

Figure 4.3 Initial mobile acquisition.

the desired level, then the mobile will continue searching for the next control channel (BCCH channel).

After the mobile has successfully synchronized to a valid BCCH, it must maintain the link and monitor the PCH. The mobile unit is also required to maintain information on the neighboring BCCHs. The information includes sync information and the average measured RF levels of at least six adjacent cell channels. This information is important for the hand-off process. In the next section we will expose the reader to essential elements of a call setup process before we discuss details of different types of calls in Sections 4.5 to 4.9.

4.4   BASIC STEPS IN THE FORMATION OF A CALL

In this section we discuss the necessary basic steps that form the foundation for complex operations that the MS and GSM systems must perform for the successful completion of a call. These operations include connection request, paging procedure, identification process, authentication, ciphering, and call clearing IMSI attach and detach [1].

4.4.1   Connection Request

To get any type of service from the system, the mobile has to request the service. This can occur in many ways. For example, the mobile unit may request a channel or can respond for the system page, IMSI attach, IMSI detach, normal and periodic location updating, call setup, or supplementary services. See Table 4.1 for details.

As shown in Figure 4.4, as a first step the mobile will request service on the Reverse Access Channel. The message consists of three bits, which include the reason for the request and a five-bit random number. If the request is not answered by the network within a set time (perhaps due to collision on the Random Access Channel), the mobile will repeat the message a predetermined number of times, til it is answered. The attached random number is the identification of the message. Upon receiving the message, BSS sends an acknowledgment with the same random number as

Table 4.1
Types of Connection Request by Mobile and Their Responses

Reason for Access	Initial Message
Paging response	RIL3-RR paging response
Location update or IMSI attach	RIL3-MM location updating request
IMSI detach	RIL3-MM IMSI detach
Call setup and supplementary services	RIL3-CC (application layer: call control management)

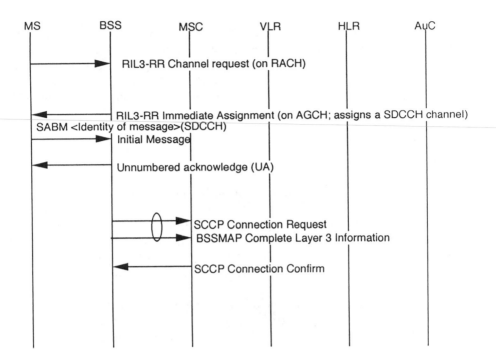

Figure 4.4 Connection request procedure.

originally received over the AGCH channel along with the channel number for SDCCH. This assigned channel is usually at 1/8 TCH (SDCCH + SACCH) rate. One-eighth TCH simply means that these channels transmit data once every eight frames; therefore, their data rate is only 1/8 of the TCH/F. It should be noted that the SACCH can also be assigned with SDCCH. This channel is also assigned when the mobile gets the assignment of a TCH. The MS at this point sends "SABM frame" as an initial message. This will allow MS to set up the link layer protocol (layer 2) over the radio path. This message contains the reason for the specific message. If the response is the result of paging, the mobile provides a "RIL3-RR paging response." If the response is the result of location updating or IMSI attach, the resulting message by mobile is "RIL3-MM location updating request." For IMSI detach the message on the application layer is "RIL3-MM IMSI detach." For call setup and supplementary services the message is RIL3-CC (Application layer: Call Control Management) and is shown in Figure 4.1. Thus, the reason for mobile response is contained as a part of the *Set Asynchronous Balanced Mode* (SABM) frame containing the setup request (CM Service Request) to the network. In response to a "SABM frame" message, BSS sends an *Unnumbered Acknowledge* (UA) message. At this point, BSS requests connection to MSC through a "SCCP connection request" message, to which the service request information is attached as a "BSSMAP complete layer 3 information" message. MSC in turn responds to BSS by a "SCCP connection confirm" message. Since one of the response messages is a result of paging, we provide details of paging in the next subsection. It should be noted that the RIL3-MM and RIL3-CC messages are between MS and MSC and pass directly through BSS. In other words, the BSS is transparent to these messages. BSSMAP messages are between BSC and MSC.

4.4.2 Paging Procedure

As discussed in the previous section, one reason for a connection request by mobile can be the result of paging by the system. The network initiates the paging procedure by broadcasting a "paging request" message on the appropriate paging subchannel. The use of a subchannel narrows the time when a group of mobile users will be looking for their page. This way they do not have to continuously monitor the PCH and thus can perform other tasks. There are three types of paging messages: Paging Request Types 1, 2, and 3 [1].

A "paging request" message includes, for each paged message by MS, an indication that defines how mobiles of different capabilities shall code the field in the "Channel Request" message. The information received in the Channel Request can be used by the network to assign a suitable channel.

A "paging request" message may include more than one MS identification. The choice of the message depends on the number of MSs to be paged and the types of identities that are used. The maximum number of paged MSs per message is four when using only TMSIs for the identification of the MSSs.

The MS is required to receive and analyze the paging messages and immediate assignment messages sent on the paging subchannel corresponding to its paging subgroup. Upon receipt of a "paging request" message, if access to the network is allowed, the addressed MS shall within 0.5 sec initiate the immediate assignment procedure by requesting a channel through the Reverse Access Channel. The establishment of the main signaling link is then initiated using an "SABM" message with the information field containing the "Paging Response" message as shown in Figure 4.5. Steps two to four in Figure 4.5 are the same as in Figure 4.4.

4.4.3 Identification Procedure

The identification procedure is used to identify the MS/SIM by its IMSI if the VLR does not recognize the TMSI sent by the MS. This can be the result of the mobile changing the MSC/VLR area from the last time it accessed the system or due to similar other reasons. If identification is required, the VLR first sends a "MAP/B Provide IMSI" message to the MSC as shown in Figure 4.6. As a result of this

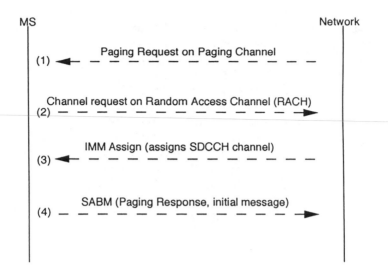

Figure 4.5 Paging procedure.

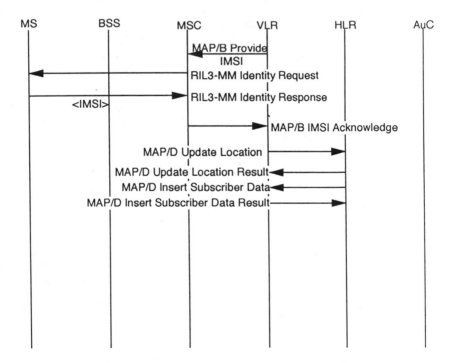

Figure 4.6 Mobile identification process.

message, MSC sends an "RIL3-MM Identity Request" message to the MS. The MS responds by returning an "RIL3-MM Identity Response" message, containing its IMSI, to the MSC. In Figure 4.6 we designate this operation by <IMSI>, meaning that the message contains the mobile IMSI. It should be noted that this is the only time that the IMSI is sent over the air unencrypted. Normally, TMSI is sent over the air, and IMSI is not required. The MSC then sends the "MAP/B IMSI acknowledge" to the VLR. If the IMSI is currently not in the VLR, then the VLR must get the individual user's file from the HLR, which identifies its IMSI. To do this, the VLR sends the HLR a "MAP/D Update Location" message. Assuming the IMSI is, in fact, registered in the HLR, the HLR responds with a "MAP/D Update Location Result" message followed by a "MAP/D Insert Subscriber Data" message containing other pertinent data needed by the VLR. The VLR acknowledges the data transfer with a "MAP/D Insert Subscriber Data result" message to HLR.

4.4.4 Authentication [1]

The authentication process may be run at each and every location update and at the initiation of every new service request. However, in some systems it may not be carried over at every location update or at every new service initiation. Thus, the frequency of authentication is mostly decided by the operator himself. The process starts at VLR. If the VLR determines that authentication is required, it sends a "MAP/D Send Parameters" message to the HLR, which relays this message to the AUC. The AUC then draws a value for the random challenge RAND and applies Algorithms A3 and A8 to generate the response SRES and the cipher key. This complete process is discussed in Section 5.3. The AUC then returns the triplet (RAND, SRES, Kc) value to the VLR in a "MAP/D Send Parameters Result" message. Actually, the AUC normally calculates and sends a few such triplets at a time for each requesting MS, so the VLR only has to request parameters from the AUC if it has no stored unused triplets for the particular MS.

The VLR then sends a "MAP/B Authenticate" message to the MSC, which in turn sends an "RIL3-MM Authentication Request" message, containing RAND, to the MS over the air. The MS calculates the required response $SRES_c$ using Algorithm A3 and authentication key K_i stored in the SIM. $SRES_c$ is returned to the MSC in a "RIL3-MM Authentication Response" message. The MSC compares $SRES_c$ with SRES (SRES response is already at MSC as it has been sent to MSC by VLR as a part of MAP/B authenticate message), and if they agree, it sends the MS an "RIL3-MM Service Accept" message. The MSC also sends the VLR a "MAP/B Authentication Complete" message. If $SRES_c$ and SRES disagree, the MSC sends the MS an "RIL3-MM Service Reject" message and terminates the request for service. The complete authentication process is shown in Figure 4.7. For further details of the authentication process, refer to Section 5.3.

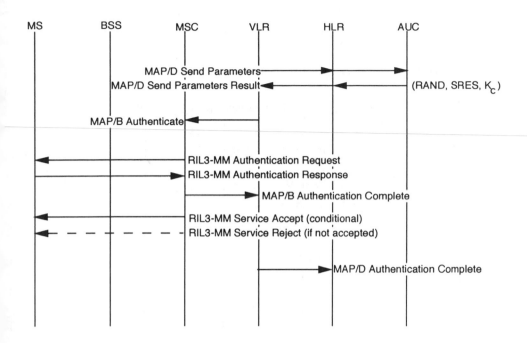

Figure 4.7 MS authentication process.

4.4.5 Ciphering

Encryption (or ciphering) is normally required for all user transactions over the RF link once the user has been authenticated by the system. This is an optional feature and may or may not be included in the particular system. Thus, the decision to switch encryption is dependent upon the operator. Ciphering begins with the VLR sending the MSC a "MAP/B Set Cipher Mode" message containing the value of K_c for use. This K_c is generated during authentication by using Algorithm A8 at AUC, as discussed in Section 5.4.

The MSC sends the new ciphering mode and ciphering key to the BSS in a "BSSMAP Cipher Mode Command" message. The BSS in turn sends an "RIL3-RR Ciphering Mode Command" message to the MS. The MS then switches to encrypted transmission and reception and sends back an "RIL3-RR Cipher Mode Complete" message in encrypted mode. After the BSS receives this message, it too switches to encrypted transmission for subsequent bursts. The BSS then sends a "BSSMAP Ciphering Mode Complete" message to the MSC to indicate that the encryption process is complete. This complete ciphering procedure is shown in Figure 4.8.

4.4.6 Call Clearing

Call clearing can be initiated by either the land-based user or the mobile. Both of these are discussed in the following subsections [1].

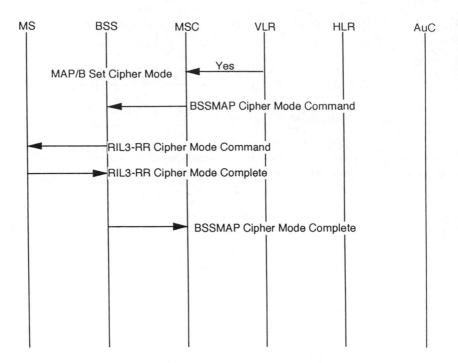

Figure 4.8 Encryption process.

4.4.6.1 *Network-Initiated (Mobile-Terminated Call)*

The release process in this case starts with a "Release" request from PSTN. Upon receiving this message the network (MSC) initiates the clearing of a call by sending a "Disconnect Message" to the MS. The MS in turn responds by sending a "Release Channel Message" in DTAP protocol to the MSC. Upon receiving the "Release Channel Message" from the MS, the network sends a "Release Complete" to the MS and sends the confirmation message for channel release to PSTN. The complete call-clearing process as initiated by land is shown in Figure 4.9(a).

4.4.6.2 *Mobile-Initiated (Network-Terminated Call)*

The MS initiates the clearing of a call by sending a "Disconnect Message" to the network. The clearing procedure applies with the exchange of the "Release and Release Complete Messages" from the network and MS, respectively.

Upon receiving the "Disconnect Message" from MS, MSC can send the "Release" message to PSTN without waiting for the "Release Complete Message" from the MS as shown in Figure 4.9(b).

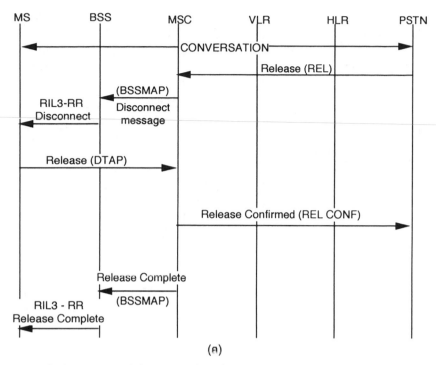

Figure 4.9 Call clearing: (a) mobile-terminated, and (b) mobile-originated.

4.4.7 IMSI Attach and Detach

The IMSI attach and detach procedures register and unregister the mobile to the system. If the mobile is attached, he will be paged in the LA where he is present. If the mobile is detached, the system will not waste its resources in paging for an incoming call [1,2]. Registering and unregistering are necessary for the system to page or not to page the mobile.

4.4.7.1 IMSI Attach Procedure

The IMSI attach procedure is used by the MS to indicate that it has reentered the active state (power on). The IMSI attach is also performed as a part of the location updating procedure. The IMSI attach procedure is a complement of the IMSI detach procedure, and it is used to indicate that the IMSI is now active in the network. IMSI attach is invoked if the attach/detach procedures are required by the network and an IMSI is activated in a MS (that is, activation of a MS with plug-in SIM or the insertion of a card in a SIM card-operated MS) within the coverage area of the

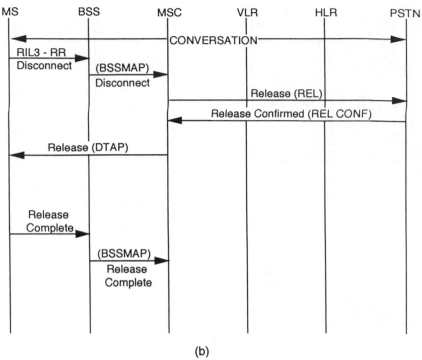

(b)

Figure 4.9 (continued).

network or when a MS with an IMSI activated outside the coverage area enters the coverage area. IMSI attached is marked in MSC/VLR with an "attached" flag. This flag is related to IMSI. The following sequence of events describes the IMSI attach procedure as shown in the Figure 4.10(a). Upon turning the power on, the MS sends an "RIL3-RR Channel Request Message" on the Reverse Access Channel to BSS. The network assigns the channel and the BSS sends an "RIL3-RR IMM Assignment Message" to MS over the AGCH (as discussed earlier) for the connection request message. (See Figure 4.4.) This message assigns the SDCCH to mobile. After the channel is assigned, MS sends an "RIL3-MM IMSI Attach" message over the SDCCH to BSS, which is forwarded to MSC and then to VLR as a MAP/B protocol message. VLR acknowledges MSC as "IMSI Attach Acknowledge" as a MAP/B protocol, which is forwarded to BSS and then to MS. MSC also sends "Clear Command" for the channel release to BSS as BSSMAP protocol, which is then forwarded to MS. Upon receiving the "RIL3-RR Disconnect" signal from MS, a "Clear Complete" message is sent to MSC as BSSMAP protocol.

4.4.7.2 IMSI Detach Procedure

Similar to IMSI attach, the IMSI detach procedure may be invoked by a MS if the MS is deactivated or if the SIM is detached from the MS. A flag (ATT) broadcasted

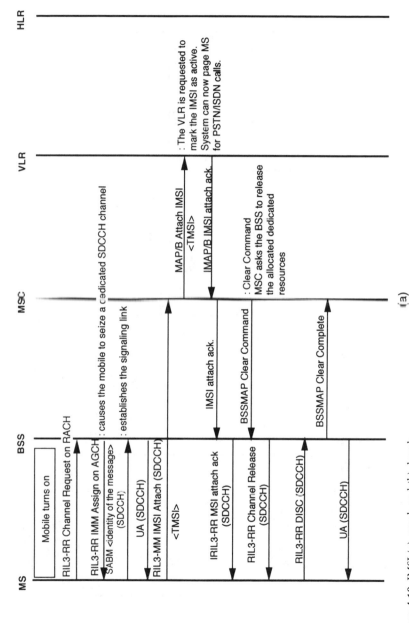

Figure 4.10 IMSI (a) attach and (b) detach.

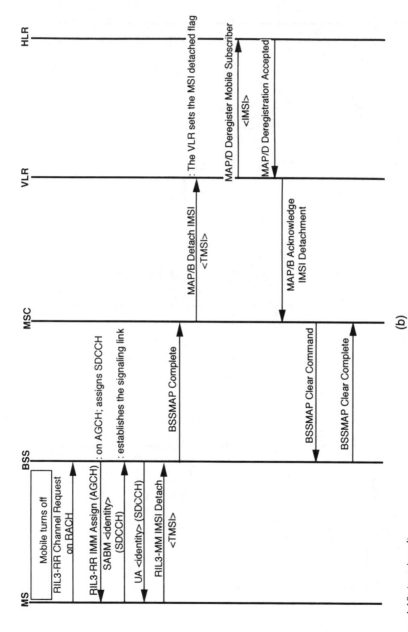

Figure 4.10 (continued).

in the "System Information Message" on the BCCH is used by the network to indicate to the MS whether the detach procedure is required. The procedure causes the MS to be declared as inactive in the network. Once IMSI detach is active, MS can neither transmit nor receive. The system will also not page MS.

The IMSI detach procedure starts with the MS sending an "RIL3-RR Channel Request" message on RACH to BSS. BSS assigns the SDCCH and notifies the channel assignment to the MS over AGCH. The MS then sends an "RIL3-MM IMSI Detach Indication" message to the BSS. The message identifies the MS (indicated here as <TMSI>) and contains an eight-bit code indicating IMSI detach. After receiving the "IMSI Detach Indication" message from MS, the BSS forwards this message in a BSSMAP complete layer 3 information message to MSC. MSC in turn updates the state of MS in the VLR with a "MAP/B Detach IMSI" message. At this stage all terminating calls to MS are rejected and the system does not page the mobile anymore. VLR forwards this message to HLR as "MAP/D Deregister Mobile Subscriber," and HLR marks MS as unregistered. HLR forwards a "MAP/B Deregistration Accepted" message to VLR, which in turn sends a "MAP/B Acknowledge IMSI Detachment" message to MSC. No response is returned to the MS as shown in the attached sequence chart, Figure 4.10(b). This is rightly so because the mobile would be switched off before the return message is sent from the BSS to MS. MSC sends a "BSSMAP Clear Command" to BSS to clear the SDCCH channel assigned to MS. The BSS acknowledges via a "BSSMAP Clear Complete" message to MSC. These commands for IMSI detach procedure are shown in Figure 4.10(b).

In the event that power is cut off from the MS, the VLR will have an implicit unregistration procedure such that if a location update or an "IMSI Detach" has been received within some predetermined time (30 min or so—it varies from system to system), the mobile is unregistered automatically. MS is also forced to register every so often (for example, say every 30 min), which is called periodic registration. If the system does not receive periodic registration it will mark the MS "detached" in its VLR. How often MS has to do the registration is indicated by the message over the BCCH. With the help of this elementary procedure, we now form a complex procedure of user location update.

4.5 LOCATION UPDATE [1]

MS location updating is performed to tell the system where to search for the MS during paging for an incoming call. If the location is known to a definite subregion of a particular PLMN, then this will reduce the number of cells where the mobile has to be paged, thereby reducing the load on the system.

The MS location is determined from the cell identification of the strongest BCCH signal received by the MS. The MS regularly measures the received signal strengths of the BCCHs for all surrounding cells at least once every 6 sec (superframe cycle). It stores at least the six strongest BCCH measurements and their identifications in the SIM, which can subsequently be used for handover decisions. MS also transmits the LA of the strongest cell to the MSC during location updating. The LA may consist of a few cells or a contiguous group of cells under the control of one BSC, as shown in Figure 4.11. However, all cells in a LA must be under the control of a single MSC in a single PLMN, as discussed in Section 2.2.

As discussed in Section 4.3, after channel acquisition, the mobile decodes the BS information over the BCCH. The BS information includes country code, network code, and *location area code* (LAC), and possibly the cell identity. If any of these pieces of information do not match with the information stored in the SIM card, the MS thinks that it has been moved into a new area and reports its location to the system so that incoming calls can be routed correctly. Also, the MS will periodically report its location if there has been no activity for a given amount of time. The time period of reporting is part of the information contained in the BCCH and may vary, due to network loading. The MS location is also updated in the system whenever the MS changes cells during conversation.

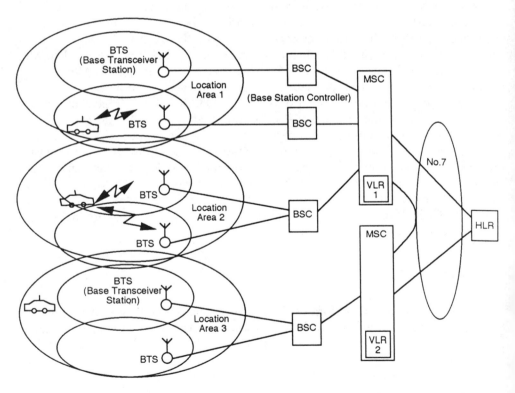

Figure 4.11 Location updating [3].

It is necessary for the system to know exactly where the subscriber unit is located. When a unit needs to be reached to receive a call, GSM pages based on the previously reported location of mobile. All idle units will receive this page and compare their ID numbers with the message to determine the intended recipient.

As explained earlier, the system must know the location of all active MSs at all times as they roam. As seen in Figure 4.11, each cell is served by one BTS. Each LA is divided into many cells, which may be served by one or more BSCs. VLR may serve one or more LAs. An inactive mobile is ignored by the system. As soon as the mobile switches its power, it retrieves its stored LAI and compares it with the one broadcast within its present cell. If they match, the mobile does not have to do anything since the subscriber is already correctly located; however, if it does not match, the mobile identifies itself by transmitting its IMSI together with the identities of the previous and present LAs. The BSS transmits this information to the associated VLR.

Each time a MS moves into a new LA, the corresponding VLR is informed. If both the present and previous areas are served by the same VLR, the MS is given a new TMSI and its location is updated in the VLR memory. On the other hand, if the mobile enters a new VLR area, its HLR, the old VLR, and the new VLR are informed. The old VLR erases the data for the mobile and the new VLR records relevant parameters needed to process calls.

The message sequence is shown in Figure 4.12. If the MS is switched on in a LA different from the previous one, or it moves across boundaries of a LA in the idle state, an "RIL3-Location Updating Request" message, which is sent from MS to BSS, is relayed to MSC. MSC in turn alerts VLR by a "MAP/B Update LA" message. The message contains the old LA that the mobile had in its storage along with its TMSI (designated here as $<LAI_o, TMSI>$). The initial process of channel request has already been discussed in connection with the request process in Section 4.4 and will not be repeated here. The process of authentication and ciphering can now start. As explained before, every GSM system may not be equipped with ciphering capability. If the system is equipped with ciphering, it will be initiated by a "Set Ciphering Mode from VLR to MSC" message as MAP/B protocol. After completing the ciphering process, the message is sent from VLR to MSC for reallocation of the TMSI if desired ("Forward new TMSI"). A "TMSI Reallocation Complete" message is sent from MS to BSS after reallocating a new TMSI. The HLR sends "MAP/D Location Update Result" to VLR, which in turn sends a "MAP/B Location Update Acknowledge" message to MSC. This message is subsequently forwarded to the MS as an "RIL3-RR Location Update Accepted" message. In the event that the HLR rejects the request, a VLR "RIL3-MM Location Update Reject" message is sent from MSC to MS (shown as dotted). Either an accept or a reject message is initiated from MSC. LA updating may not be accepted because of the following reasons.

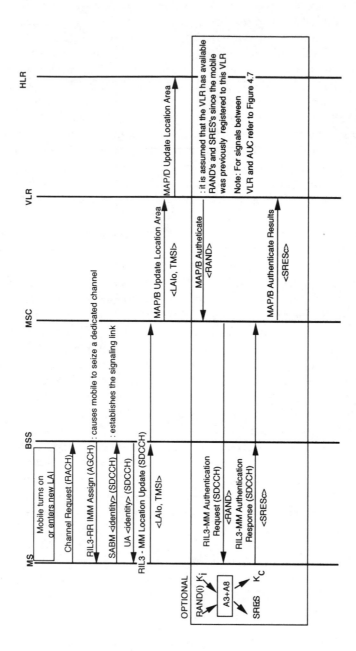

Figure 4.12 Location update process.

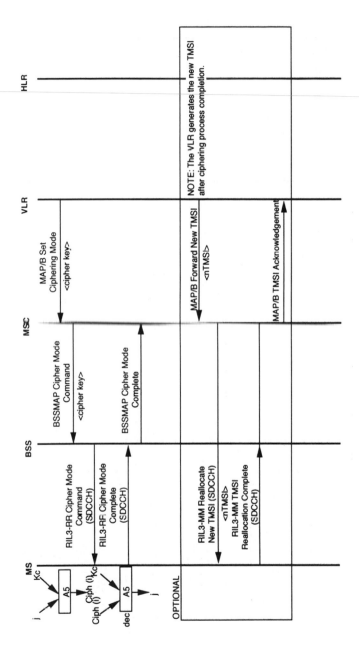

Figure 4.12 (continued).

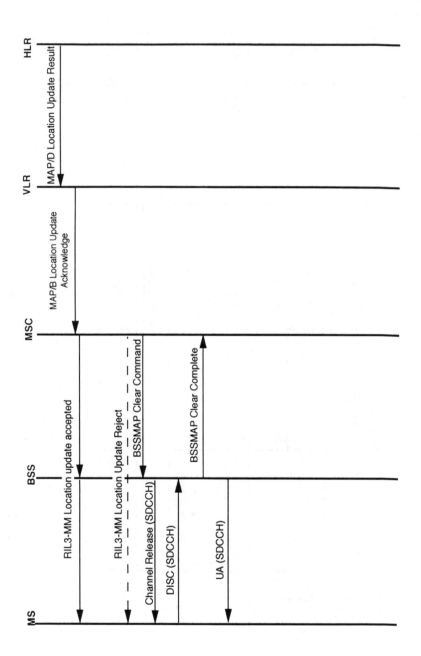

Figure 4.12 (continued).

- Unknown subscriber;
- Unknown LA;
- Roaming not allowed;
- System failure.

After receiving a location update accept or reject message, the MSC asks the BSS to release the allocated dedicated resource by sending "BSSMAP Clear Command" to BSS, which is then forwarded to MSC as a "BSSMAP Clear Complete" message, which completes the location updating process.

Thus far, we have discussed all the basic steps in the formulation of a call. In the next section we shall discuss the details of complex calling processes.

4.6 MS-PSTN CALL

Before we discuss this topic, let us address a side issue connected with the call setup scenario that is known as *Off-Air Call Setup* (OACSU). In the basic call set-up scenario, the transmission path between the MSC and the MS is fully operational for the requested service before the MSC is aware that both parties are ready to converse [1]. Thus the TCH is assigned right after the call initiation process without waiting for an MS alert signal from the system. A possible variant of OACSU is that the allocation of the suitable RR is delayed as much as possible in an attempt to save RRs. The distinction between the two cases is shown in Figure 4.13(a,b). In the OACSU case, the system assigns a TCH only after the alert signal is initiated by the mobile; TCH assignment occurs before the alert signal initiation for calls without OACSU. As a matter of fact, the assignment of a TCH can wait in the OACSU case until the distant party (that is, the party connected to PSTN/ISDN in this case) answers. This maximizes the traffic-carrying capacity of the system. Given this brief introduction of OACSU, let us discuss the complete process of a mobile unit originating a call to PSTN/ISDN subscriber. It should be noted that both these cases are the same, except for the stage when the TCH is assigned by the system.

Let us assume that the mobile is active and idle (IMSI attached) and that the user of the MS wants to set up a call. The subscriber dials the PSTN subscriber's number and starts the call originating process by pressing the "send" button. At this moment the mobile sends the "Channel Request" message over the Reverse Access Channel, which is followed by a response over AGCH by BSS ,which identifies the SDCCH allocated to mobile. The MS initiates an immediate assignment and sends a SABM frame containing the set-up request (CM Service Request) to the network. The same message is returned as a UA-frame over the SDCCH. When MS is connected to MSC, the MSC sends a "MAP/B Service Request" message to VLR. At this stage, the network may initiate authentication and subsequently starts the cipher mode. TMSI, if required, may also be assigned. After TMSI allocation the

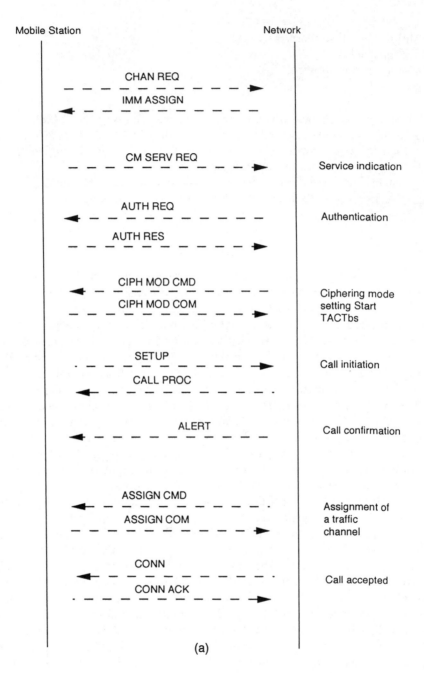

Figure 4.13 Mobile originating call (a) with OACSU (late assignment) and (b) without OACSU.

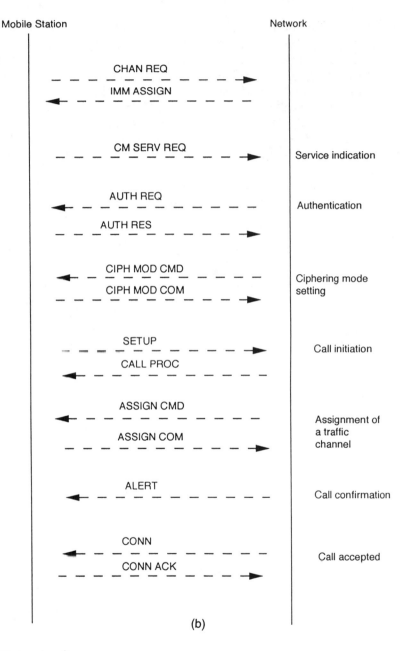

(b)

Figure 4.13 (continued).

MS will initiate call establishment by sending the "RIL3-CC Setup" message to the network. The message contains an MS-ISDN number, the number of the called party, and the type of service requested. Upon receiving this message, MSC sends to VLR a "MAP/B Send Call Setup Information" message, which contains the information of what is desired. If the originating MS is authorized for the desired services (such as, type of call and destination), the VLR responds with a "MAP/B Call Complete" message listing the MS capabilities, subscribed services, and parameters needed by the network to properly allocate a channel for the call. The MSC then sends the MS an "RIL3-CC Call Proceeding" message to MS. MSC also assigns the TCH through an "RIL3-CC Assignment Command." After receiving the "RIL3-CC Assignment Complete" message from MS, MSC sends a "TUP/ISUP *Initial Address Mobile*" message (IAM) to the PSTN/ISDN switching center. The switching center is now responsible for properly routing the call. When the called party receives this message, it rings to alert the subscriber and returns a "TUP/ISUP Address Complete" message to the PSTN/ISDN, which in turn alerts the MSC by sending "TUP/ISUP *Address Complete*" Message (ACM). The MSC then sends an "RIL3-CC Alerting" message to the MS, informing that the called party has been alerted.

When the called party answers by taking his receiver off the hook or pressing the appropriate key on his terminal, the terminal sends the PSTN or ISDN switching center a "TUP/ISUP Answer Signal" message as appropriate. The switching center relays this message to the GSM MSC, which informs the MS with an "RIL3-CC Connect" message. The MS responds with an "RIL3-CC Connect Acknowledgment" message. The MSC then establishes the two-way communication connection between the GSM MS and the called land-line party. The conversation now starts as shown in Figure 4.14.

The system may also ask for an IMEI check, which is initiated by MSC to BSS and then to MS. After receiving the IMEI response from MS, a "Check IMEI" message is sent to HLR. HLR responds to MSC by sending a "Check IMEI Response" message. It should be noted that this is an optional feature that may or may not be exercised by an operator.

4.7 PSTN-MS CALL

Mobile terminating call establishment is initiated by the network by sending the "Paging Request" message. Upon receiving this message the MS initiates the immediate assignment procedure and responds to the network by sending the "Paging Response" message. Details of this are illustrated in Figure 4.15.

For an MS-terminated call from a PSTN/ISDN caller to a GMS PLMN, the caller first places his call with a "TUP/ISUP Initial Address Mobile" message to the PSTN/ISDN switching center in his network. This message contains the MS-ISDN telephone number (<MSISDN>) and the type of service required. The PSTN/ISDN

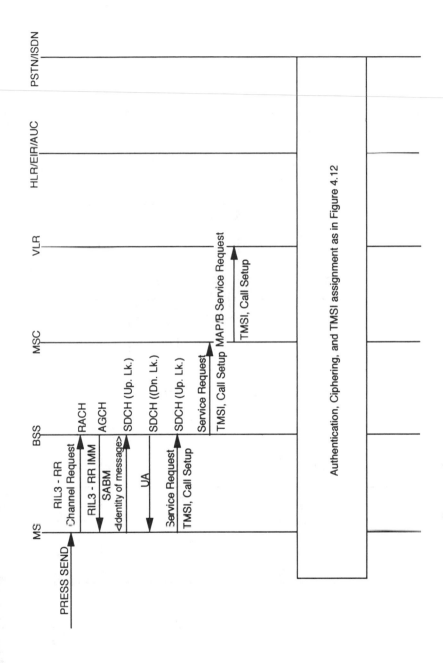

Figure 4.14 MS originating calling procedure.

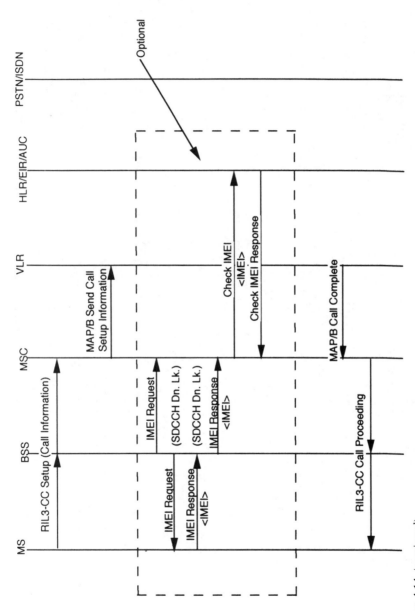

Figure 4.14 (continued).

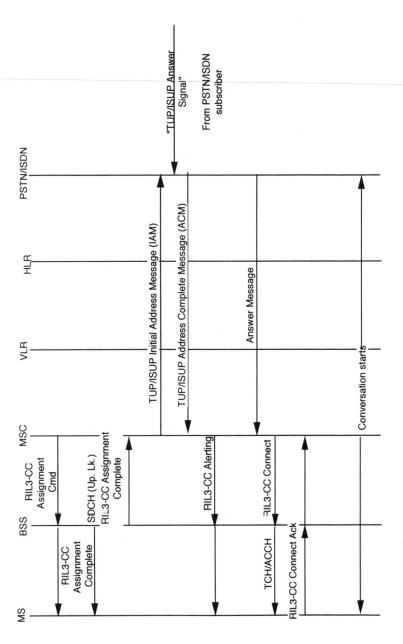

Figure 4.14 (continued).

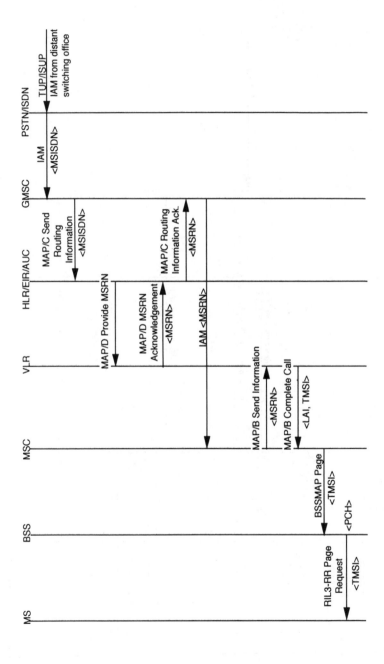

Figure 4.15 MS terminating calling sequence.

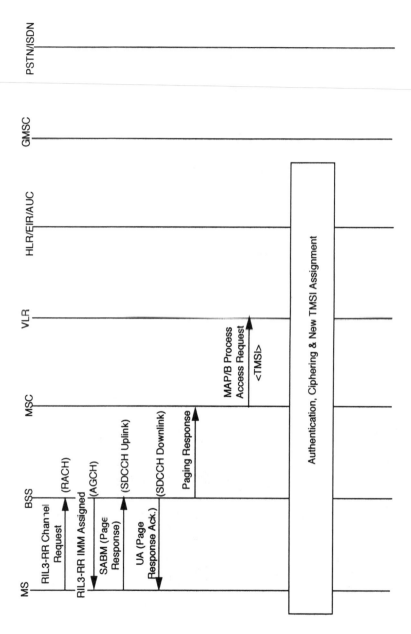

Figure 4.15 (continued).

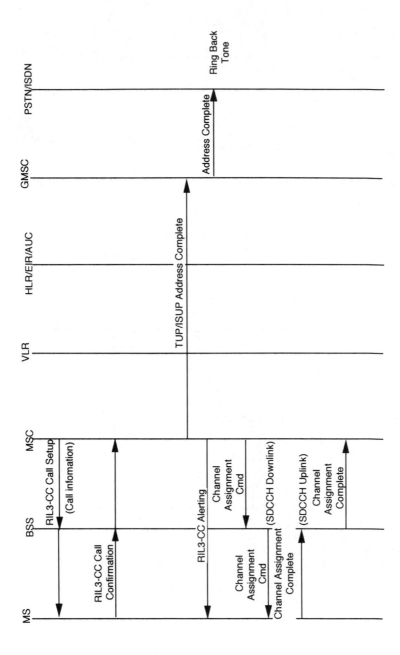

Figure 4.15 (continued).

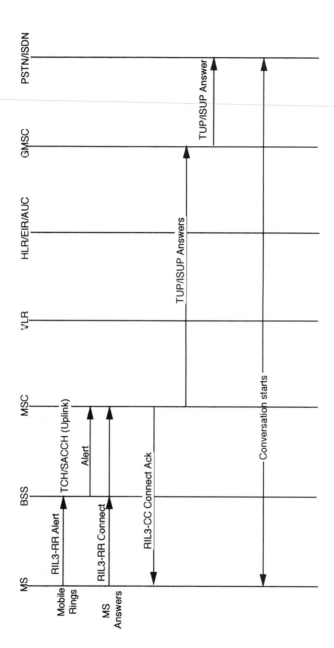

Figure 4.15 (continued).

switching center recognizes the GSM PLMN code in the IAM message and routes this message to the GMSC. The GMSC recognizes the user HLR designation in the MS-ISDN message and sends a "MAP/C Send Routing Information" message to the HLR to find out where the MS is presently located. The HLR knows in which VLR LA the MS is presently located and sends to VLR a "MAP/D Provide MSRN" message to obtain the MSRN. The VLR provides in response a "MAP/D MSRN Acknowledgment" message. The HLR then sends to the GMSC a "MAP/C Routing Information Acknowledgment" message identifying to the GSM system where the MS is and its MSRN. Since the GMSC now knows the whereabouts of mobile, it sends the IAM message to the proper MSC where the mobile is presently located.

At this stage, the MSC sends to the VLR a "MAP/B Send Information" message requesting the called MS's capabilities, subscribed services, and parameters needed by the network to properly allocate a channel for the call. Provided that the called MS is capable and authorized to receive the call, the VLR sends back to the MSC a "MAP/B Complete Call" message, giving the information needed to properly assign a channel for the call. The LAI and the mobile TMSI are parts of the "MAP/B Complete Call" message. The MSC uses this information to route a "BSSMAP Page" message to all BSS cells in that LA. The BSS in turn sends "RIL3-RR Page Request" for the MS.

Upon recognizing its TMSI in the paging message, the terminating MS proceeds through the previously described process of channel request through the Reverse Access Channel to gain assignment of a SDCCH and get connected to the MSC. After the MS is connected to the MSC, the MSC issues a "MAP/B Process Access Request" message to the VLR. If the TMSI, sent by the MS, does not agree with that stored in the VLR, the VLR may request MS go through the previously described identification procedure. The MS may also be required to go through the authentication and ciphering processes, as previously described.

Next, the MSC sends the MS an "RIL3-CC Call Setup" message giving the phone numbers of both the calling (PSTN or ISDN number) and called (MS-ISDN number) parties and the type of service requested. The MS performs a comparability check and responds to MSC with an "RIL3-CC Call Confirmation" message that is forwarded to MSC. At this stage MSC sends an TUP/ISUP Address Complete" message to GMSC that in turn is sent to the PSTN/ISDN switching center. At this time the calling party gets the ring tone. At the same time the MS is assigned a TCH by the previously described channel assignment procedure. Now MS rings or buzzes to alert the subscriber of the incoming call and the MS sends an "RIL3-CC Alerting" message to the MSC to inform that the subscriber has been alerted. When the mobile subscriber answers the alert by pressing the Answer Key on the MS subscriber set, the MS sends an "RIL3-CC Connect" message to the MSC. The MSC then responds with an "RIL3-CC Connect Acknowledge" message to the MS and simultaneously sends a "TUP/ISUP Answer" message to the GMSC. The GMSC and the PSTN/ISDN switching center relay this message to the originating PSTN/ISDN terminal.

After the complete two-way communication channel is established between the fixed and mobile terminals, the conversation starts.

4.8 MS-MS CALL

The establishment of a MS to MS call within the same BSC control area is the most basic GSM call operation, shown in Figure 4.16. The protocol begins with the originating MS going through the Connection Request Procedure, as discussed in Section 4.4.1, to request and be assigned a channel (SDCCH channel) and send an initial message requesting a call. At the end of the connection request, the MSC sends a "MAP/B Access Request" message to the VLR [1,3].

If the VLR does not recognize the TMSI sent by the MS, then the MS must go through the identification procedure as described in Section 4.4.3. This process is shown as optional (dotted) in Figure 4.16. The authentication process may also be run either with TMSI or IMSI received by the VLR. After authentication, the VLR acknowledges the process access request with a "MAP/B Access Request Accepted" message to the MSC. The system may then go through the ciphering procedure as discussed in Section 4.4.5. The MS may also have to go through the TMSI reallocation procedure as if the mobile has just changed its LA.

At this point, the originating MS sends an "RIL3-CC Setup" message giving the phone numbers (MS-ISDN) of both the calling (itself) and called parties and the type of service desired. The MSC in response sends to the MS an "RIL3-CC Call Proceeding" message.

Next, the MSC checks the MS-ISDN phone number of the called party and, in this particular example, discovers that the called mobile is registered in the local HLR (since the called mobile is within the same BSC). The MSC then queries the HLR for the called MS's location through a "MAP/C Send Routing Information" message. The HLR responds with a "MAP/C Send Routing Information Result" message. From this message, the MSC finds that the called MS is located locally and is also listed in the local VLR. The MSC then queries the VLR, with a "MAP/B Send Information for Incoming Call Setup" message requesting the called MS's capabilities, subscribed services, and required parameters by the network to properly allocate a channel for the call. If the called MS is capable and authorized to receive the call, the VLR sends to the MSC a "MAP/B Complete Call" message giving the information needed to properly assign a channel for the call.

Since the called mobile belongs to the same LA, the MSC uses this information to route a "BSSMAP Paging" message to all BSSs controlling cells in that LA. BSS then pages the called MS over their paging channels with an "RIL3-RR Paging Request" message on the PAGCH.

Upon recognizing its TMSI in the paging message, the terminating MS proceeds through the same connection request procedure, identification, authentication, ciphering, and TMSI reallocation procedure as discussed for the originating MS.

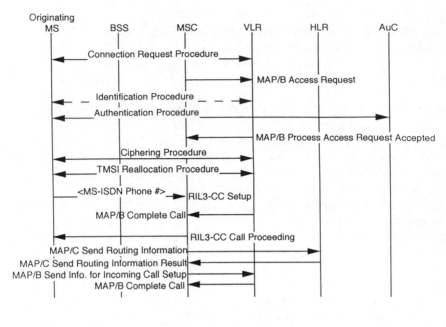

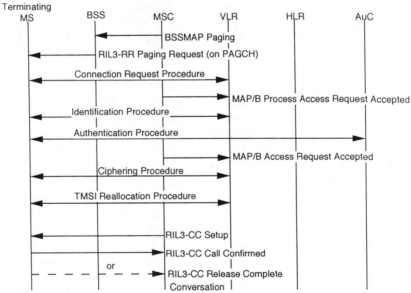

Figure 4.16 MS to MS call within the same BSC.

The MSC then sends the terminating MS an "RIL3-CC Setup" message providing details of the call, including the calling party and the connection services required. If the MS chooses not to accept the call, he rejects the setup with an "RIL3-CC Release Complete" message to the MSC. On the other hand, if the called MS chooses to accept the call, it responds with an "RIL3-CC Call Confirmed" message. Conversation now starts between the calling and called MSs.

4.9 CALL HANDOVER [1,3,4]

Handover is defined as the process of automatically transferring a call in progress to a different cell to avoid adverse effects of user movements, such as the loss of received signal strength. There are three purposes for handover. First, rescue the call, which imposes urgency for call transfer. If the call is not transferred immediately, it will be lost. The reason for call loss may be the excessive cochannel interference within the cell. This is the most common purpose. The second purpose may be to reduce the interference that a call in one cell may cause to calls in another cell. The third purpose is to ease traffic congestion by moving some calls from a highly congested cell to a cell that is not so congested.

There are also three cases of handover, all of which must be treated somewhat differently. First, there is handover from one radio channel to another of the same or different BTS (same BSC). Second, there is handover between channels of different BSCs under the control of the same MSC. Third, there is handover between channels under the control of different MSCs in the same PLMN.

These three hand-off processes are as shown in Figure 4.17. Assume that the initial mobile call is established through MSC (a) to BSC (1), through link 1a to cell 1. If the measurements by the mobile and the BSC indicate that the adjoining cell 2 is a better choice, the BSC selects link 1b and a new voice channel and informs the mobile to tune to this new channel. The new speech path becomes MSC (a) to BSC (1), through link 1b to cell 2. The initial link 1a is released. Further movement of the mobile may bring the mobile to cell 3, which is controlled by a different BS but still within the same MSC. The process is the same, but the new path becomes MSC (a) to BSC (2), through link 2a to cell 3. Finally, the mobile may come to a cell that is controlled by a different BSC and MSC, for example MSC (b). In this case MSC (a) will extend the fixed telephone link to the new MSC (b), but retain supervision of the call. The MSC (a), in this case, is known as the anchor MSC. The MSC (a) delegates activation and deactivation to MSC (b) during future handover to a new MSC. The voice channel path in this case becomes MSC (a) to MSC (b) to BSC (3), through link 3a or 3b to the mobile. When the MS enters a cell that depends on another MSC to MSC link (like link 4), a new transit link is prepared by MSC, in this case link 4 (from MSC [a]), and replaces the old link 3 between itself and the old MSC (b).

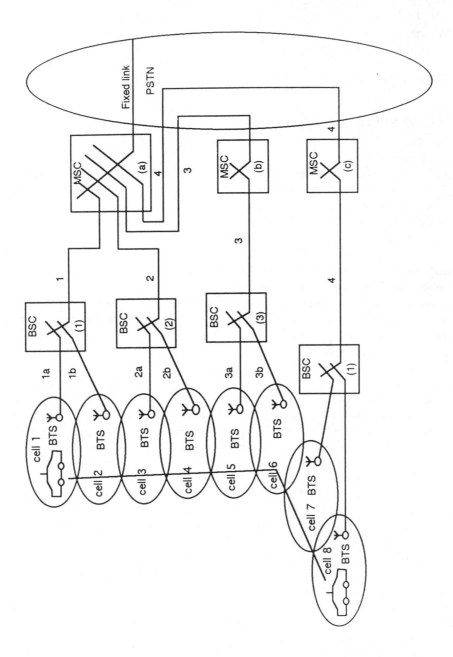

Figure 4.17 Mobile handover [3].

There are two modes of handover: synchronous and asynchronous. In synchronous handover, the old and new cells are synchronized so that their TDMA timeslots start at exactly the same time. This permits the MS to compute the timing advance for the new cell without any help. The difference in timing advance between the old and new cells is then just twice the difference in arrival times of signals within the same slot to the MS from the two different BTSs. In asynchronous handover, the old and new cells are unsynchronized, so the MS cannot independently correct the timing advance in this way. The timing advance must then be initialized at both the MS and the new BTS during handover. The MS sends the new BTS access bursts with a zero timing advance, and the BTS then determines the required timing advance from the round-trip time delay of the message. The BTS then sends the required timing advance to MS.

We shall now provide the detailed protocols for the simplest of all the cases, namely, the handover between two radio channels belonging to the same BSC. Details of this are shown in Figure 4.18.

As stated, there are two cases: handover between channels on the same BTS (same cell) and between channels on a different BTS (different cell). In both cases BTSs are controlled by the same BSC. Handover to another cell is the usual case, but handover to another channel in the same cell might also take place in order to mitigate interference with other cells. Thus, we shall discuss here the handover to different cells controlled by the same BSC. In this diagram, the MS is shown at both ends, indicating its connections to the old and new BTSs.

With a call in progress, the BSC may determine that a change of channel or cell (BTS-2) is necessary. The BSC requests a new channel from BTS-2 with an "RSM Channel Activation" message. BTS-2 allocates a channel, if available, and responds with an "RSM Channel Activation Acknowledge" message to the BSC. The BSC then sends an "RIL3-RR Handover Command" message to the MS on the FACCH, via BTS-1 (mobile is currently talking through BTS-1), assigning the new channel, its characteristics, the power level to use, and whether to use synchronous or asynchronous handover. The message will also include the assignment of a new SACCH. Upon receiving this message, the MS initiates the release of the old channel and the connection to the new one.

Two procedures are possible depending on whether the old and new cells are synchronized or not. In the synchronous mode, after switching to the new channels, the MS sends to the new BTS, in successive assigned multiframe slots on the FACCH, four "RIL3-RR Handover Access" messages as shown in Figure 4.18. It then activates the new channels in both directions. When it has received sufficient "Handover Access" messages, the new BTS may also (optional) send an "RSM Handover Detection" message to the BSC.

In the asynchronous mode, the MS starts sending a continuous stream of "RIL3-RR Handover Access" messages to the new BTS until it receives in response an "RIL3-RR Physical Information" message giving the timing advance to apply.

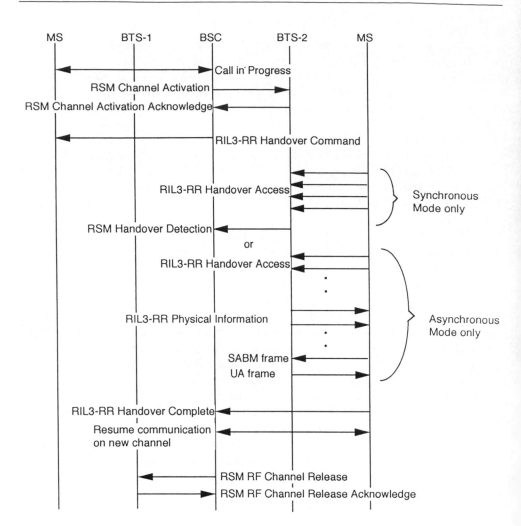

Figure 4.18 Handover between radio channels of the same BSC.

This handover message is repeated several times to ensure that the message is received by BTS-2. After starting the "Handover Access" messages, the MS activates the new channels in receiving mode only. It does not activate the transmit side until it receives the timing advance in the "Physical Information" message. The MS then sets up the link layer of the radio link by sending a SABM frame (an X.25 packet switched protocol) answered by an UA frame as discussed earlier.

After the lower layer connections are successfully established, the MS sends an "RIL3-RR Handover Complete" message to the BSC over the new FACCH.

Communication over the newly established channel is then resumed through BTS-2. The BSC also directs BTS-1 to release the old channels by sending an "RSM RF Channel Release" message with acknowledgment from BTS-1.

4.10 SUMMARY AND CONCLUSIONS

Three important states of a mobile—namely, MS turned off, MS turned on but idle, and MS busy—have been fully explored in this chapter. All the basic steps used in forming a call were covered in Section 4.4, on which we built complex protocols for MS location update, MS-PSTN call, PSTN-MS call, and MS-MS call. A thorough understanding of this chapter will enable the reader to easily grasp the operational aspects of the complex GSM system.

Problems

4.1 Name three distinct states of the mobile. What functions must mobile perform in these states?

4.2 Why is initialization necessary for mobile after the power is turned on?

4.3 Name three distinct paging types. What are the main distinctions between them?

4.5 Define functions performed within the following procedures.
  1. Identification
  2. Encryption and ciphering
  3. Call clearing
  4. IMSI attach and detach
  5. Location update

4.6 Why it is necessary for the mobile to register in the system? Can one classify registration as a special case of location update?

4.7 What are various factors for which the location update reject message may be sent from MSC to mobile? Elaborate on each of those reasons.

4.8 Explain the concept of "Off-Air Call Setup." What are the advantages of this scheme? Is there any advantage of call establishment without OACSU as explained in Section 4.6 and Figure 4.13(a)? Compare the advantages and disadvantages of both cases.

4.9 What are different cases of handovers? Draw the signal and response diagram for all the cases and show their differences clearly.

References

[1] ETSI/GSM Section 4.08, "Mobile Radio Interface Layer 3 Specification,"April 1993.
[2] Mouly, M., et al., *The GSM System for Mobile Communications,* Michel Mouly and Marie Bernadette Pautet, 1992, Section 2.3. 49, Rue Louise Bruneau, F-91120 Palaiseau, France.
[3] Mehrotra, A., *Cellular Radio Analog and Digital Systems,* Norwood, MA: Artech House, 1994.
[4] ETSI/GSM Section 04.08, "European Digital Cellular Telecommunication System (Phase 2) Mobile Radio Interface Layer 3 Specification."

CHAPTER 5

SECURITY MANAGEMENT

5.1 INTRODUCTION

At an early stage in the development of the Pan-European mobile radio system GSM, it was realized that security was an important issue that needed to be addressed. As a result, a security expert group (MOU-SG) was set up (as discussed in Section 1.2) to integrate features into the system to protect against perceived threats. It was apparent that the weakest part of the system was the radio path, as this could be easily eavesdropped with radio equipment. There was also a need to authenticate users of the system so that the resources would not be misused by nonsubscribers.

It was easy to see that the PLMN needed a higher level of protection than traditional telecommunication networks. Therefore, to protect the system against the two aforementioned cases, the following parts of the system have been reinforced and provided with various security features. The objective of all this was to make the radio path as secure as the fixed telephone network. The operators of the system wished to ensure that they could issue bills to the right people and that their services were not compromised, whereas the users required some privacy against traffic being overheard.

There are four basic security services provided by GSM.

- Anonymity: TMSI assignment;
- Authentication;

- Signaling data and voice protection against eavesdropping: encryption;
- User's SIM module and ME ID.

Anonymity is the use of a temporary identifier for the MS user. A user's IMSI has to be protected over the air interface. GSM does this by temporarily assigning a user an ID that is known as a TMSI. Both the permanent and temporary IDs are stored in the SIM. When a mobile first switches on his radio set in a new MSC/VLR area, the real identity (IMSI) is used and a temporary identification (TMSI) is issued. From then on the TMSI is used for all communication between mobile and the system. Authentication is used to identify the user (or holder of an SC) to the network operator. It uses a technique that can be described as a Challenge and Response. A simple representation of authentication is shown in Figure 5.1. Encryption against eavesdropping can be provided using the following process. A random number is generated by the network and sent to the mobile. The mobile uses the random number as the input to the plain text for encryption and, using a secret key unique to the mobile, transforms this into an output (cipher text) that is sent back to the network.

The network can check that the mobile really has the right secret key by performing the same process and comparing the responses. In this process a series of bits are transformed by mathematical or logical functions into another series of bits. The number of transformations is determined by the key so that an exhaustive search of all the possible keys must be made. The encryption process is illustrated in Figure 5.2, where the input is Plaintext and the output is Ciphertext.

Both authentication and user confidentiality processes involve the use of encipherment algorithms. The following algorithms are used in GSM.

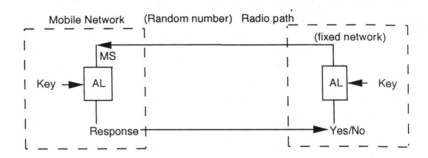

AL: Authentication Algorithm

Figure 5.1 Authentication process.

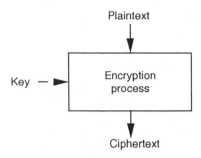

Figure 5.2 Encryption process.

A3: Used for subscriber authentication;
A5: Used for ciphering/deciphering. This algorithm is standardized throughout all GSM networks;
A8: Used for cipher key generation. This algorithm is defined by the PLMN.

These security features are either implemented as supplementary services that can be selected by the subscriber or as network functions involved in the provision of one or several telecommunication services

To a large extent, the use of these three algorithms is interrelated. The GSM encipherment algorithms are not available for general distribution, except to those companies directly involved in its implementation. The use of encipherment is one of several important contributions of GSM.

The objective of this chapter is to clearly outline the security features adopted in GSM, which include: (1) TMSI instead of its permanently attached number, called IMSI; (2) authentication; (3) ciphering; and (4) an *equipment ID* (EID), which assures that no stolen or unauthorized ME is used in the system.

5.2 TEMPORARY MOBILE SUBSCRIBER IDENTIFICATION

When a MS moves to a dedicated channel in a new MSC/VLR area, there is some period during which the network does not know the identity of the subscriber. This has a major consequence; specifically, all the signaling exchanges up to and including the first message carrying a nonambiguous subscriber identity must be sent in clear. A third party could, at this stage, listen to this identity and know the whereabouts of the mobile.

A mobile subscriber of a GSM PLMN is identified by a personal, permanently attached number called an IMSI. This number is meaningful only within the GSM premises; it is composed of a country code, a PLMN code, a HLR code, and a serial number in the data base of the HLR. Several other numbers are attached to a mobile subscriber, either permanently (for instance, the ISDN directory number) or temporarily (for instance, address to forward incoming calls, TMSI). The location registers of the system infrastructure maintain the correspondence between all these numbers. Details of IMSI, TMSI, and MSISDN have been covered in Section 2.3.4.

A subscriber identity confidentiality procedure is required for the protection of IMSI. Each time a MS requests one of the system's location updating procedures, call attempt or the service activation, MSC/VLR allocates to an IMSI a new TMSI and transmits it to MS with the order to substitute TMSI for IMSI for all future communications within the GSM system. From then on, the signaling exchange between MSC/VLR and MS makes use of only TMSI so that the identity of the subscriber (IMSI) is not again sent over the radio path. Thus, the subscriber identity confidentially feature means that the IMSI is not made available or disclosed to unauthorized individuals, entities, or processes. This feature protects the privacy and the identities of the subscribers who are using GSM/PLMN resources. It also prevents the location of a mobile subscriber from being traced by listening to the signaling exchanges on the radio path. Thus, the subscriber identity confidentiality (TMSI) is used whenever IMSI is requested over the radio path. Only in cases of location updating failure or when the MS has no TMSI available does the system need to refer to IMSI. This process of assigning new TMSI (TMSI-n) as a result of a location update is shown in Figure 5.3. As seen from the figure, the process starts with the mobile request for a location update, which goes to VLR. After proper authentication and cipher mode activation, the VLR provides location update acknowledgment with the new TMSI (TMSI-n). After the mobile receives new TMSI, it activates the TMSI acknowledgment process to BSS, which is finally forwarded to VLR by MSC, thereby ending the new TMSI assignment process.

Thus, protection is obtained by using an identity alias (the TMSI), which is used instead of the subscriber's permanent identity (the IMSI) whenever the system wants to address the subscriber. This alias must be agreed upon beforehand between the MS and the network during protected (ciphered) signaling procedures. The translation between TMSI and IMSI is performed at MSC and also within MS. Both these identities are carried transparently through BSS as far as possible. The reader should note that both the VLR and the HLR will associate TMSI with their corresponding IMSI in their data bases. The TMSI is a temporary number that is associated to a given mobile subscriber only within the localized area and for a certain period of time. The localization of each mobile subscriber within the cellular networks is defined by the LAI, together with the TMSI. The TMSI and LAI are automatically updated in both the MS and the location registers of the system, through a location updating procedure, in order to manage the mobility of the subscriber. Also, as the TMSI is only local to VLR, after changing the VLR area a MS may come up with an unknown TMSI (this will happen as the mobile moves to a different MSC/VLR combination). Therefore, the VLR is able to request the MS to provide its IMSI, which is globally unique.

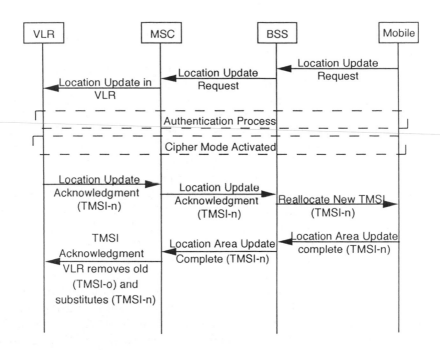

Figure 5.3 TMSI assignment process as a result of location update.

Thus, to ensure the confidentiality of the mobile subscriber identity, the use of the IMSI on the radio path is replaced as far as possible by the TMSI except under unusual circumstances when the system may ask the user for his IMSI.

5.3 AUTHENTICATION

The authentication feature ensures to a very high level of probability that the user is who they claim to be [1–4]. The purpose of the authentication is to protect the network against unauthorized use. It also enables the protection of the GSM PLMN. Subscriber authentication is performed at each registration, at each call set-up attempt (mobile originating or terminated), and before performing some supplementary services such as activation or deactivation of the mobile (that is, IMSI attach and detach). Authentication is not mandatory prior to IMSI attach and detach procedures. The frequency with which a particular PLMN applies the authentication procedure to its own subscribers is its own responsibility. However, a PLMN shall apply the authentication procedure to visiting subscribers as often as this feature is applied to those subscribers in their home PLMN. The authentication procedure is performed after the subscriber identity (IMSI/TMSI) is known by the network and before the channel is encrypted. We discussed this feature in considerable detail in Chapter 4.

A very simple authentication method is the use of a password (or a PIN code). Since listening once to this personal code by outside intruders is enough, the level of protection achieved by such a method is very low in a radio environment. However, GSM does make use of a PIN code when the mobile uses his SIM; this PIN code is checked locally by the SIM itself, without transmission on the radio interface. As stated, to gain access to the system, the user must first enter the correct PIN on the MS keyboard. The MS compares this PIN with the PIN stored in the SIM. If the two PINs agree, user access to the MS is granted. However, many users will choose to disable the PIN entry requirement after the first registration in order to speed up the process to access the GSM system, even though this increases their vulnerability, since one may be charged for calls made as a result of a lost or stolen SIM card. In addition, GSM uses a more sophisticated method, which consists of asking a question that only the right subscriber equipment (in this case, the SIM) may answer. The heart of this method is that a large number of such questions exist, and it is unlikely that the question can be answered correctly by the wrong MS. The generic process of authentication is shown in Figure 5.4.

The identity claimed by a mobile subscriber who asks for a service from his MS is ascertained by an authentication mechanism. The authentication algorithm (called A3 in the GSM specifications) computes both at the MS and at the AUC, a SRES from a RAND, using an individual secret key K_i attached to the mobile subscriber. The RAND whose value is drawn between 0 and $2^{128} - 1$ is used to generate the response by mobile as well as by the fixed part of the network. It should be noted that the authentication process is carried out both at the mobile as well as at the MSC simultaneously. BSS remains transparent to this process. It should also be noted that the mobile only receives the RAND over the radio path and in turn returns the SRES to the network. Thus, an air interface mobile identification is not disclosed.

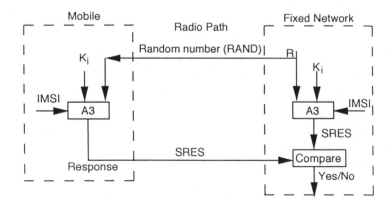

Figure 5.4 Generic authentication process.

At subscription time, the subscriber authentication key, K_i, is allocated to the subscriber together with its IMSI. The K_i is stored in the AUC and is used to generate a triplet within the GSM system (K_c, SRES, RAND). As stated, the same K_i is also stored at the mobile in the SIM. In the AUC, the following steps are carried out in order to produce one triplet. A nonpredictable RAND is produced. RAND and K_i are used to calculate the SRES and the *Ciphering Key* (K_c) using two different algorithms (A3, A8). This triplet RAND, SRES, and K_c are then delivered to the HLR. This is shown in Figure 5.5. Thus, for each subscriber the triplets are stored automatically in HLR and, upon request, delivered to the MSC/VLR combination.

The AUC begins the authentication and cipher key generation procedures after receiving an identification of the subscriber from the MSC/VLR. The AUC first queries the HLR for the subscribers authentication key, K_i. It then generates a 128-bit RAND for use as a challenge to be sent to the MS for verification of the MS's authenticity. RAND is also used by the AUC, with K_i in the algorithm A3 for authentication, to calculate the expected correct response, SRES, from the MS. RAND and K_i are also used in the AUC to calculate the K_c with algorithm A8. SRES is a 32-bit number, and K_c is a 64-bit number.

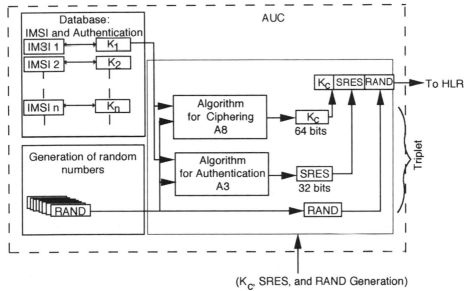

(K_c, SRES, and RAND Generation)

IMSI: International Mobile Subscriber Identity
K_c: Ciphering Key
K_n: Subscriber Authentication Key
SRES: Signed Response
RAND: Random Number

Figure 5.5 Generation of K_c, SRES, and RAND at AUC.

The values of RAND, SRES, and K_c are transmitted to the MSC/VLR for interaction with the MS. Algorithm A3 and A8 are not fully standardized by GSM and may be specified at the direction of PLMN operators. Different PLMNs may use different and proprietary versions of these algorithms. Also, to protect the secrecy of the user the K_i is not sent to the MSC/VLR. Based on the discretion of the PLMN operator, K_i can be of any format and length. The MSC/VLR forwards the value of RAND to the MS, which also has the correct K_i and algorithm A3, which are both stored in its SIM. The SIM then uses RAND and K_c in these algorithms to calculate the authentication response $SRES_c$ and K_c. The MS sends the calculated response $SRES_c$ back to the MSC/VLR, which compares $SRES_c$ with the value SRES received from the HLR/AUC. If $SRES_c$ and SRES agree, subscriber access to the system is granted and K_c is transferred to the BTS for use in encrypting and decrypting messages to and from the MS. If $SRES_c$ (computed SRES at the mobile) and SRES disagree, subscriber access to the system is denied. In summary, VLR initiates authentication toward the MS and checks the authentication result. The complete authentication process is shown in Figure 5.6.

Algorithm A3 is a high-grade, one-way function for which it is easy to derive SRES from K_i and RAND but very difficult to derive Ki from the transmitted values of RAND and the calculated value of $SRES_c$. The SIM is protected against the unauthorized readout of algorithms A3 and A8.

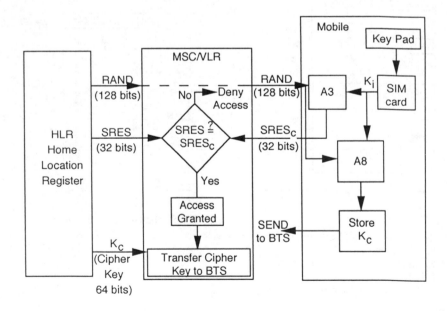

Figure 5.6 Authentication process in GSM system.

5.4 ENCRYPTION

Obtaining a good protection against unauthorized listening is not an easy matter with analog transmission, but digital transmission brings an excellent level of protection by using digital cryptographic methods [3,5,6]. The confidentiality feature of physical connections (physical radio channels) means that the user information and signaling exchanged between BTS and MS is not made available or disclosed to unauthorized individual entities or processes. The purpose of this feature is to ensure the privacy of the user information (voice and nonvoice) as well as the user-related signaling elements. All speech and data is ciphered, and all associated signaling information is protected. The ciphering algorithm is synchronized with the TDMA clock and adds very little complexity to the MS. The cipher key is obtained as a side product of the authentication procedure and differs from call to call. This has been taken advantage of in GSM, where a single encryption algorithm is used to protect all of the transmitted data in dedicated mode, whether it is user information (speech, data, etc.), user-related signaling (for example, the messages carrying the called phone numbers), or even system-related signaling (such as the messages carrying radio measurement results to prepare for handover).

Encryption is a process by which a series of bits is transformed by mathematical or logical functions into another series of bits. The number of transformations is determined by the key, so an exhaustive search for all possible keys must be tried. The confidentiality of the information elements carried on the radio path (signaling and user data) is ensured by systematic encryption. The ciphering/deciphering algorithm (called A5) uses a cipher key K_c that is generated during the authentication procedures. K_c is computed from the RAND by an algorithm (called A8) driven by K_i. Algorithm A8 is common to all GSMs. Figure 5.6 shows the process of generating K_c. For the authentication procedure, when SRES is being calculated at the mobile, K_c is also calculated using another algorithm (A8). This key setting takes place in the fixed system as well as in the MS. At the ciphering start command (from VLR to BSS), K_c is used by the MS and the BTS in order to cipher and decipher the bit stream that is sent over the radio path. In addition to the authentication procedures, a key setting may be initiated by the network as often as the network operator wishes. The command to use the encryption key is sent over the logical channel SDCCH as soon as the identity of the mobile subscriber is known by the network. K_c must be agreed upon by the MS and the network prior to the start of encryption. The choice of GSM is to compute K_c independently from the effective start of encryption during the authentication process. K_c is then stored in a nonvolatile memory inside the SIM so as to be remembered even after a switched-off phase. This key is also stored in the visited MSC/VLR on the network side and is ready to be used for the start of encryption. It should be noted that the actual encryption/ decryption of user data (for example, speech) takes place within the MS and the BSS. For this purpose the encryption key is downloaded from the MSC to the BTS via BSC. After authentication, the transmission is ciphered, and K_c is used for ciphering/deciphering. This is shown in Figure 5.7.

It is worth noting that the length of the significant part of key K_c, which is the output by algorithm A8, is fixed by the group of signatories of the GSM MOU and may be less than the maximum allowed 64 bits. In that case, the significant bits are complemented with zeros, so the format always uses the full 64 bits. As far as algorithm A5 is concerned, all patterns of 64 bits are possible and meaningful. This mechanism allows the level of security to be increased in the future if needed, without any change to algorithm A5 and (therefore without any change of the ME) by increasing the number of significant digits within the limits of 64 bits. Data flow on the radio path is obtained by bit by bit binary addition of the user data flow

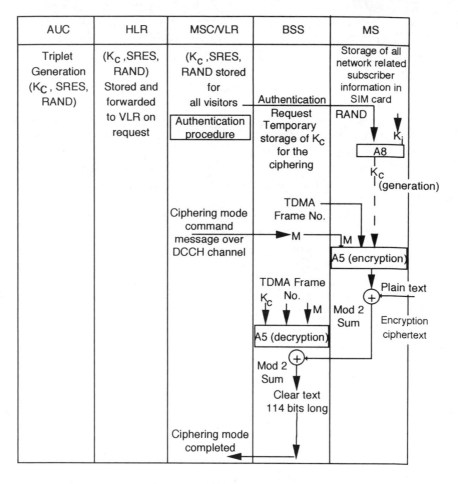

Figure 5.7 Sequential steps for encryption and decryption process.

and ciphering bit stream generated by GSM algorithm A5 using K_c. This exact process of encryption/decryption at the mobile and BTS is shown in Figure 5.8. Code words S_1 and S_2 for downlinks and uplinks are changed at every frame: S_1 when modulo 2 is added with Plaintext outputs Ciphertext; and, on the other side, the Ciphertext when modulo 2 is added with S_1 outputs Plaintext. The ciphering/deciphering function is placed on the transmission chain between the interleave and the modulator. Since algorithms A3 and A8 are always running together, in most cases these two are implemented as a single algorithm. Algorithm A3 is standardized in the whole of GSM.

SDCCH encipherment begins under the control of the network after the completion of the authentication procedure (if any) or after key K_c has been made available at the MS and BTS. No information elements for which protection is needed must be sent before the completing the initiation of the ciphering and deciphering processes.

When a TCH is allocated for user data transmission, the key used is that set during the preceding DCCH session (call set-up). Enciphering and deciphering processes start immediately. The complete ciphering process is shown in Figure 5.9, which also shows the start of ciphering after the user authentication is complete and a new TMSI has been assigned as a result of a location update.

When a handover occurs, the necessary information (for example, K_c, initialization data, etc.) is transmitted within the system infrastructure to enable the communication to proceed from the old BS to the new BS. The key, K_c, remains unchanged as a result of handover.

5.5 CHARACTERISTICS OF THE SIM

For the user's own security, the SIM must handle a four-digit PIN code. The mobile subscriber has the option, by acting on his SIM, to change his PIN as often as

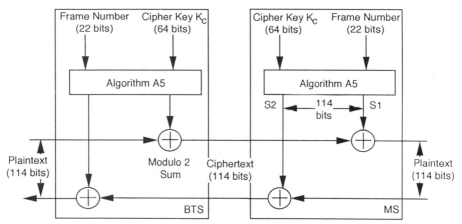

Figure 5.8 Encryption/decryption process.

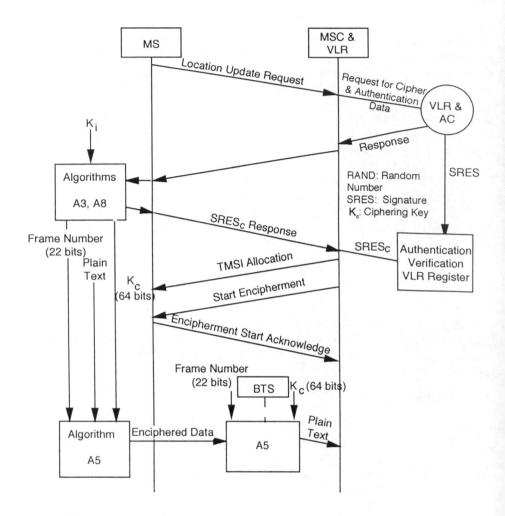

Figure 5.9 GSM encryption process.

he wants and even has the capability to disable the PIN verification in certain circumstances. The SIM enables the mobile subscriber to access to the GSM system as long as he is not blocked. Blocking the SIM is the SIM's self-initiated procedure and occurs only after three consecutive false presentations of the PIN code. The SIM can be unblocked by using a personal unblocking key [2,7,8].

In 1987, GSM decided that all the information elements contained in a MS that are related to the mobile subscriber must be stored and operated within a specific module, called SIM [9]. The remaining part of the MS, called the ME, is intended to contain all the mechanisms and devices that are needed to access the

GSM services (for instance, the radio equipment) but are not specific to a given subscriber. Moreover, the SIM must be removable from the ME, and the SIM-ME interface must be GSM standardized. These features allow the ME to be a universal equipment operable by different mobile subscribers, in turn, each using his own SIM. From the network operator's point of view, such an implementation provides a great flexibility in subscription management. In summary, the complete MS consists of the following.

SIM (subscriber specific data) + ME (equipment for accessing GSM network)

Here the subscriber's specific data is stored in the module or in the plastic card while the ME is just the means to access the services provided by GSM system. The Subscriber Identity Module Expert Group was created by GSM with the tasks of elaborating upon the SIM—ME interface specification and treating all the relevant fields. SIM is implemented in two forms—either as a part of an IC card or as a plug in SIM—and can be removed from ME whenever desired by the mobile. The removability of the SIM means that it is inserted in the ME whenever the mobile subscriber wants to use it and can be removed when the MS is unattended. The first option is that the SIM is part of an ISO standardized IC card. If this card is for multiservice, then the SIM is one of the supported applications called a GSM application. The ME shall implement a *Card Accepting Device* (CAD) that will accept a SIM card. The second option is the plug-in SIM, which is a small dedicated module implemented on a cut-out IC card. The size of the mechanical interface with ME are fully standardized by GSM.

In order to allow the mobile subscriber to operate his SIM in different places with possibly different MEs, the SIM must also contain the current values of the temporary data, namely, TMSI, LAI, and K_c. It is worth noting that the technology adopted by manufacturers for the SIM realization must be such that frequent updating of the stored data are made possible. That is, the SIM should be upgradable from time to time.

The SIM storage capability may provide facilities to memorize and manage additional information elements related to the mobile subscriber in association with GSM services or MS features. Some of these storage capabilities are as follows.

- Storage of short messages and associated parameters;
- Management of an abbreviated dialing numbers list;
- Implementation of a fixed dialing numbers list;
- Memorization of various bearer capability configuration parameters of terminals;
- Memorization of the advice of charge information given by the network;
- Management of a list of preferred PLMNs for connection;
- Implementation of the MS feature of barring for outgoing calls.

SIM is also used to control or to ease the access of the MS to the network by having the following capabilities.

- Storage of a list of BCCH frequencies;
- Storage of network access control parameters;
- Storage of PLMN where a previous location updating was refused;
- Storage of certain location updating failure causes.

The SIM life covers the whole period from the very beginning when it is manufactured, passing by the personalization phase when it is allocated to a mobile subscriber, and until the moment it is put out of service. For specification purposes, GSM only distinguishes two phases during the SIM life. The GSM network operation phase is when the SIM is allocated to a given subscriber and operated in association with a ME in order to access the GSM services. All operations performed during this phase need to be fully standardized by GSM because their implementation in any ME must work with every SIM.

Second, the GSM administrative management phase covers all the various operations needed for the establishment and the continuity of the SIM capability to access the GSM system. Manufacturing, service provider operations, and personalization (when the SIM is loaded with IMSI or K_i, for example) are part of this phase. The SIM does not interface with a GSM ME for these operations. Consequently, the opinion that prevails in GSM is that most of the events of this phase mainly concern the commercial relations to be established between the PLMN network operator and the customers.

5.6 EQUIPMENT IDENTIFICATION

The administrative use of the IMEI enables the operator to check the ME identity at call setup [1–10]. The purpose of this feature is to make sure that no stolen or unauthorized ME is used in the system.

An equipment Identification procedure consists of the following steps. First, the MSC/VLR requests the IMEI from the MS and sends it to a stand-alone entity called the EIR. Then, upon reception of the IMEI at the AUC, the EIR makes use of three possible defined lists.

- A white list containing all number series of all equipment identities that have been allocated in the different participating GSM countries—obviously right equipment;
- A black list containing all equipment identities that are considered to be barred—may be the result of stolen equipment;
- A gray list containing (at the operator's decision) faulty or nonapproved ME—under observation but not barred from service.

Though the GSM specification recommends using the equipment ID at each and every call, the frequency of identification really depends on the individual operator. The system operator can make a decision in this regard. The equipment identification process starts with MSC/VLR requesting MS for its IMEI. In response, the MS sends its ID, which if positively checked by the EIR allows the mobile to proceed further with the call. Mobile is not allowed to continue with the call if the equipment ID does not match with the stored value of the ID in the register. The complete equipment identification process is shown in Figure 5.10. As shown in Figure 5.10, an IMEI request is initiated by an MSC/VLR combination as a result of MS requesting for the call setup. Upon receiving the IMEI request, MS sends the equipment identification to MSC/VLR, which is subsequently checked against the stored values in the EIR.

5.7 CONCLUSIONS

In this chapter we discussed the four security services provided by the GSM system. They are anonymity, authentication, encryption, and the positive identification of

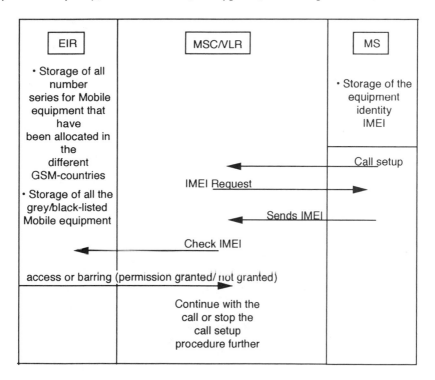

Figure 5.10 Equipment identification process.

ME before providing service. Authentication ensures that the network is accessed by the legitimate subscribers. Radio path is protected due to ciphering. Mobile identity is not easy to detect due to the use of TMSI instead of IMSI. An equipment ID ensures that the mobile is using the correct brand of transceivers.

Problems

5.1 Why do you think that PLMN needs increased protection against eavesdropping compared to a regular telephone system? State the main objectives of the operator and the subscribers in this area.

5.2 Describe the following terms.
1. Challenge;
2. Response;
3. Anonymity;
4. Authentication;
5. Encryption;
6. TMSI;
7. IMSI;
8. LAI.

5.3 State various advantages of assigning TMSI to a subscriber. Figure 5.3 shows that the VLR sends a new TMSI assignment message only after authentication and ciphering processes are complete. Can VLR send this TMSI assignment message before completion of authentication and activation of the cipher mode?

5.4 Narrate conditions under which the system will be forced to ask mobile for its IMSI.

5.5 Why is it absolutely essential for the operator to have authentication of the visiting subscriber?

5.6 For what reasons is the PIN number used? What is its main purpose?

5.7 Briefly describe two different types of SIM implementation in GSM. Provide the respective advantages and disadvantages of these two implementations. List important items stored in a SIM.

5.8 Why do you think it is essential to standardize the SIM-ME interface?

5.9 Explain four different types of security services provided by GSM.

5.10 Why do you think the ciphering key K_c must differ from one call to another?

References

[1] Maloberti, A., et al., "Radio Subsystem Functions and Elements," GSM Seminar, Budapest, October 1990, Session 3.1.
[2] GSM Recommendation 09.02, "Mobile Application Part (MAP) Specification."
[3] Siegmund, M. R., et al., An Introduction to GSM, Norwood, MA: Artech House, 1995, Section 3.8, pp. 44–48.

[4] Dechaux, C., et al., "What are GSM and DCS," *Electrical Communication*, 1993, pp 118–127.

[5] Brune, U. J., "The Mobile Application Part Protocol," *GSM Seminar*, Budapest, October 1990, Session 4.2.

[6] Mehrotra, A., *Cellular Radio Analog and Digital Systems*, Norwood, MA: Artech House, Section 7.5.2.4, pp. 305–309.

[7] Pautet, M., et al., "GSM Protocol Architecture: Radio Sub-System Signaling," *IEEE VT Conf.*, 1991, pp. 326–332.

[8] Mouly, M., et al., *The GSM System for Mobile Communications*, Michel Mouly and Marie-Bernadette Pautet, 1992, Section 7.2, pp. 477–492. 49, Rue Louise Bruneau, F-91120 Palaiseau, France.

[9] ETSI/GSM specification Vol. 2.17, Section 3, January 1993.

[10] ETSI/GSM specification Vol. 3.20, Section 3, January 1993.

CHAPTER 6
▼▼▼

TECHNICAL DETAILS OF GSM

6.1 INTRODUCTION

The GSM system is digital and based on narrowband TDMA technology, which means that each radio frequency will carry a number of simultaneous calls. In 1987 all European countries within CEPT agreed on using 200-kHz carrier separation, which is to be time-divided between eight users. The system is a large step from first-generation analog cellular systems. It provides international roaming and many new services, such as encoded speech within Europe. The choice of digital TDMA improves capacity and reduces the cost of BS sites compared to the first-generation system based on FDMA technology.

The frequency band used for the uplink is anywhere from 890 MHz to 915 MHz (from MS to cell site) and anywhere from 935 MHz to 960 MHz (from cell site to MS) for the downlink. The bandwidth for GSM systems is 25 MHz, which provides 124 carriers, each having a bandwidth of 200 kHz. With eight users per carrier, there are about 1,000 actual speech or data channels. The number of channels will double to 2,000 with the introduction of a half rate speech coder.

With 200 kHz of carrier spacing and this data rate, the spectral efficiency of the system is 1.35 bps/Hz (270.8/200). During an effort to maximize the spectrum efficiency of the GSM system, it was found that a significant increase in spectrum efficiency can be achieved by utilizing voice activated transmission. The basic principle, DTx, is to switch on the transmitter only for those periods when there is active speech to transmit. By doing so, the average interference on the air is reduced, thus allowing a smaller frequency reuse cluster size. Mobile power requirements are also reduced.

With 270.8 Kbps divided among eight users in GSM, the per-user data rate is 33.85 Kbps. A speech coder is a regular pulse excitation with a long-term predictor. For full rate speech the data rate is 13 Kbps. The speech coding algorithm chosen by GSM, the *Regular Pulse Excitation—Long-Term Prediction LPC* (RPE-LTP), where 260 bits are produced every 20 ms, results in a bit rate of 13 Kbps. The output bits from the speech coder are divided into three classes, depending on the sensitivity of the bit errors, and they are coded differently. These bits are distributed as 78 bits of side information (for example, filter coefficients, gain, and pitch information) and 182 bits of residual bit stream [1,2].

The channel coder for speech is based on a convolutional half rate code having the constraint length of five. A puncturing scheme is applied is such that the bits from the speech coder are protected differently depending on their sensitivity. The side information is left uncoded, while for 182 residual bits both block and convolutional coding is used. An interleaver collects the speech information generated over 40 ms ($2 \times 456 = 912$ bits) and interleaves over eight frames, which correspond to ≈ 40 ms. The interleaving distance of eight is found adequate to randomize the burst errors due to multipath. In other words, errors that are clustered in the sequence of 912 channel bits tend to be randomly dispersed in the bit stream presented to the decoder at the receiving end. With an interleaving duration of 40 ms and an interleave distance of eight, the frame length is theoretically limited to 5.0 ms. Thus, the actual frame length chosen is 4.6 ms.

The modulation method in GSM is GMSK, which facilitates the use of narrow bandwidth and allows for both coherent and noncoherent detection capabilities. In GMSK, the rectangular pulses are passed through a Gaussian filter prior to their passing through a modulator. The normalized pre-Gaussian bandwidth is kept at 0.3 (BT product), which corresponds to a baseband filter bandwidth of 81.25 kHz for an aggregate data rate of 270.8 Kbps. The BT product is a compromise between adequate BER and low-level sidelobes necessary for satisfying the adjacent channel interference requirements.

With a bit interval of 3.7 μs, the GSM signal will encounter significant intersymbol interference in the mobile radio path due to multipath (multipath minimal delay spread nearly equal to 3 μs to 6 μs in urban areas). As a consequence, an adaptive equalizer is necessary [1,2]. Different types of equalizers are allowed in the GSM system. Both *Decision Feed Back* (DFE) and Viterbi equalizers have been used.

The delay in a GSM network is limited to about 80 ms in order to avoid speaker annoyance. To maintain an acceptable speech quality during the call duration, a hand-off process has been implemented in the GSM network. This will allow the system to maintain a certain quality of signal. So that a user's signal, as a result of their location at varying distances from the BS, does not overlap at the cell site, the BS instructs a user to advance their timing appropriately.

Figure 6.1 provides the block diagram of the mobile radio system. Here, the user speech data at 104 Kbps is encoded at 13 Kbps, and then forward error correction coding is applied. In order to mitigate channel burst error, user data is interleaved over eight frames. GMSK allows "class C" amplification at the mobile. A class C amplifier can be used for a constant envelope signal and provides saturated operation with maximum efficiency. In view of Figure 6.1 and this discussion, the following topics are discussed in this chapter.

- Spectral efficiency;
- Speech coding algorithm;
- Block and convolutional coding;
- Digital modulation;
- Channel characterization and equalization;
- Network delay requirements;
- Discontinuous transmission;
- Timing advance;
- Hand-off mechanism.

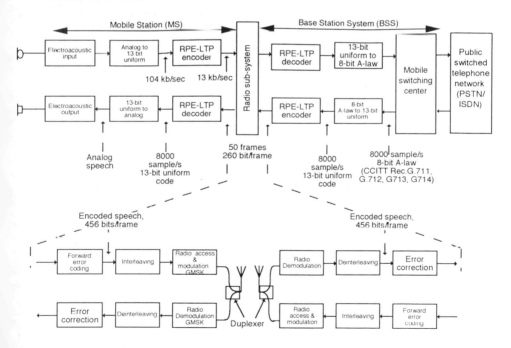

Figure 6.1 Block diagram of GSM radio system [1].

6.2 SPECTRAL EFFICIENCY OF GSM SYSTEM

The overall spectral efficiency of the system can be calculated by multiplying the modulation and the multiple access efficiencies separately [1]. In this section, we will discuss both of these efficiencies and then apply these equations to the evaluation of the spectral efficiency of the GSM system. The measure of spectral efficiency with respect to modulation is defined as

$$\eta_m = \frac{\text{Total number of channels available within the system}}{(\text{Total available bandwidth}) \cdot (\text{Cluster area})} \tag{6.1}$$

where η_m denotes the modulation efficiency in Channels/MHz/Km^2 and is computed

$$\eta_m = \frac{B_t/B_c}{B_t(NA)} = \frac{1}{B_c NA} \tag{6.2a}$$

Here B_t denotes the total bandwidth available to the system in MHz, B_c is the voice channel bandwidth or channel spacing in MHz, N represents the number of cells per cluster(cluster size), and A is the area of a cell in square kilometers.

From (6.2a) we observe that the spectral efficiency is independent of B_t and only depends upon the channel bandwidth B_c and cluster area NA. We call this the modulation efficiency since the channel bandwidth B_c is a function of modulation. By decreasing the channel bandwidth, the modulation efficiency of the system can be increased provided that it does not force an increase in the cluster area NA. Assuming that the cell area of the system is controlled by geographical constraints (propagation related) rather than the modulation, (6.2a) can be rewritten as [1]

$$\eta_m \propto \frac{1}{NB_c} \tag{6.2b}$$

Thus, the relative efficiency of the system x with respect to the system y can be written as

$$\eta_r = \frac{(\eta_m)_x}{(\eta_m)_y} \tag{6.3a}$$

or

$$\eta_r = \frac{(B_c)_y \cdot N_y}{(B_c)_x \cdot N_x} \tag{6.3b}$$

Dropping the subscript c for convenience,

$$\eta_r = \frac{B_y \cdot N_y}{B_x \cdot N_x} \tag{6.3c}$$

The number of cells in a cluster N depends upon the tolerance of a given modulation format to interference from the nearby cells reusing the same channel. Assuming that the channel ambient noise is insignificant compared to cochannel interference, then, N can be expressed in terms of D/R or C/I as follows [1]

$$\frac{C}{I} = \frac{1}{6}\left(\frac{D}{R}\right)^\alpha = \frac{1}{6}(3N)^{\alpha/2} \tag{6.4}$$

where C/I is the carrier to cochannel interference, D/R is the cochannel reuse distance, N is the cluster size, and α is the propagation exponent.

Here only the first tier of six cochannel interferers are considered, which adequately represents the practical systems. Assuming the fourth-power propagation law in the urban environment, (6.3c) can be expressed in terms of C/I requirements of the two systems as

$$\eta_r = \frac{B_y\sqrt{(C/I)_y}}{B_x\sqrt{(C/I)_r}} \tag{6.5}$$

which demonstrates the interesting property that the relative system efficiency decreases according to increases in the square root of C/I, but directly in proportion to the reduction in channel bandwidth. Let us examine some alternate expression for spectral efficiency before we provide some examples.

6.2.1 Some Alternate Measure of Spectral Efficiency

Based upon the discussion of the previous section, one measure of spectral efficiency is Channels/MHz/Km^2 and the other measure is Erlangs/MHz/Km^2. Following the definition of an Erlang as the quantity of traffic on a voice channel, or a group of channels per unit time, one can relate the two definitions. From (6.1)

$$\eta_m \text{ (Erlangs/MHz/Km}^2) = \frac{\text{Traffic offered by } (B_t/B_c) \text{ channels}}{B_t \cdot (NA)} \tag{6.6}$$

which when re-arranged yields

$$\eta_m \text{ (Erlang/MHz/Km}^2) = \frac{\text{Traffic offered by } [(B_t/B_c)/N] \text{ channels}}{B_t \cdot A} \tag{6.7}$$

Trunking efficiency can be included in (6.7) to represent the total amount of carried traffic through the system

$$\eta \ (\text{Erlangs/MHz/Km}^2) = \eta_t x \eta_m = \frac{\eta_t[\text{Traffic carried by } [B_t/B_c)/N] \text{ channels}]}{B_t \cdot A}$$

$$(6.8)$$

where η_t is the trunking efficiency, which provides the measure of how many of the total number of calls received by the system are carried through the system. Essentially, there are two traffic formulas: Erlang-B and Erlang-C [1]. Erlang-B is a pure loss or a blocking formula in which blocked calls are cleared. That is to say, if a call arrives when all channels are busy, the call is immediately cleared from the system. On the other hand, in Erlang-C, a call arriving when all the trunks are busy will wait within the system indefinitely until the channel becomes free. The following observations can be made about the prior equations.

- The C/I ratio, which determines the quality of the received signal, is a function of cluster size N.
- The relationship between the total number of available channels and the traffic carried is nonlinear; that is, for a given percentage increase in the number of channels, the amount of additional traffic carried exceeds the increase in the number of channels.
- Knowing the average traffic per user (Erlangs/user) during the peak hours and the Erlang/MHz/Km^2, the capacity in terms of Users/MHz/Km^2 can be derived.

6.2.2 Multiple Access Efficiency of FDMA and TDMA Systems

The objective of the multiple access techniques is to combine signals from different sources on to a common transmission medium in such a way that, at the destinations, the different channels can be separated without mutual interference. In other words, multiple access systems permit many users to share a common medium in the most efficient manner [1]. There are three basic types of multiple access techniques: (1) FDMA, (2) TDMA, and (3) *Code Division Multiple Access* (CDMA). In FDMA, users share the radio spectrum in the frequency domain. The user is allocated part of the frequency band, which is used throughout the conversation. In TDMA the users share the radio spectrum in the time domain. An individual user is allocated a timeslot during which they have access to the whole frequency band allocated for the system (wideband TDMA) or only part of the band (narrowband TDMA) [1,3]. The CDMA combines FDMA and TDMA techniques. For CDMA based on spread spectrum, each user is assigned a unique pseudorandom user code and thus can access the frequency-time domain uniquely. All the three *multiple access* (MA) techniques should have an efficiency of unity provided the signals transmitted by users are orthogonal to each other. However, this is difficult to attain. In FDMA, the number of channels for speech is reduced due to necessary guard bands between channels in order to reduce the filter roll-off requirements. In TDMA, the efficiency is reduced due to inclusion of guard time and synchronization sequence. Similarly, the efficiency of a CDMA system is reduced due to the nonorthogonality of the codes. Mathematically, the MA efficiency factor η is defined as

$$\eta = (\text{MA efficiency in time domain}) \cdot (\text{MA efficiency in frequency domain}) \quad (6.A)$$
$$= \eta_t \cdot \eta_f$$

We now evaluate η for FDMA and TDMA systems.

6.2.3 FDMA

$$\eta_{\text{MA,FDMA}} = \eta_f \cdot \eta_t = \eta_f = \frac{B_c M_a}{B_t} \leq 1 \quad (6.9)$$

where M_a is the total number of voice channels available to the system, B_c is the voice channel bandwidth or channel spacing, and B_t is the total bandwidth available to the system in megahertz. Obviously, its efficiency in the time domain is unity as an individual user occupies the channel for 100% of the time. Alternately, (6.9) can also be expressed as

$$\eta_{\text{MA,FDMA}} = \frac{[\text{Voice channel interms of bits/s}]M_a}{[\text{Total BW interms of bits/s}]} \quad (6.10)$$

Figure 6.2 represents the layout a typical FDMA/TDMA system. Here, the available frequency band is divided among many carriers, each having a certain bandwidth. In general, the efficiency is evaluated in consideration of the guard band, which is required in most communication systems. However, in cellular systems a guard band between channels is not included as the adjacent channel is never allocated to the same cell.

6.2.4 TDMA

We define the TDMA efficiency for wideband and narrowband separately. The efficiency of the wideband TDMA is expressed as

$$\eta_{\text{MA,TDMA}} = \frac{\tau \cdot M_t}{T} \leq 1 \quad (6.11)$$

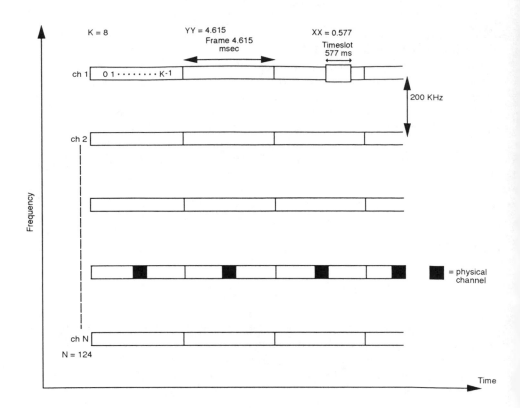

Figure 6.2 Layout of a typical FDMA/TDMA system.

where t denotes the timeslot duration for speech or data transmission in seconds, T is the frame duration in seconds, and M_t is the number of timeslots for voice transmission in a frame. Obviously, its efficiency in the frequency domain is unity. It is assumed that the total available band is shared by all users. Normally, in a wideband TDMA system, the number of slots or users is ≥ 10. A GSM system uses the narrowband TDMA schemes where the total band is split into number of subbands, each using TDMA techniques. For narrowband TDMA systems, the frequency domain efficiency is not unity, as the individual user channel does not occupy the whole frequency band available to the system. Narrowband efficiency is defined as

$$\eta_{\text{MA,TDMA}} = \eta_t \eta_f = \left(\frac{\tau \cdot M_t}{T} \right) \cdot \left(\frac{B_u \cdot M_u}{B_t} \right) \leq 1 \qquad (6.12)$$

where B_u denotes the bandwidth that an individual uses during his timeslot and M_u is the number of users who share the same timeslot in the system but have access to different frequency bands.

In (6.12) the first factor is the wideband TDMA efficiency, assuming that the FDMA efficiency is unity; similarly, the second factor is the FDMA efficiency, assuming that the TDMA efficiency is unity. We now provide examples of the European Digital Cellular System (GSM), which we will use to evaluate the overall spectral efficiency of the system.

Example 1

The European Digital Cellular System has an individual user data rate of 33.85 Kbps (270.8/8) in which the speech with error protection has the rate of 22.8 Kbps (456 bits/20 ms). The data rate of the SACCH is 0.95 Kbps (114 bits of data transmitted over one multiframe period of $4.615 \times 26 = 120$ ms). Find the overhead data rate and the TDMA efficiency, assuming that the system contains 125 channels of 200-kHz bandwidth each and with total system bandwidth of 25 MHz. The frame duration is 4.62 ms divided equally between eight slots.

$$\text{Overhead data rate} = (33.85 - 22.8 - 0.95) = 10.1 \text{ Kbps} \qquad (6.13a)$$

Thus, the slot duration allocated to speech is

$$\tau = \frac{(33.85 - 10.1)}{33.85} \frac{4.62}{8} = 0.405 \text{ ms} \qquad (6.13b)$$

where $t = 4.62$ ms is the frame duration and $M_t = 8$ are the number of channels. Also, $B_u = 200$, $M_u = 125$, and $B_t = 25$ MHz. Substituting these values in (6.12), the efficiency of GSM is

$$\left(\frac{0.405 \times 8}{4.62}\right)\left(\frac{200 \times 124}{25000}\right) = 69.6\% \approx 70\% \qquad (6.13c)$$

Thus, the overhead portion of the frame is $(1 - \eta_{\text{TDMA}})$ or 30.0%.

Example 2

We now apply the previous definitions to estimate the spectral efficiency of a GSM system with the following assumptions. The required C/I for a digital system is claimed to be 6 dB lower than the present analog system. Thus, the following parameters are assumed for the analysis:

Total spectrum = 25 MHz;
Knowledge channel bandwidth = 25 kHz;
Channel bandwidth/user = 25 kHz;

Digital channel bandwidth = 200 kHz;
Cell repeat pattern, N = 4;
Number of TCHs (125 × 8) = 1,000;
Number of channels/cell = 250;
Traffic (Erlang/Cell, Erlang table with 2% blocking probability) = 235.8;
Offered traffic in Erlang/cell = 231.1.

Thus the spectral efficiency of the present analog system is

$$= \frac{231.1}{25 \times 2.6 \times R^2} = \frac{3.55}{R^2} \text{ Erlang/MHz/Km}^2 \qquad (6.14a)$$

Cell areas of the analog and digital cellular are related as [1]

$$\frac{A_{\text{Dig}}}{A_{\text{Analog}}} = \sqrt{\frac{[C/I]_{\text{Dig}}}{[C/I]_{\text{Analog}}}} \qquad (6.14b)$$

Thus the ratio of digital to analog area is

$$\sqrt{10^{-0.6}} = 0.5 \qquad (6.14c)$$

Thus, the spectral efficiency of Phase 1 of the European digital system including the effect of multiple access efficiency is

$$\frac{231.1 \times 0.7}{25 \times 2.6 \times R^2 \times 0.5} = \frac{4.97}{R^2} \text{ Erlang/MHz/Km}^2 \qquad (6.14d)$$

Similarly, with half rate speech coding, it can be shown that the spectral efficiency of the digital system becomes $\cong (10.0)/R^2$ Erlang/MHz/Km^2. Thus, the relative efficiency (Phase 2) with respect to the present analog system is ≈ 2.8. Plots of the European analog, digital phase 1, and digital phase 2 cellular system's capacity is shown in Figure 6.3.

6.3 SPEECH CODING

The selection of the most appropriate speech code for the GSM system was based on the number of proposals and fairly extensive subjective testing in various languages and different operating conditions [4–6]. Initially over 20 different proposals from nine European countries were made. Before the international formal listening tests began, this number was reduced by national tests to six coders from six countries.

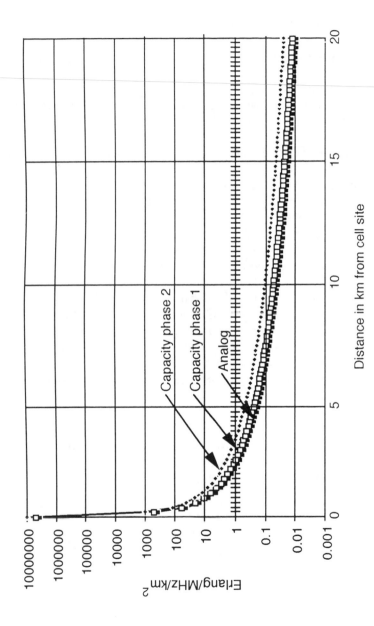

Figure 6.3 Cellular system capacity in Europe.

After initial evaluations, two subband coders were withdrawn. The remaining coders (four out of six), as shown below, were two different pulse-excited coders and two subband coders for final evaluation and selection.

- *RPE-LPC:* Regular-Pulse Excitation—Linear Predictive Coding by Germany (Philips).
- *MPE-LTP:* Multipulse Excitation—Long-Term Prediction by France (IBM). The particular speech codec implementation used a 13.2 Kbps transmission rate and 2.8 Kbps *Forward Error Correction* (FEC) coding. Thus, the gross bit rate was 16 Kbps.
- *SBC-APCM:* Subband coding relies on dividing the baseband speech signal into a number of subbands and then coarsely encoding the amplitude of the signal contained within each subband. Subband coding with block adaptive PCM in 14 subbands by Sweden (Ellemtel). This codec used *quadrature mirror filters* (QMF) to split the input signal into 16 subbands of 250-Hz bandwidth each, of which the two highest bands were not transmitted. The gross transmission rate of the subband signals was 10 Kbps; the side information was 3 Kbps, which was further protected by 3 Kbps FEC, resulting in a total data rate of 16 Kbps.
- *SBC-ADPCM:* Subband coding with adaptive differential PCM in six subbands by England (British Telecom Research) with adaptive delta PCM. In this scheme the speech input signal was split into eight subbands, out of which only six were transmitted. The bit allocation of the subbands was fixed, hence no side information was transmitted, which made the scheme more robust under noisy conditions; hence no FEC protection was equipped. The data rate was 15 Kbps only.

Tests for evaluation and final selection were conducted in seven languages, three input levels (12 dB, 22 dB, and 32 dB below overload), the effects of bit error rates (0., 1:1000, and 1:100), transcoding, and two forms of environmental noise. The average *Mean Opinion Score* (MOS) over a five-point scale (Table 6.1), bit rate

Table 6.1
Subjective Opinion Scales

Quality Scale	MOS	Listening Effort Required
Excellent	5	Complete relaxation possible; no effort
Good	4	Attention necessary; no appreciable effort
Fair	3	Moderate effort
Poor	2	Considerable effort
Bad	1	No meaning understood with feasible effort

(prior to channel coding), and complexity in terms of *Millions Of Operations Per Second* (MOPS) are shown in Table 6.2. The analog system is included for comparison only.

From the above table RPE-LTP has the best quality in terms of MOS value. Thus, this was adapted by GSM. The net bit rate of 14.77 Kbps was higher than the desired rate. This codec was redesigned to include the long-term prediction loop of the French codec, which resulted in a reduction of the speech rate to 13 Kbps. We shall now address the study of a RPE-LTP speech decoder in detail.

6.3.1 The RPE-LTP Speech Encoder

The simplified diagram of the RPE-LTP encoder is shown in Figure 6.4, and the complete details are given in Figure 6.5. We first describe the overall coding process through Figure 6.4 followed by details of coding through Figure 6.5.

In Figure 6.4 the input speech frame, consisting of 160 signal samples (uniform 13-bit PCM samples), is first preprocessed to produce an offset-free signal, which is then subjected to a first-order pre-emphasis filter [4,5,7,8]. The 160 samples obtained are then analyzed to determine the coefficients for the short-term analysis filter (LPC analysis). These parameters are then used to filtering the same 160 samples. The result is 160 samples of the short-term residual signal. The filter parameters, termed reflection coefficients, as shown in Figure 6.4, are transformed to *log-area-ratios* (LARs) before transmission.

For LTP analysis, the speech frame is divided into four subframes with 40 samples of the short-term residual signal in each. Each subframe is processed blockwise by the subsequent functional elements. Before processing each subblock of 40 short-term residual samples, the parameters of the long-term analysis filter, the LTP lag and the LTP gain, are estimated and updated in the LTP analysis block on the basis of the current subblock of the present and a stored sequence of the 120 previous reconstructed short-term residual samples.

Table 6.2
Comparison Chart for Four Different Codecs for Evaluation

Codec Type	MOS (1 = Bad, 5 = Excellent)	Number of Bands	Net Bit Rate (Kbps)	Gross Bit Rate (Kbps)	MOPS
RPE-LPC	3.54	1	14.77	16	1.5
MPE-LTP	3.27	1	13.20	16	4.9
SBC-APCM	3.14	16	13.0	16	1.5
SBC-ADPCM	2.92	8	15.0	16	1.9
Analog FM	1.95	–	–	–	–

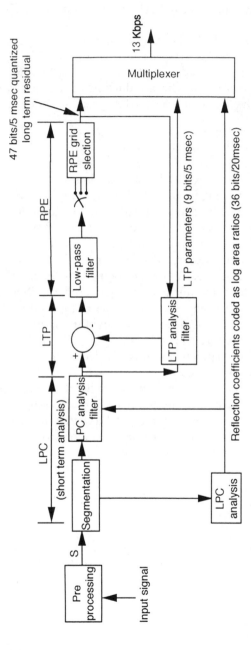

Figure 6.4 Simplified block diagram of RPE-LTP encoder.

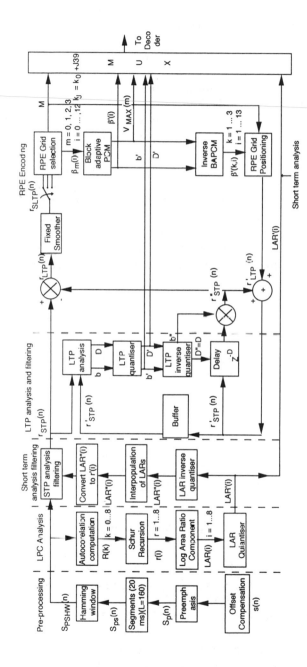

Figure 6.5 Block diagram of the RPE-LTP encoder.

A block of 40 long-term residual signal samples is obtained by subtracting 40 estimates of the short-term residual signal from the short-term residual signal itself. The resulting block of 40 long-term residual samples is fed to the regular pulse excitation analysis, which performs the basic compression function of the algorithm. As a result of the RPE-analysis, the block of 40 input long-term residual samples is represented by one of four candidate subsequences of 13 pulses each. The selected subsequence is identified by the RPE grid position (M) [4–10].

The detailed description of the encoder now follows. The elements of the speech encoder consists of the following functional parts: (1) preprocessing, (2) LPC analysis, (3) short-term analysis and filtering, (4) LTP analysis and filtering, and (5) RPE encoding. We now describe functions of each of the blocks.

During preprocessing, the offset compensation is applied in order to prevent a DC component being translated into an annoying side tone by the process of high-frequency regeneration in the decoder. A first-order FIR preemphasis filter is used for numerical reasons. Preemphasis can be deployed to increase the numerical precision in computations by emphasizing the high frequency, which is normally at a lower power. Here, a single-pole filter with the transfer function of the following form is used

$$H(z) = 1 - c_1 z^{-1} \qquad (6.15)$$

where the coefficient $c_1 = 0.9$. The pre-emphasized speech is segmented into blocks of 160 samples. In the autocorrelation method, the preprocessed speech samples are windowed by the Hamming window, represented as

$$W(n) = 0.54 - 0.46 \cos\frac{2\pi n}{N} \qquad (6.16)$$

The windowed signal, represented as $S_{PSHW}(n)$, contains N samples. The Hamming window has a tapering effect toward the edge of the block, while it has no effect at the middle of the range. The output of the Hamming window is represented as

$$S_{PSHW}(n) = K S_{PS}(n)\left(0.54 - 0.46 \cos\frac{2\pi n}{N}\right) \qquad (6.17)$$

The value of the constant $K = 1.5863$ and is chosen to keep the same power at the output of the Hamming window as well as at the input. The $S_{PS}(n)$ represents the pre-emphasized, segmented speech.

In LPC analysis, the overall objective for the LPC analysis is to generate the excitation sequence (LAR'(i)) for the decoder. The signal is also passed through an inverse filter that represents speech organs. For each segment of $L = 160$ samples, nine autocorrelation coefficients $R(k)$ are computed from $S_{PSHW}(n)$ by

$$R(k) = \sum_{n=0}^{159} S_{PSHW}(n)S_{PSHW}(n - k) \qquad k = 0, 1, \ldots, 8 \qquad (6.18)$$

A new LPC analysis is performed for each segment of the computed autocorrelation coefficients $R(k)$ by calculating eight reflection coefficients $r(i)$ using the Schur recursion algorithm. Due to the better quantization characteristics, the reflection coefficients are converted into LARs using the following equation

$$LAR(i) = \log_{10}\left(\frac{1 + r(i)}{1 - r(i)}\right) \qquad i = 1, \ldots, 8 \qquad (6.19)$$

A piecewise linear approximation with five segments is utilized in practical application as

$$LAR'(i) = \begin{cases} r(i), & \text{if } |r(i)| < 0.675 \\ \text{sign}[r(i)][2|r(i)| - 0.675] & \text{if } 0.675 < |r(i)| < 0.95 \\ \text{sign}[r(i)\text{sign}[r(i)][8|r(i)| - 6.375] & \text{if } 0.975 < |r(i)| < 1.0 \end{cases} \qquad (6.20)$$

The quantized form of LAR, $LAR'(i)$, is sent to the decoder and locally reused for short-term analysis and filtering subblock. Since various $LAR'(i)$, $i = 1 \ldots 8$ filter parameters have different dynamic ranges, 6, 5, 4, and 3 bits are allocated to the first, second, third, and fourth pairs of LARs, respectively.

In short-term analysis filtering, the eight coefficients of the short-term analysis filter are preprocessed as follows. First, the quantized and coded LARs are decoded [4–10]. Then the most recent and previous sets of LAR coefficients are interpolated linearly within a transition period of 5 ms to avoid spurious transients. The spurious may occur if the filter coefficients are changed abruptly. Finally, the interpolated LARs are reconverted into the reflection coefficients $r(i)$ of the FIR lattice filter, which is used to compute STP residual $r_{STP}(n)$. This constitutes the analogy to the acoustic tube model of the human speech production. The computation cycle outlined so far is repeated every 20 ms and produces 160 samples of the prediction error signal $r_{STP}(n)$.

In LTP analysis and filtering, the LTP loop is used to compute the estimate $r''_{STP}(n)$ of the residual signal $r_{STP}(n)$ from the reconstructed excitation signal $r_{LTP}(n)$. The LTP filter is characterized by the gain b and the delay D according to the equation

$$r''_{STP}(k) = b''r'_{STP}(k - D) \qquad (6.21)$$

where b'' denotes the quantized versions of b. The parameters b and D are calculated every 5 ms (40 samples). For each subsegment of 40 samples of the residual $r_{STP}(n)$, beginning with $r_{STP}(k_o)$, the cross-correlation functions $R_j(L)$ are calculated according to the equation

$$\sum_{i=0}^{39} r_{STP(k_j+i)} r''_{STP}(k_j + i - L) \qquad k_j = K_o + j40, \ j = 0, 1, 2, 3; \ L = 40, \ldots, 120$$

$$(6.22)$$

We note that $j = 0, \ldots, 3$ are subsegment numbers. K_o corresponds to the first value of the current frame. Optimum delay is calculated by determining the maximum value of $R(L)$, that is, where the currently processed subsegment is the most similar to its previous history. This is most probable at the pitch periodicity or by its multiple. Thus,

$$R(n) = \text{MAX}[R(l); \ l = 40, \ldots, 120] \qquad (6.23)$$

The LTP gain b for the jth subsegment is calculated as

$$b = \frac{R(N)}{\displaystyle\sum_{i=0}^{39} r^2_{STP}(K_j + i - N)} \qquad (6.24)$$

Once the parameters b and D are found they are quantized to find b' and D'. The LTP parameters b' and D' are encoded with two and seven bits. With these computed parameters, the LTP residual $r_{LTP}(n)$ is calculated as the difference of the STP residual $r_{STP}(n)$ and its estimate $r''_{STP}(n)$, which has been computed by the help of the locally decoded LTP parameters (b'', D) as follows

$$r_{LTP}(n) = r_{STP}(n) - r''_{STP}(n) \qquad (6.25)$$

$$r''_{STP} = b'' r'_{STP}(n - D) \qquad (6.26)$$

Here $r'_{STP}(n - D)$ represents an already known segment of the past history of $r'_{STP}(n)$ that is stored in the buffer. Finally, the content of the buffer is updated by using the locally decoded LTP residual $r'_{LTP}(n)$ and the estimated STP residual to from $r''_{STP}(n)$

$$r'_{STP}(n) = r'_{LTP}(n) + r''_{STP}(n) \qquad (6.27)$$

For RPE encoding, a FIR block filter algorithm is applied to each subsegment of 40 samples of the residual signal $r_{LTP}(n)$. Only the 40 middle samples of convolution of the 40 input samples with the 11-tap impulse response is calculated. The filtered signal $r_{SLTP}(n)$ is down-sampled by a ratio of 3, resulting in four candidate sequences $b_m(i)$ of length 13

$$\beta_m(i) = \beta(k_j + m + 3i) \quad \begin{array}{l} m = 0, 1, 2, 3; \\ i = 0, 1, \ldots, 12; \\ k_j = K_o + j39 \end{array} \quad (6.28)$$

where k_j defines the beginning of the *j*th subsegment and m denotes the phase of the decimation grid. The optimum candidate sequence $\beta_m(i)$ selected is that with the maximum value of the energy. Thus the solution of the RPE-approximation of the prediction error signal $r_{LTP}(n)$ requires energy calculations and can be interpreted as a generalization of the sample rate decimation process in a baseband RELP-coder. Finally, the selected RPE-sequence $\beta_m(i)$ consisting of 13 samples is quantized by block *adaptive* PCM (APCM). The block maximum is coded logarithmically with six bits, the normalized samples are quantized with three bits, and the grid position m is coded with two bits. The summary of the overall bit allocation resulting in a raw data rate of 13 Kbps is given in Table 6.3.

The decoding process is the inverse of the encoder where the parameters m, $\beta'_m(i)$, and $V_{max}(m)$ are decoded and used to reconstruct the excitation signal of the long-term synthesis filter. The sample rate of the denormalized RPE samples are increased by a factor of three by inserting zero samples and placing the nonzero samples in the correct temporal grid position M.

6.4 BLOCK AND CONVOLUTIONAL CODING

Coding aims at improving transmission quality when the signal encounters disturbances (such as significant noise when the reception level is low, interferences,

Table 6.3
Total Bit Allocation for RPE-LTP Speech Encoder

Parameter	Number of Bits
LPC coefficients LAR(i) = 9	36
LTP coefficients $b' = 4$	8
LTP delays $D' = 4$	28
RPE grids $m = 4$	8
Block maxima $V_{max}(m) = 4$	24
RPE samples $\beta'_m(i) = 52$	156
Total number of bits/frame	260
Frame duration (ms)	20
Bit rate (Kbps)	13

multipath propagation, and Doppler shift). It results, however, in an increased number of bits. Coding consists of adding to the source data some redundant information calculated from this source information. Decoding makes use of this redundancy to detect the presence of errors or estimate the most probable emitted bits given those received. Errors are detected when the transmitted redundancy is different from that calculated, with the received data.

Error control codes can be divided into two categories, namely, block and convolutional codes. In the case of block codes, the data is segmented into blocks of k message or information bits, and each block can represent any one of 2^k distinct messages. The encoder adds $(n - k)$ bits and forms a block n-bits long, which is called code bits or channel symbols. These $(n - k)$ additional bits are known as redundant bits, parity bits, or check bits and carry no information. The code is referred as (n, k) code. The check bits in the code block are dependent only on the information bits in the message block, hence the word block codes. The ratio $(n - k)/k$ within a block is called the redundancy of the code, and the ratio of the data bits to the total number of bits, k/n, is called the code rate. The code rate is the portion of a code bit that constitutes information. Thus for rate 3/4 code, there are three bits of information for every four bits of coded word. The redundancy in this case is 25% and the bandwidth expansion is 4/3. On the other hand, for half rate code, each code bit carries a half bit of information; both the redundancy and the bandwidth expansion is 100%. The Japanese cellular system uses (43, 31) and (11, 7) BCH codes for forward and reverse channels. Thus, the code rates are 31/43 and 7/11 and the redundancies are 28% and 36%, respectively. The block diagram of coder, decoder including a channel modulator and demodulator is shown in Figure 6.6.

Besides the code rate k/n, an important parameter of a code word is its weight, which is simply the number of nonzero elements that it contains. When all the M code words have equal weight, the code is called a fixed-weight or a constant-weight code. Let us consider two code words C_j and C_i in an (n, k) block code. The distance between two code words is defined as the measure of the difference between them or the number of positions where they differ. This measure is called Hamming distance $d(j, i)$. Clearly, $d(j, i)$ for $j \neq i$ satisfies the condition $0 < d(j, i) < n$. The smallest distance value of $d(j, i)$ among all code words is called the minimum distance of the code and is denoted $d_{min}(j, i)$. The minimum distance is important because it represents the weakest link in a chain and provides the minimum capability, which provides the strength of the code. The Hamming distance for (40:28) and (48:36) used in the forward and reverse channels of the first-generation cellular system in the United States has a minimum distance of 5.

Example

If two codes j, i are represented as

$$j = 1\ 0\ 0\ 1\ 1\ 0\ 1\ 0\ 1\ 1\ 1$$
$$i = 1\ 1\ 0\ 0\ 1\ 1\ 0\ 1\ 0\ 1\ 0$$

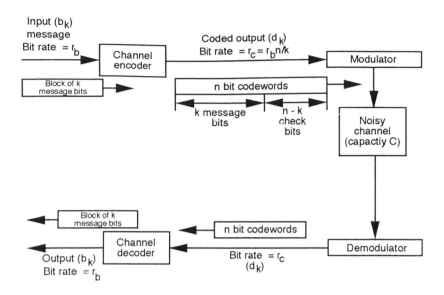

Figure 6.6 Block diagram of general coder/decoder [1].

(thus $d(j, i) = 7$), by the property of modulo-2 addition, we conclude that the sum of two binary vectors is another vector whose binary ones are located in those positions in which the two vectors differ. Thus in the above example

$$j + i = 0\ 1\ 0\ 1\ 0\ 1\ 1\ 1\ 1\ 0\ 1$$

From this example one can note that the Hamming distance between two code vectors is equal to the Hamming weight of their sum; that is, $d(j, i) = W(j + i) = 7$.

Block codes are often used when we have block-orientated signaling, such as in GSM mobile radio where the speech data is sent in blocks of 20 ms. Block code is often used to detect errors when an *automatic repeat request* (ARQ) is implemented. When we detect an error, we ask for a retransmission.

Convolutional coding is associated with error correction when the ARQ facility is not available. When we digitize speech and transmit it, we cannot retransmit as this would lead to intolerable delays. In convolutional coding, the block of coded digits generated by the coder depends not only on the digits in the current message block that is shifted into the coder but also on bits in preceding message locks. Figure 6.7 provides the basic block diagram of a convolutional encoder that is identified by constraint length K and the coding rate m/n.

In GSM both block and convolutional coding methods are used. First, some of the information bits are block coded, building a block of information along with parity (check) bits. All the block-coded bits are then passed through convolutional

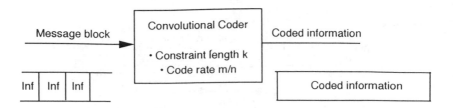

Figure 6.7 Convolutional coding.

code to form the final coded bits. The two steps apply to both speech and data, though the coding schemes are a little different. The reason for this "double" coding is that we want to correct errors if we can (achieved by convolutional coding); after this we can detect (block coding) whether the information is damaged beyond repair and, if so, ignore the information. In summary, codes used in GSM are

- *Convolutional codes:* These codes are used for error correction purposes. They achieve high efficiency when they are combined with a likelihood estimation scheme. The discussion of likelihood estimation theory is beyond this book, and the reader is referred to an excellent treatment of this by Viterbi and Omura [11].
- *Fire code:* This code is used to detect and correct "bursty" errors, that is, errors that are clustered together. It is used in concatenation after a block convolutional code, for which residual errors often come out in groups.
- *Parity code:* Simple parity codes are used for error detection.

We now investigate coding schemes as applied to GSM system for speech, data, and signaling applications.

6.4.1 Full Rate Speech Coding

To understand how the basic information is compiled into a frame, consider the speech coding mechanism shown in Figures 6.8(a,b). Basic speech is sensed by a coder for 20-ms segments and produces 260 bits at the output. Thus the output data rate of the speech coder is 13 Kbps [1,11]. Every 20 ms the channel coder releases 456 bits. The residual data consisting of 182 bits and 78 bits of side information when passed through the half rate convolutional encoder provides 456 bits of coded data. The residual data consisting of 182 bits and 78 bits of side information when passed through the half rate convolutional encoder provides 456 bits of coded data. The speech blocks are grouped into three classes of sensitivity to errors depending on their importance to the intelligibility of the speech. In descending order these are

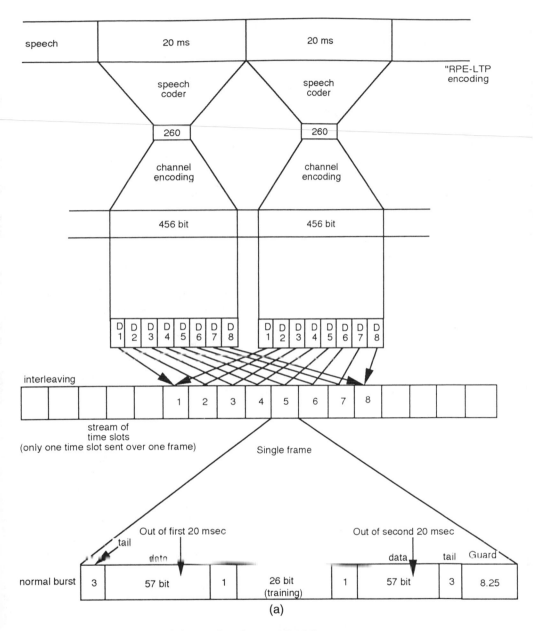

Figure 6.8 (a) Digital mobile and (b) speech coding in GSM [1].

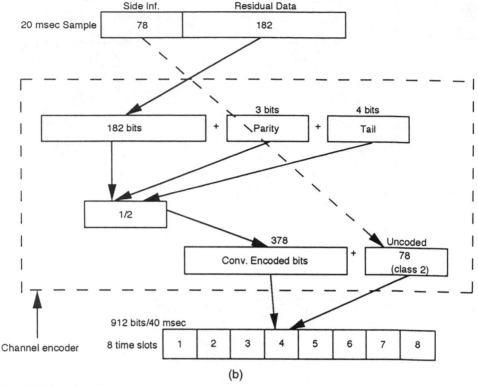

Figure 6.8 (continued).

- *Class 1a:* Three parity bits are derived from the 50 class 1a bits. Transmission errors within these bits are catastrophic to speech intelligibility; therefore, the speech decoder is able to detect uncorrectable errors within the class 1a bits. If there are class 1a bits in errors, the whole block is usually ignored.
- *Class 1b:* The 132 class 1b bits are not parity checked but are fed together with the class 1a and parity bits to a convolutional encoder. Four tail bits are added first and then $r = 1/2$, $K = 5$ convolutional code provides an output of 378 bits.
- *Class 2:* The 78 least sensitive bits are not protected at all.

Before convolutional coding, three bits of parity are added to class 1a bits. The generating polynomial is

$$G(D) = 1 + D + D^3 \tag{6.28}$$

The hardware implementation is shown in Figure 6.9. The operation is such that for the first 50 clocks the switch remains closed and for the last 51 to 53 clocks the switch is kept open.

Convolutional coding adds redundant bits in such a way that the decoder can, within limits, detect errors and correct them. This code is applied to both the class 1b and class 1a bits (including the parity bits generated from the class 1a bits). In order for a code to be able to correct errors, a certain number of additional bits have to be added. The added bits are called redundancy bits. The convolutional code employed here uses a rate of $r = 1/2$ and a delay of $K = 5$. This means that five consecutive bits are used to calculate the redundancy bits and that for each data bit an additional redundant bit is added. Before the information bits are encoded, four bits are added. These bits are all set to zero and used to reset the convolutional code. Since we always use five bits to calculate the appropriate redundancy bits (in GSM application), we need the trailing four zeros for the last data bit. The block diagram of the convolutional encoder used for speech is shown in Figure 6.10. A 40-ms segment of speech is interleaved with a distance of eight to combat errors in the multipath surroundings. This distribution of 2×456 bits over eight frames is shown in Figure 6.8(a). Thus a total of 114 bits of speech data per user is sent per frame, and the total information is sent over eight frames. Since a total of 456 bits are generated in 20 ms, the user data rate is 456/0.02 or 22.8 Kbps, which includes 13 Kbps of raw data plus 9.8 Kbps of parity, tail bits, channel coding, and training sequence.

The resulting 456-bit block is then transmitted using an interleaving scheme with a depth of eight. This means that eight frames are used to transmit these bits. In real life, bit errors often occur in bursts due to the fact that long fading dips affect several consecutive bits. Unfortunately, channel coding is most effective in detecting and correcting single errors and not when errors are in burst. To deal with

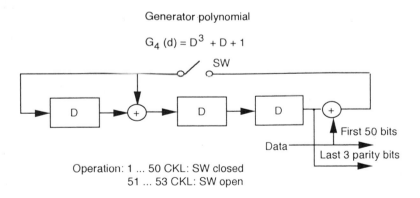

Generator polynomial

$$G_4(d) = D^3 + D + 1$$

Operation: 1 ... 50 CKL: SW closed
51 ... 53 CKL: SW open

Figure 6.9 Systematic block encoder (53, 50).

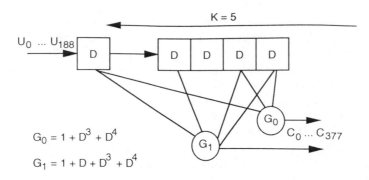

Figure 6.10 Convolutional encoder for speech.

this problem we want to find a way of separating consecutive bits of a message so that these are sent in a nonconsecutive way. This is done by interleaving, for which we provide details Section 6.4.5.

The example below shows the code generation for convolutional code. When the machine starts, four flush bits are added to initialize the coder. The speech bits come in the input port and a single bit delay is provided with each cycle of clock.

Example

Bit stream (input)	1 0 1 1 0 1 0 0 1 1 0 1 1
Adding of four 0 bits	1 0 1 1 0 1 0 0 1 1 0 1 1 0 0 0 0
(Four flush bits)	
Delay of one bit (D^1)	0 1 0 1 1 0 1 0 0 1 1 0 1 1 0 0 0 0
Delay of two bit (D^2)	0 0 1 0 1 1 0 1 0 0 1 1 0 1 1 0 0 0 0
Delay of three bit (D^3)	0 0 0 1 0 1 1 0 1 0 0 1 1 0 1 1 0 0 0 0
Delay of four bit (D^4)	0 0 0 0 1 0 1 1 0 1 0 0 1 1 0 1 1 0 0 0 0
1st stage $(1 + D^3 + D^4)$	1 0 1 0 1 0 0 1 0 0 0 0 1 1 1 0 1 0
2nd stage $(1 + D^1 + D^3 + D^4)$	1 1 1 1 0 0 1 1 0 1 1 0 0 1 1 1 0 0

Since the convolutional coder provides two bits at the output corresponding to a single bit at the input, the coder output is

"1110111001001011001010000111111100100"

Figure 6.11 shows the complete channel-coding scheme for all the speech bits. It is interesting to note that the class 2 bits, which are least important, are not protected at all. In this scheme, 189 bits enter the convolutional coder, $2 \times 189 = 378$ bits come out, and 78 class 2 bits are added to the 378 bits to yield 456 bits.

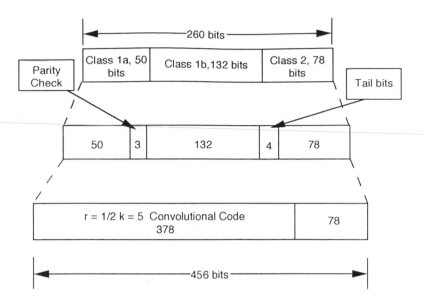

Figure 6.11 Full rate channel coding.

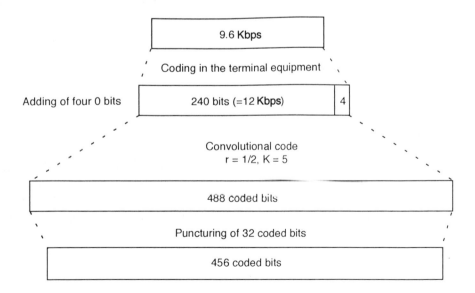

Figure 6.12 Coding for 9.6-Kbps data rate.

This is exactly 4 × 114, and 114 is the number of coded bits within one burst, or 8 × 57, which is the number of bits included into eight subblocks.

6.4.2 Data Coding at 9.6 Kbps and 2.4 Kbps

The data TCHs require a higher net rate (net rate means the bit rate before coding bits have been added) than their actual transmission rate. For example, the 9.6-Kbps service will require 12 Kbps because status signals (such as the RS-232 DTR "Data Terminal Ready") have to be transmitted [1,12,13]. The user's bit stream is divided into four blocks of 60 bits each, for a total of 240 bits that are coded together in a convolutional code. In contrast to the coding of the speech data, a block code is not applied prior to the convolutional coding because the error detection is performed within the terminal equipment. As one must always do with convolutional codes, four zero bits are added to the 240 data bits to reset the decoder as shown previously. The parameters for the convolutional code are the same as those used for speech coding ($r = 1/2$, $K = 5$). The convolutional code accepts, then, 244 bits and outputs 488 coded bits. Since 488 bits do not fit easily in the 456 bits-per-block scheme that we discussed under speech data, there is an excess of $488 - 456 = 32$ bits. We puncture the 488 bits according to a certain rule to reduce their number by 32. The 32 punctured bits are not transmitted. Since convolutional encoding is used here for data, an additional transmission delay can be allowed. This is unlike speech where an unacceptably long delay will cause customer dissatisfaction. Thus, interleaving depths up to 19 (19 bursts to all the data in one block) are used. See Figure 6.12.

6.4.3 Channel Rate of 2.4 Kbps

The 2.4-Kbps data rate uses the same interleaving scheme as the common TCH and provides a coded output. There is, of course, a change in the convolutional code parameters; specifically, it uses $r = 1/6$ and $K = 5$. The data rate coming from the terminal equipment is 3.6 Kbps, which already includes the terminal equipment's error correction coding. The customer's coded data are divided into 72-bit blocks to which, again, four zero bits are added. The convolutional code transforms the $72 + 4 = 76$ input bits into $76 \times 6 = 456$ coded output bits, which are mapped onto eight subblocks in the same way, as in speech. Figure 6.13 shows the implementation scheme.

6.4.4 Control Channel Signaling Data

The diagram shows the principle of the error protection for the signaling data. This scheme is used for all the logical signaling channels, except for the data portions in

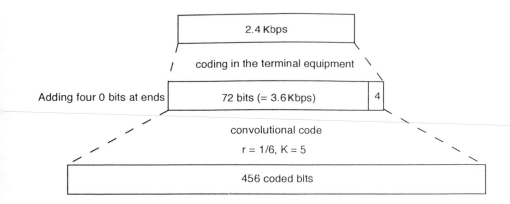

Figure 6.13 Coding for 2.4-Kbps data channel.

the SCH and the RACH. The diagram applies to SCH and RACH, but with different numbers.

Recalling the OSI Model, the radio subsystem (the physical layer, that is, layer 1) receives from the data link layer (layer 2) blocks of 184 bits. These are first protected with a cyclic block code of a Fire Codes class, which are particularly suitable for the detection and correction of burst errors; it uses 40 parity bits. Before the convolutional encoding, four tail bits (flush bits) are added to ensure equal protection for the last bits in the block. They assist the decoder to error correct the last coded symbols received.

The output from the encoding process for each of 184 bits of signaling data is 456 bits, exactly the same as for full speech. However, the interleaving depth is four instead of eight (it takes four bursts to transmit the full block of 456 bits). Figure 6.14 shows the coding scheme for control channel signaling data.

Note that FACCH, because it steals speech bursts from a subscriber channel, experiences the same kind of interleaving as the data that it replaces (interleaving depth of eight). A summary for speech, data, and control channels is shown in Figure 6.15.

6.4.5 Interleaving

In order to combat the effects of error due to interference and noise, error correction techniques are used [1]. The redundancy introduced due to error-correcting codes increases the data rate. For example, the raw data rate, due to speech coding, is only 260 bits over a period of 20 ms, as shown in Figure 6.11. However, after block and channel encoding, the number of bits are increased to 456 bits, resulting in a data rate of 22.8 Kbps. Error-correcting codes are better at correcting randomly

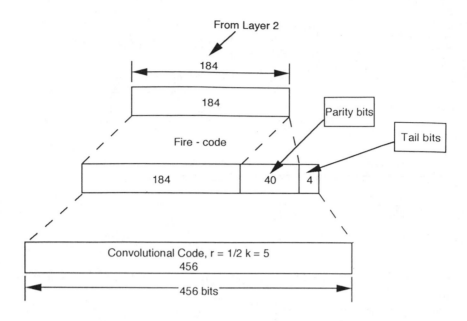

Figure 6.14 Coding for control channels.

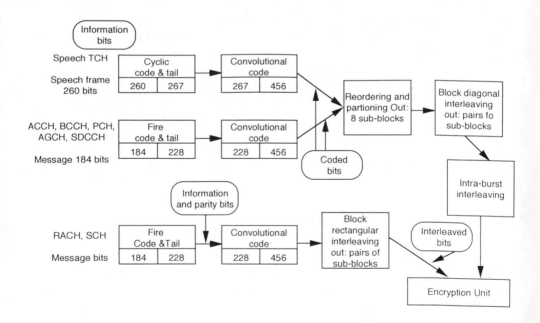

Figure 6.15 Summary coding for a GSM system.

distributed errors but does not work well when the errors occur in bursts. For bursty errors, interleaving of data is recommended since in reality, bit errors often occur in bursts. This is due to the fact that long fading dips affect several consecutive bits. To deal with this problem we want to find a way of separating consecutive bits of a message so that these are sent in a nonconsecutive way. This is done by interleaving, which is the process of distributing data bits in a different order in which they are generated.

After encoding the logical channel, the next step is to build its bits stream into bursts that can then be transmitted within a TDMA frame structure. It is at this stage that the process of interleaving is carried out. That is, the interleaving process spreads the content of one block of data either across four timeslots for the most control channels as explained, eight timeslots for full rate speech, and up to nineteen timeslots for data TCHs. This process is important, because it protects the data against interference, noise, or physical interruption of the radio path. Under adverse mobile surroundings normally 10% to 20% bursts are destroyed or corrupted. As an example, whole bursts are regularly lost on radio channels if the mobile passes through a tunnel or if just about any other type of interference occurs. The purpose of interleaving is to ensure that only some of the data from each traffic block is contained within each burst. By this means, when a burst is not correctly received, the loss does not affect the overall transmission quality because the error-correction techniques are able to interpolate for the missing data. It is interleaving that is largely responsible for the robustness of the GSM data, enabling it to withstand significant noise and interference and to maintain the quality of service presented to the subscriber. Since the fades experienced by the mobile radio system occur at a much slower rate than the 270 Kbps transmission rate of GSM, errors will tend to occur in bursts, which may destroy an entire TDMA burst. To overcome this, the bits in each message block are interleaved over eight bursts, which reduces the average errors per block to a rate manageable by the FEC.

The basic interleaving process is shown in Figure 6.16. The 456 bits are subdivided into the eight subblocks in the following way. Bit number 0 goes into subblock 1, bit number 1 goes into subblock 2, and so on until all eight subblocks are used up. Bit number 8 ends up in subblock number 1 again. The first four subblocks are put into the even-numbered bits of four consecutive bursts, and the second four subblocks are put into the odd numbered bits of the next four consecutive bursts. First, the 456-bit encoded speech message block is read into an 8-column by 57-row matrix RAM, filling each row in turn. The bits are then read out of the RAM by column, forming eight subblocks of 57 bits each. Note that adjacent bits in the code word are placed into different subblocks. As each burst contains 114 traffic-carrying bits, it is in fact shared by two speech blocks. Each block will share four bursts with the block preceding it and four with the block that succeeds it. A burst will then be transmitted in the designated timeslot of eight consecutive TDMA frames, thus providing the interleaving depth of eight.

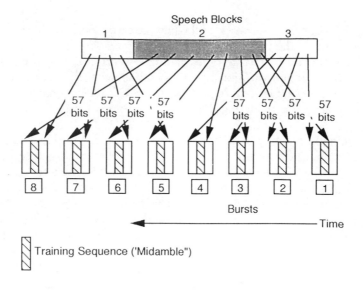

Figure 6.16   Basic interleaving process.

Using a technique called diagonal interleaving, the subblocks are then distributed over eight successive bursts, all in their assigned timeslot in successive TDMA frames. Each burst can hold 114 message bits, which are shared between bits from adjacent 456-bit blocks. Each timeslot is shared by two subblocks from different blocks. The bits from subblocks 0 to 3 use even bit positions, while the bits from subblocks 4 to 7 use odd bit positions. The first four bursts are thus shared with the preceding message block, and the last four bursts are shared with the ensuing message block. Of course, this entire process must be undone in the receiver. Figure 6.17 shows the exact process.

6.5 DIGITAL MODULATION

In this section we first discuss the requirements pertinent to cellular radio as well as those especially chosen by GSM standards. General requirements for modulation for digital cellular systems are as follows [1].

- Efficient/compact modulation;
- Good BER performance;
- Efficient use of mobile dc power;
- Applicability to the cellular environment;
- Implementation ease.

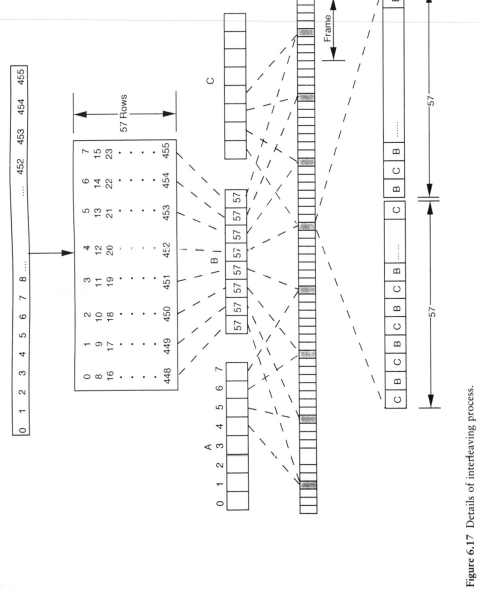

Figure 6.17 Details of interleaving process.

Channel separation, which is commonly used as a measure of modulation efficiency, should be made as narrow as possible. Assuming that the RF spectrum in the adjacent channels cannot be overlapped with each other, the required channel separation f_s is given by

$$f_s = B + 2\Delta f + \Delta D \qquad (6.29)$$

where B is the transmission bandwidth occupied by the RF signal power spectrum, Δf is the carrier frequency drift in each transmitter, and ΔD is the doppler shift. Since the transmission bandwidth, $B = R_d/m$, where R_d denotes the channel data rate and n is the transmission efficiency determined by the digital modulation method, (6.29) can be written as [1]

$$f_s = (R_d/m) + 2\Delta f + \Delta D \qquad (6.30)$$

Thus, to make the channel separation narrower in the digital system for voice transmission systems, it is necessary to develop the following: (1) Reduce the encoded speech channel data rate; (2) Narrow the band-efficient digital modulation, and raise the value of m; (3) Lower the oscillator drift, that is, stabilized carrier frequency. Of course, nothing can be done for the doppler that is a function of the speed of the vehicle. The carrier frequency drift is given as the product of the RF carrier frequency and the frequency stability of local oscillators. Considering that the mobile radio unit has to be necessarily simplified, miniaturized, and economized, it will not be easy to realize a local oscillator that has a frequency stability of less than 10^{-6}/year unless any special frequency stabilization technique is adopted. Thus, assuming the frequency stability of the two adjacent channel oscillators are limited to $2 * 10^{-6}$/year, the value of $\Delta f \leq 2$ kHz at 900 MHz.

For a cellular radio application, where the transmission is over the multipath surroundings, the value of m lies between 1 bit/Hz and 2 bits/Hz. Choosing m above 2 bits/Hz degrades the BER performance of the system as the constellation comes closer. Using the prior value of Δf and the transmission efficiency m of 1 and 2, the required channel separation, f_s, versus the channel data rate, R_d, is plotted in Figure 6.18.

Thus, for a 25-kHz channel bandwidth, the maximum data rate can be 22 Kbps to 45 Kbps for transmission efficiencies lying between 1 bps/Hz to 2 bps/Hz. Let us note that the data rates for GSM are ≈ 34 Kbps/25 kHz channel bandwidth and this lies in this range.

The recommendations for the output RF spectrum mask due to the GMSK modulation are shown in Figure 6.19. At the nominal bandwidth of 200 kHz the spectrum must decay by 30 dB with respect to its value at the carrier frequency. The specified relative power levels for frequency offsets from the carrier equal to or greater than 400 kHz depend on the output power level of the transmitter and the

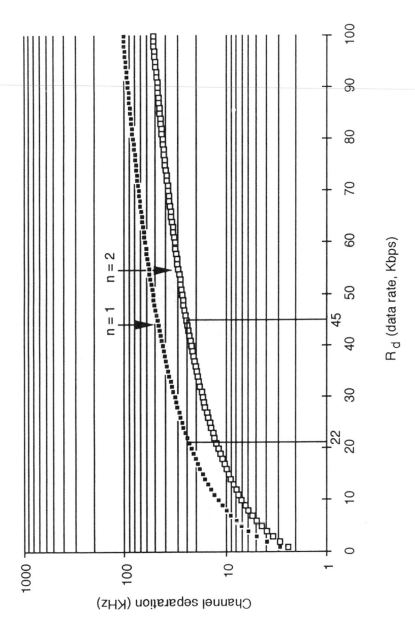

Figure 6.18 Channel separation versus data rate in Kbps.

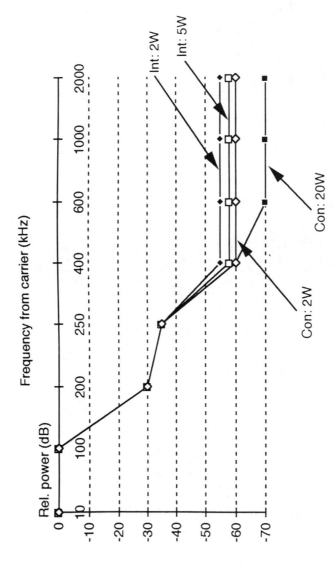

Figure 6.19 GMSK RF spectrum output mask.

type of MS. For transmitter output power levels below 43 dBm (20W), the specifications allow slightly higher levels in the spectrum at points up to 600 kHz to 1.8 MHz away from the carrier. The example of curve Continuous: 2W represents a transmitted power of 2W at the antenna connector. Figure 6.19 demonstrates that for 10-dB lower transmitted power (2W), a 10-dB higher modulation spectral mask is acceptable when compared with a 20 W transmitter. Higher levels at points 400 kHz to 1.8 MHz away from the carrier are allowed also for equipment with integral antennas (for example, portable sets) operating at a power below 37 dbm (5W), as represented by the curves Intermittent: 5W and Int: 2W, respectively.

The switching transients caused by the transmission of bursts of RF energy widen the output spectrum. The RF spectrum due to the switching transients is required to be 23, 26, 32, and 36 dbm down relative to the level specified by the modulation mask at frequencies of 400 kHz, 600 kHz, 1200 kHz, and 1800 kHz measured from the carrier frequency respectively. The switching transients can be reduced by ramping the output power up and down when transmitting a burst instead of just keying the transmitter on and off. The information transmitted in the burst must not be affected by the process of power ramping, which is performed at the beginning and end of the timeslot using the mask illustrated in Figures 6.20(a, b). The timeslot in the figure corresponds to a duration of 156.25 bits, that of the burst length. In a normal burst, frequency correction burst, or synchronization burst a GP of 8.25 bit periods is inserted between adjacent slots. The remaining 148 bit periods form the active part of the bursts. In an access burst the GP after the burst is 68.25-bits long, leaving an active part of 88 bit periods.

The useful part of a burst in all cases is one bit period shorter than the active part and begins halfway through the first bit period. During that part of the burst when information is transmitted, the amplitude of the modulated RF signal must stay approximately constant. The power control of the transmitted signal exemplified by the ramping of the transmitted power occurs during the GPs. From the above figure one can observe that approximately 70-dB power, ramp-up during 28 μs corresponding to 7.6 bit intervals, while ramp-down takes place in 18 μs, that is, 4.9 bit intervals. When bursts are transmitted at the same frequency in consecutive timeslots, that is, no frequency hopping is used, power ramping between the slots is not required and the signal transmitted in the guard times between the active slots may not be a modulated signal. In this case, the recommended time masks apply for the beginning and the end of the series of consecutive bursts. We now discuss two modulation schemes, namely, MSK and its modified form, GMSK, adapted for the GSM system, given the following review and elaboration of the general requirements for digital cellular systems.

With regard to good BER performance, generally, a higher level of modulation will increase the spectral efficiency. Thus, $m = 4$ should be better than $m = 2$. However, a higher level of modulation increases the BER as the signal constellation points come closer for the same power level. Thus, higher level schemes must be

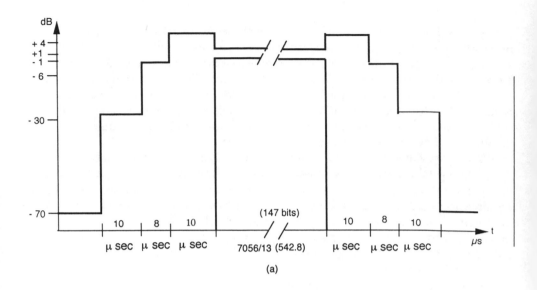

(a)

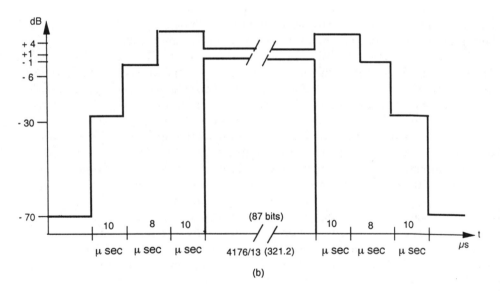

(b)

Figure 6.20 Power ramping time-masks for (a) normal burst and (b) access burst.

traded off with increased BER. On the other hand, the modulation scheme should be such that the BER requirements are met at lower values of the *C/I* ratio or in the presence of a high interfering power. The low required *C/I* will permit cochannel cells with lower geographical separation and, hence, a higher traffic-carrying capacity.

Consider the efficient use of mobile dc power. In cellular radio, the mobile is always power limited, especially those with handheld radios. Constant envelope modulation will facilitate the efficient utilization of available dc power using a class C power amplifier. In view of this, the MSK and GMSK are good choices.

The applicability to the cellular environment is important because regardless which modulation we choose, it must not only be able to withstand the severe multipath fading but also perform satisfactorily; that is, the BER should at least be $< 10^{-2}$ or better. This is the minimum acceptable BER for speech communication.

Finally, it is absolutely essential that the chosen modulation scheme should be easily implemented, preferably using large scale integrated circuits.

6.5.1 MSK Modulation

As previously discussed, due to the limited availability of power at the mobile, a higher order modulation, such as 8-PSK cannot be adopted for cellular radio application [14,15]. At the same time we cannot live with QPSK due to the difficult filtering requirements limiting the adjacent channel leakage. This leads us to choose a modified OQPSK modulation known as MSK. MSK is an adaptation of OQPSK, in which the modulating pulses are sinusoidal instead of rectangular. Thus, MSK can be generated and coherently detected as OQPSK with sinusoidal pulses. Having a constant envelope property allows the power amplifier to work at saturation without significant distortion. MSK is also a special case of coherent FSK modulation, known as *Fast Frequency Shift Keying* (FFSK), where the minimum spacing between high and low tones is 0.5 * data rate. This is the minimum required tone spacing that allows the two frequency states to be orthogonal to each other. MSK derives its name from the fact that this is the minimum frequency spacing for a coherently orthogonal sinusoid.

Amplitude versus the time graph of MSK results in one complete cycle of the lower shift frequency and one and a half cycles of the higher shift frequency over a symbol period. In actuality, the difference between the two frequencies will be a half cycle over *n* cycles of carrier. Here, just for making the diagram clear, one cycle is shown as mark frequency and one and a half cycle shown as space frequency. The *I/Q* diagram, time waveform, and phase function for MSK are shown in Figure 6.21(a–c).

Since MSK is a special case of OQPSK with sinusoidal pulse weighing, the signal can be defined as

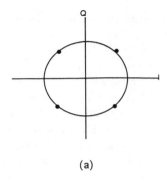

(a)

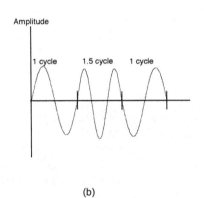

(b)

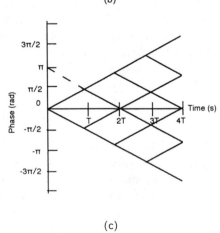

(c)

Figure 6.21 (a) *I/Q* diagram, (b) signal time waveform, and (c) phase function of the FFSK.

$$s(t) = a_I(t)\cos\left(\frac{\pi t}{2T}\right)\cos 2\pi f_c t + a_Q(t)\sin\left(\frac{\pi t}{2T}\right)\sin 2\pi f_c t \qquad (6.31a)$$

which can alternatively be expressed as

$$s(t) = \cos(2\pi f_c t + b_k(t)\frac{\pi t}{2T} + \Phi_k) \qquad (6.31b)$$

where

$$b_k \begin{cases} = +1 & \text{for } a_I a_Q = -1 \\ = -1 & \text{for } a_I a_Q = +1 \end{cases}$$

and

$$\Phi_k \begin{cases} = 0 & a_I = 1 \\ = \pi & a_I = -1 \end{cases}$$

The various component waveforms of (6.31a) are shown in Figure 6.22. From the figure and (6.31a, b), the following conclusions can be drawn for MSK modulation.

- It has a constant envelope.
- During each bit interval, the phase of a MSK carrier is shifted linearly with time by $\pm \pi/2$.
- $b_k = +1$ corresponds to a higher frequency f_H, and $b_k = -1$ corresponds to a lower frequency f_L.
- A high tone f_H contains one more half cycle of carrier than does the low tone f_L.
- Carrier frequency f_c is never transmitted, only f_L or f_H are present in the modulated signal.
- For a modulation bit rate of R_b

$$f_H = f_c + 0.25 R_b \quad \text{(when } b_k = +1) $$
$$f_L = f_c - 0.25 R_b \quad \text{(when } b_k = -1)$$

coherent detection allows the orthogonal detection of a MSK tone with a minimum tone spacing, $\Delta = f_H - f_L = 0.5 R_b$.

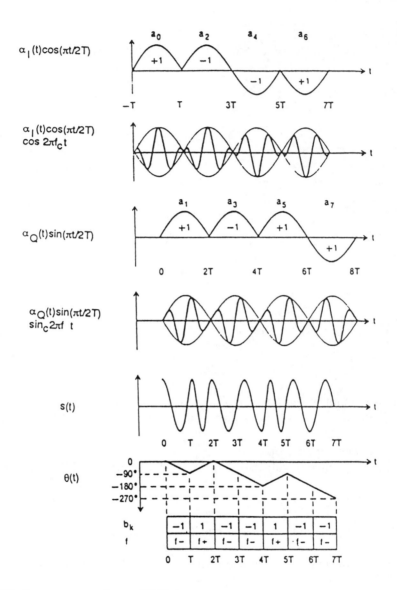

Figure 6.22 Component waveforms of (6.31a).

The modulation index is 0.5, as shown

$$m_f = \Delta/R_b = (f_H - f_L)/R_b = 0.5 \qquad (6.32)$$

One method of MSK realization (serial realization) is shown in Figure 6.23. Equation (6.31b) can be written as

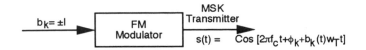

Figure 6.23 One form of MSK generation.

$$s(t) = \cos[2\pi f_c t + \Phi_k + b_k(t)\omega_T t] \qquad (6.33)$$

where $\omega_T = 2\pi(0.25/T) = 0.5\pi R_b$ or $f_T = R_b/4$. Here, $f_H = f_c + f_T$ and $f_L = f_c - f_T$. The corresponding receiver structure is shown in Figure 6.24.

The power spectra of QPSK, OQPSK, and MSK can be found by taking the Fourier transform of the symbol shaping function. The $p(t)$ for QPSK and OQPSK is given by

$$p(t) = \begin{cases} 1/\sqrt{2} & |t| \leq T \\ 0 & \text{elsewhere} \end{cases} \qquad (6.34a)$$

and for MSK by

$$p(t) = \cos \pi t/2T \qquad |t| \leq T \qquad (6.34b)$$

The normalized spectral density, $G(f)$, for QPSK and OQPSK is given by

$$\frac{G(f)}{T} = 2\left(\frac{\sin 2\pi f T}{2\pi f T}\right)^2 \qquad (6.35a)$$

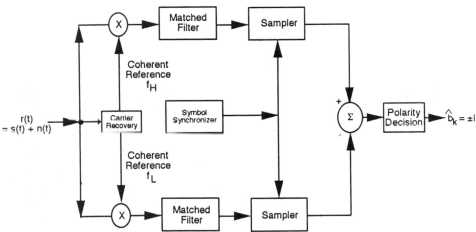

Figure 6.24 MSK receiver (serial form).

and for MSK by

$$\frac{G(f)}{T} = \frac{16}{\pi^2}\left(\frac{\cos 2\pi fT}{1 - 16f^2T^2}\right)^2 \qquad (6.35b)$$

The spectra are sketched in Figure 6.25. The difference in rates of falloff of these spectra can be explained on the basis of the smoothness of the pulse shape p(t). The smoother the pulse shape, the faster the drop of spectral tails to zero. Thus, MSK that has a smoother pulse has lower sidelobes than QPSK and OQPSK. FFSK is phase continuous and the QPSK discontinuous, which corresponds to a slope of 12 dB/octave and 6 dB/octave, respectively. Unfortunately, this performance improvement is somewhat offset by an increase in the main lobe width with the first zero crossing at a normalized bandwidth of 1.5 (two-sided). Thus, MSK is not suitable for narrowband application.

6.5.2 GMSK Modulation

As discussed previously, MSK is an improvement over QPSK (in terms of out-of-band power) because it makes the phase change linear and limits it to $\pm \pi/2$ over a bit interval. A low side lobe in the output power spectral density represents the effect of this linear phase change [16]. Unfortunately, the main lobe is even wider than QPSK/OQPSK cases (Figure 6.25). A low side lobe definitely helps to control

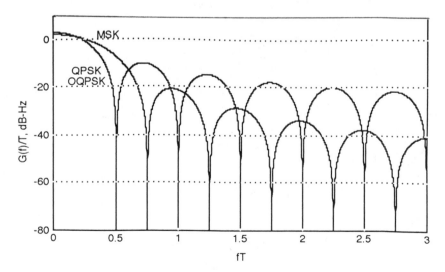

Figure 6.25 Spectral density of QPSK, OQPSK, and MSK.

adjacent channel interference. Both the side lobe power level and the width of the main lobe can be reduced by introducing a baseband Gaussian-shaped filtering of rectangular pulses before modulation. This controlled amount of ISI further reduces the phase discontinuities in the carrier. As a result, rolloff of the main lobe of the transmitter spectral density is increased, and the side lobe level is reduced [14,15].

The power spectrum density plot as a function of normalized frequency difference from the carrier center frequency $(f - f_c)T$, with the normalized bandwidth of the baseband Gaussian filter B_bT as a parameter, is shown in Figure 6.26 [1]. Here B_bT is the 3-dB bandwidth of the Gaussian filter and T is the data pulse width. A normalized bandwidth value of ∞ represents a normal MSK power spectrum density. For a normalized frequency difference of 1.5 (representing $(f - f_c)$ = 200 kHz, R_d = 270.8 Kbps, $B_{IF}T_s$ = 200 × 2/270.8 ≈ 1.5), and with B_bT = 0.24, the power spectrum density is shown to have a value below −80 dBc.

A plot of the spurious radiated power in the adjacent channel to the desired channel power, with normalized frequency separation f_sT as abscissa and B_bT as a parameter, is shown in Figure 6.27. With f_s = 200 kHz and T = 1/(270.8*10³) sec,

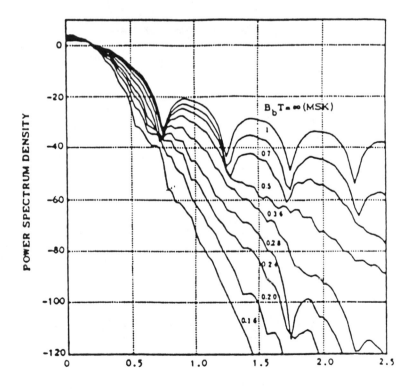

Figure 6.26 Normalized frequency difference versus PSDF of GMSK.

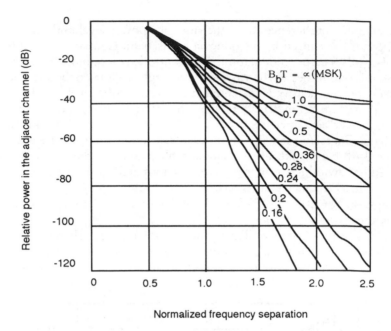

Figure 6.27 Relative power radiated in the adjacent channel.

which provides normalized frequency separation of 1.5, the power in the adjacent channel is below 60 dB for $B_bT = 0.24$. Thus, a GMSK with $B_bT = 0.24$ (maximum value) can be adapted as a digital modulation scheme where the ratio of spurious radiated power in the adjacent channel to that of the desired channel will be about −60 dB.

A GMSK has been adapted by the European GSM system. For the GSM system, a normalized bandwidth of B_bT is 0.3, with a total channel data rate of 270.8 Kbps (if the bandwidth = 0.3 * 271 kHz). A choice of $B_bT = 0.3$ appears to be a compromise between BER and out-of-band interference. As B_bT drops below 0.3, the BER exponentially increases due to a drastic reduction in signal power.

6.5.2.1 GMSK Representation

A MSK is a continuous phase and a constant amplitude modulation represented by

$$s(t) = A \cos[2\pi f_0 t + \Phi(t, \alpha)] \qquad nT \le t \le (n + 1)T \qquad (6.36)$$

The information is contained in the phase term $\Phi(t, a)$. For n bits, the total accumulated phase is given by

$$\Phi(t, \alpha) = 2\eta h \sum_{i=0}^{n} \alpha_i q(t - iT) \qquad (6.37)$$

where a_i represents data symbols $\{\pm 1\}$ and h is the modulation index defined as the Δf/bit rate. Thus,

$$s(t) = A \cos[2\pi f_0 t + 2\pi h \sum_{i=0}^{n} \alpha_i q(t - iT)] \qquad (6.38)$$

The baseband pulse response $g(t)$ is related to the baseband phase response $q(t)$ by an integral given by [14–16]

$$q(t) = \int_{-\infty}^{t} g(\tau) \, d\tau \qquad (6.39)$$

The pulse response $g(t)$ can be found via the impulse response of the Gaussian filter. The premodulation Gaussian filter has a transfer function of the form

$$H(f) = Ke\left[(-f/B)^2 \frac{\ln 2}{2} \right] \qquad (6.40)$$

where B is the filter 3-dB bandwidth point and K is a constant. Taking the inverse Fourier transform, the impulse response is given by

$$h(t) = K\sqrt{\frac{2\pi}{\ln 2}} Be^{-2\pi^2 B^2 t^2/\ln 2} \qquad (6.41)$$

Considering the filter response $g(t)$ to a unit rectangular pulse of width T, centered at the origin, we write

$$g(t) - K\sqrt{\frac{2\pi}{\ln 2}} B \int_{t-T/2}^{t+T/2} e^{-2\pi^2 B^2 x^2/\ln 2} \, dx \qquad (6.42a)$$

$$= K/2[\text{erf}\sqrt{\frac{2}{\ln 2}} \pi B(t - T/2)\} + \text{erf}\{\sqrt{\frac{2}{\ln 2}} \pi B(t + T/2)\}] \quad t > 0 \quad (6.42b)$$

where

$$\text{erf}(y) = \frac{2}{\sqrt{\pi}} \int_{0}^{y} e^{-u^2} \, du \qquad (6.42c)$$

and $\text{erf}(y) = \text{erf}(-y)$. From the above, $g(t) = g(-t)$. Thus, by knowing the pulse response $g(t)$, the phase response $q(t)$ can be found by applying (6.39).

6.5.2.2 GMSK Modulator

The most straightforward way of implementing a GMSK modulator is to transmit the data stream through a Gaussian low pass filter and apply the resultant waveform to a *voltage-controlled oscillator* (VCO) as shown in Figure 6.28. The output of the VCO is then a frequency modulated signal with a Gaussian response. The premodulation filter should have narrow bandwidth and sharp cutoff as well as low overshoot impulse response while preserving the filter output area to assure a $\pi/2$ phase shift at the modulator output at the end of every bit interval.

In order to solve the phase-ambiguity problem in the acquisition of a coherent local reference, differential encoding is included in the transmitter with a corresponding differential decoding in the receiver. The modified GSM transmitter block diagram is shown in Figure 6.29.

The main disadvantage of a modulator is the instability of VCO. As a result, the signal produced by this is not suitable for coherent demodulation. It is difficult to keep the center frequency within the allowable value under the restriction of maintaining the linearity and the sensitivity for the required FM modulation.

An alternate realization of a GMSK modulator by Murota and Hirade is shown in Figure 6.30, where the data input first phase shifts the BPSK modulator by $\pi/2$ before passing through a phase-locked loop. Unfortunately, VSLI implementation of the above modulator is difficult to realize.

6.5.3 Demodulation

Three different techniques can be used for the demodulation of GMSK signals. They are (1) differential detection, (2) coherent detection, and (3) FM discriminator detection.

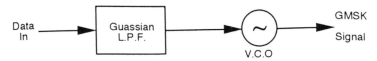

Figure 6.28 Simple GMSK modulator.

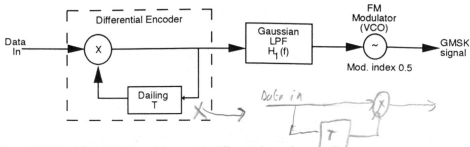

Figure 6.29 Modified GMSK modulator with differential encoder.

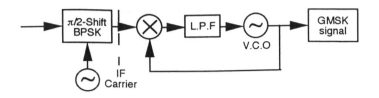

Figure 6.30 PLL GMSK modulator.

Since differential detection does not require an absolute phase reference in the receiver, this is a preferred choice for signal recovery in multipath surroundings. Both 1-bit and 2-bit delay versions have been proposed in the literature. Receiver performance in the two-bit delayed version is superior to that of the 1-bit differential detection because the collected energy over a 2-bit interval is larger. Coherent detection is similar to that of detecting straight MSK signals. The FM discriminator (noncoherent demodulation) does not account for phase and, thus, may not be very desirable.

The analysis of one-bit and two-bit differential detectors has been completed by Simon and Wang. BER curves for one-bit and two-bit detection with different values of $B_b T$ are shown in Figure 6.31. As seen from this curve, BER degrades substantially below $B_b T$ of 0.32. On the other hand, performance improvement is marginal above this value.

6.6 CHANNEL CHARACTERIZATION

For a complete understanding of the need for channel equalization, a thorough understanding of the channel model that includes multipath fading, time dispersion, correlation and coherence bandwidth, and doppler effects must be achieved. Thus, the objective of this section is to characterize these channel parameters before discussing the study of the equalizer. Both *Minimum Mean Square* (MMSE) and *Decision Feedback* (DFB) equalizers are briefly discussed.

Receiver sensitivity means that we are talking about the weakest input signal required for a specified output. Say that we have to receive P watts to be able to detect the information sent from the Tx-antenna. Then, if the signal falls below P watts, the information is lost. Thus, it is clear that we cannot plan our system according to the global mean value of the signal strength only. We must take precautions against the fading, and therefore we introduce the term fading margin. If we want to have an interruption-free transmission path, the global mean value has to be as many decibels above the receiver sensitivity as the strongest (deepest) fading dip gives rise to, say Y dB. We then have a fading margin of Y dB.

Reflections from objects far away from the receive antenna, that is, *time dispersion*, cause *Intersymbol Interference* (ISI). The signals became spread out in time

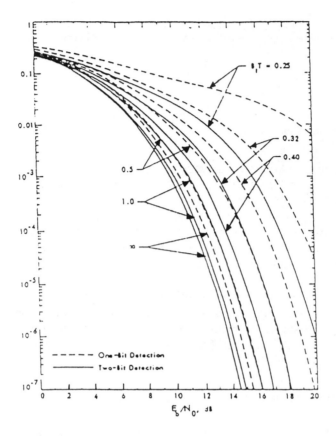

Figure 6.31 Performance of one- and two-bit differential detection of GMSK; receiving $B_r T$ is optimized for each $B_t T$. (For the two-bit detection case, the detection threshold is optimized for each E_b/N_o.)

and adjacent symbols interfere with each other. This causes problems when we try to decide which information has been sent. An example of this is shown in Figure 6.32. The sequence "1", "0" is sent from the BS. If the reflected signal arrives exactly one bit time after the direct signal, then the receiver will detect a "1" from the reflected wave at the time as it detects a "0" from the direct wave. The symbol "1" interferes with the symbol "0".

In GSM the bit rate is 270 Kbps, which leads to 3.7 ms/bit. One bit therefore corresponds to 1.11 km, so if we have a reflection from 1 km behind the MS the reflected signal will have a 2-km longer path than the direct one. This means that the reflection will mix a signal coming two bit-times later than the wanted signal, resulting in interference with the direct path signal.

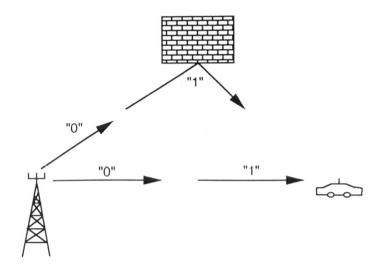

Figure 6.32 Time dispersion.

The problem of correlation between two signals at cellular radio arises for the determination of channel bandwidth. *Correlation and the coherence bandwidths* are related to each other through the time delay distribution of the received signals. Different path lengths of the received signal give rise to different propagation time delay. The presence of different time delays in the received waves causes the statistical properties of the two signals at different frequencies to become essentially independent for a large frequency separation. The frequency separation Δw at which the correlation between signals drops below a certain value (say less than 0.7) is known as a coherence bandwidth. If one transmits a signal whose bandwidth exceeds the coherence bandwidth of the channel, it is not possible to correct the amplitude and phase distortions of the received signal by means of a single complex correction factor (equalizer) applied over the whole bandwidth. For digital data transmission having signal bandwidth greater than the channel coherence bandwidth, the error probabilities can be expected to increase rapidly as the ratio of the required signal bandwidth to the channel bandwidth increases. Additionally, since the irreducible error probability for the channel is directly related to the coherence bandwidth, it will not be possible to control errors. We will discuss later the covariance of two signals and prove that it is simply the characteristic function of the time delays suffered by the component plane waves that compose the cellular radio field.

Assuming the coherence bandwidth to be the frequency separation at which the normalized value of the correlation coefficient attains a value less than 0.7, the nominal value of coherence bandwidth Δf is given by $1/(2\pi T)$, which corresponds to a correlation value of about 0.7. Thus for New York City, at 900 MHz where

$T \approx 5$ μs, the coherence bandwidth is approximately 31 kHz; while in suburban areas where T can be as small as 0.25 μs, the coherence bandwidth is 20 times greater or roughly 0.6 MHz [1,17,18].

If the transmitted signal's bandwidth is narrow compared to the channel's coherence bandwidth (B_c), all transmitted frequency components encounter nearly identical propagation delays; that is, the so-called narrowband condition is met and the signal is subjected to nonfrequency-selective or flat envelope fading. When the signal bandwidth is increased, for example, to accommodate several TDMA time slots as in the GSM system, the channel becomes more dispersive, resulting in intersymbol interference.

The doppler shift is expressed as

$$f_0 = f_m \cos \alpha \qquad (6.43)$$

where α is the angle of incidence and uniformly distributed and $f_m = V/\lambda$ = maximum value of doppler frequency f_m at $\alpha = 0$. This doppler frequency f_m is directly related to the phase change Δf caused by the change in path length $\Delta \lambda$ between signals. We note that the doppler shift is bounded to $\pm f_m$, which in general is far smaller than the carrier frequency f_c. It can be seen that component waves arriving from ahead of the vehicle experience a positive doppler shift while those that arrive from behind the vehicle have negative doppler. Thus each component of the received signal is shifted by different values of doppler frequency. For example, a vehicle traveling at the speed of 55 mph and receiving signal at a carrier of 850 MHz will introduce a maximum doppler shift of $V/\lambda = 69.3$ Hz.

Since the doppler shift in this case comes out to be approximately 70 Hz, the receiver, IF bandwidth must also exceed this value also. In order to provide exactly specified, identical test conditions for different implementations of the GSM system, in particular for various Viterbi equalizers, a set of 12-tap and 6-tap typical channel impulse responses were defined, some of which are depicted in Figure 6.33. The Doppler spectrum choices are either Rician or the classical Rayleigh. For the classical Rayleigh model the doppler power spectrum is given by

$$D(f) = \frac{1}{2\pi f_m}[1 - (f/f_m)^2]^{-1/2} \qquad -f_m < f < f_m \qquad (6.44)$$

where $f_m = v_m/\lambda$ is the doppler spread, v_m is the mobile vehicle velocity, and λ is the wavelength of the carrier frequency. The Rician spectrum is the sum of the classical doppler spectrum and one direct path, weighted so that the total multipath power is equal to that of a direct path, and is given by

$$D(f) = \frac{0.41}{2\pi f_m}[1 - (f/f_m)^2]^{-1/2} + 0.91\delta(f - 0.7f_m) \qquad -f_m < f < f_m \qquad (6.45)$$

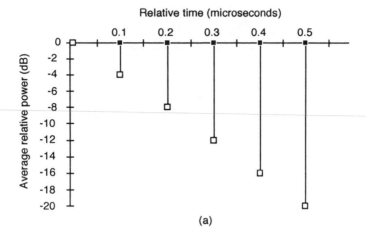

(a)

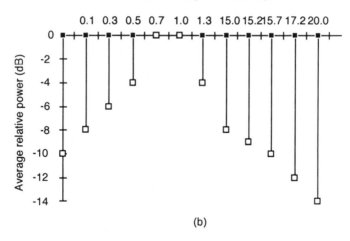

(b)

Figure 6.33 Typical GSM specified channel impulse responses: (a) rural environment, (b) hilly environment, (c) urban environment, and (d) equalizer test profile.

The model for the GSM system test is defined with both 12 and 6 taps. The configuration with six taps is particularly used for multipath simulation on an interfering signal. The *Rural Area* (RA) response is the least hostile amongst all standardized responses, decaying fast within a one-bit interval, and in terms of BER performance it behaves as a single-path nondispersive channel, where no *Viterbi Equalizer* (V) is required. The *Hilly Terrain* (H) model has a short-delay section due

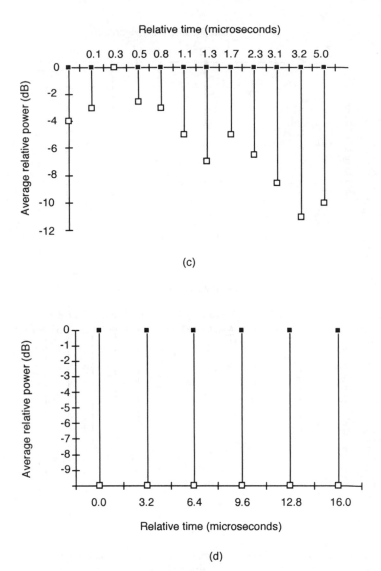

Figure 6.33 (continued).

to local reflection and a long-delay part around 15 μs due to distant reflections; therefore, it can be considered as a two- or three-path model, providing useful diversity gain, when using a Viterbi Equalizer. The *Typical Urban* (T) impulse response spreads over a delay interval of 5 ms, which is almost equal to two 3.69-ms bit intervals duration, and therefore may result in a serious ISI problem.

The last standardized impulse response is to test the Viterbi Equalizer performance and is constituted by six equidistant unit-amplitude impulses representing six equal-powered independent Rayleigh-fading paths with a delay-spread up to 16 μs. With these impulse responses in mind, the required channel is simulated by summing the appropriately delayed and weighted received signal components. Figure 6.33(a,b) show plots for these cases. Having discussed the characteristics of channel, let us now take up the study of the equalizer used in a GSM system.

6.6.1 Channel Equalization

We look upon the channel as a channel. The filter is excited by a known signal, and then we compare the output with the (known) input The comparison gives us the impulse response of the filter.

The impulse response of a filter is the output when the input is a very short pulse, in principle as shown as *a* and *b* in Figure 6.34. In *a* we have no reflections, and in *b* we have the case of one reflection. If we know how long and how strong the reflections are, we can take this into account when the received burst is detected. This is done by the equalizer, which creates a model of the transmission channel, that is, the air interface, and in addition calculates the most probable receiver sequence.

In a GSM system, data is transmitted in bursts, which are placed within timeslots. In the middle of the burst, a training sequence of a known pattern and with good autocorrelations properties is placed. This training sequence is used by the equalizer to create the channel model, which changes all the time but during one burst it is considered to be constant. If we know the impulse response of a filter, we can always find the output for any input signal.

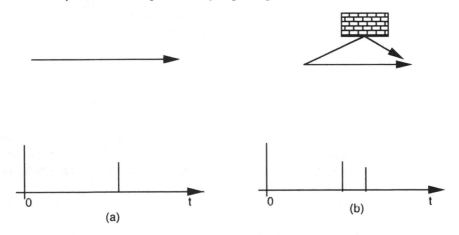

Figure 6.34 Impulse response (a) without reflection and (b) with one reflection.

Now we have created a channel model. What we do now is to generate all possible sequences and feed them through our channel model. The input sequence that gives an output sequence almost like that received is regarded as the originally sent sequence.

6.6.1.1 Use of Training Sequence in Equalizer

As shown in Figure 6.35 the training sequence has been inserted in the middle of the burst in order to minimize its maximum distance with a useful bit and is therefore sometimes called "midamble."

Eight different training sequences, as shown below, have been specified. Consider the case of two similar interfering signals arriving at the receiver at almost the same time. If their training sequences are the same, there is no way to distinguish the contribution of each to the received signal. The situation is much clearer when the two training sequences differ and are as little correlated as possible. Distinct training sequences will therefore be allocated to channels using the same frequencies in cells that are close enough so that they do not interfere.

Figure 6.35 shows the autocorrelation function of one of these eight training sequences calculated between the central 16 bits and the whole 26-bit sequence. All eight sequences share the central correlation peak surrounded by five "0" on each side.

Training Sequence Code (TSC)	Actual Bit Configuration of Training Sequence
0	00100101110000100010010111
1	00101101110111100010110111
2	01000011101110100100001110
3	01000111101101000100001110
4	00011010111001000001101011
5	01001110101100000100111010
6	10100111110110001010011111
7	11101111000100101110111100

By correlating the stored training sequence with the received sequence, the channel impulse response can be measured. This impulse response is used for other parts of the frame. In actuality, the Viterbi equalizer is used, which in turn reduces the amount of calculation required to predict the likely sequence of data. The Viterbi equalizer is shown in Figure 6.36.

If we know how long and how strong the reflections are, we can take this into account when the received burst is detected. This is done by the equalizer, which creates a model of the transmission channel, that is, the air interface, and in addition calculates the most probable receiver sequence.

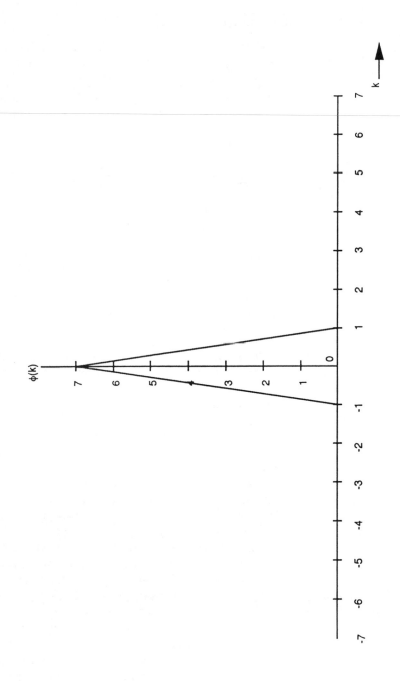

Figure 6.35 Autocorrelation of training sequence.

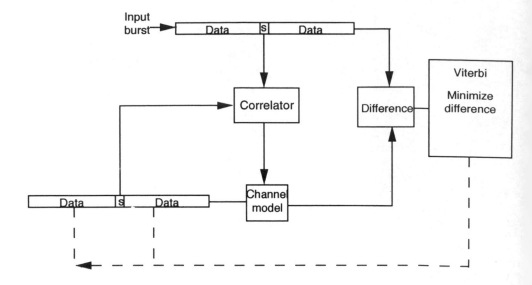

Figure 6.36 Viterbi equalizer.

As discussed, due to the multipath the MS is dispersed in times. Due to the smearing of signals in time, a radio signal introduces ISI. To decode the signal reliably, an estimate of the channel transfer function is required, which can then be used by an equalizer to compensate. Conceptually, the equalizer takes the different time-dispersed components, weighs them according to the channel characteristics, and sums them after inserting the appropriate delay between components so that the replica of the transmitted signal is restored. The problem in cellular radio becomes more complex due to the dynamic nature of the channel; as the mobile moves through multipath surroundings, the equalizer must continually adapt to the changed channel characteristics that vary at the rate proportional to the wavelength of the signal. If the operational mode is TDMA, this will necessitate measuring the channel impulse response from one slot to another. A constant training sequence may be included in every data burst to make this fast and repeated evaluation. The equalizer knows the transmitted training sequence; it also knows what it has really received and thus can make an estimate of the channel transfer function. If the multipath is severe, it may even be necessary that the equalizer be adaptive during the individual burst of transmission. Thus, an adaptive equalizer continuously updates the transfer function estimate, making sure that the decision error does not increase too much during the channel transmission. In GSM two types of nonlinear equalizer, one based on *Maximum Likelihood Sequence Estimation* (MLSE) and the other based on DFE, have been proposed as candidate solutions for this application. The use of nonlinear equalization offers significant advantages under severe channel distortions, since it avoids the noise enhancement that would otherwise arise a if linear equalizer were used. Both these equalizers are used to adapt to the dynamic changes of the channel and operate under two different modes of operations: (1) training mode, which enables a proper setting of the equalizer parameters by processing the preamble at the beginning of each timeslot, and (2) the tracking mode, during which the equalizer parameters are continuously adjusted to match the channel variations during the timeslot [12,19].

6.6.1.2 *Maximum Likelihood Sequence Estimation Equalization*

The block diagram of the Viterbi (MLSE) adaptive receiver is shown in Figure 6.37 [1]. The MLSE principally scans all possible data sequences that could have been transmitted by a digital data source, computes the corresponding receiver input sequences, compares them with the actual input sequence received by computing a weighing parameters known as metric, and selects the data sequence that shows the highest a posteriori probability (highest metric) to be that actually transmitted. The process of metric computation is the same as used in Viterbi decoding in convolutional codes. The MLSE algorithm is almost unrealizable due to the large number of computations even for relatively short binary sequences. The MLSE technique has become realizable after recognizing that the Viterbi algorithm can effectively be used not only to decode convolutionally coded sequences but also to deal with the unintentional coding produced due to ISI. The MLSE receiver considered here consists of *Matched Filter* (MF) with N taps and a Viterbi processor. Here, the received signal is sampled and each sample is filtered through a transversal filter to approximate the

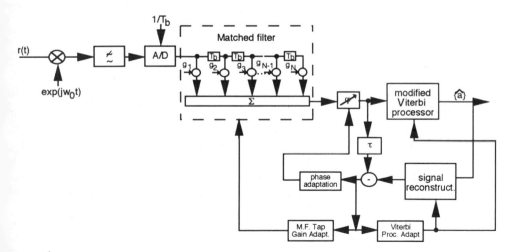

Figure 6.37 Block diagram of the Viterbi (MLSE) adaptive receiver.

matched filter response. The use of MF makes the receiver insensitive to the carrier, clock phases provided the MF coefficients are correctly adjusted, and the time span of the MF is long enough to include all the channel impulse response. The choice of the number of taps, N, is important as it relates directly to the maximum echo delays that are expected to be equalized and also to the complexity of the Viterbi processor.

It has been found that a choice of N between 5 and 7 is satisfactory under most practical conditions (five taps as in Fig. 6.37). The *Channel Impulse Response* (CIR) can be derived by correlating the timeslot with the modulating sequence. This facilitates the setup of the MF tap gains, which are the complex conjugate of the channel impulse response. The complexity and the processing effort depends exponentially on the bit periods during which CIR has to be computed. If the processing is excessive, the CIR has to be truncated, which in some cases will degrade the system performance. The output of the MF equalizer is fed to the Viterbi processor. The metric computation in the Viterbi processor does not require any squaring operation. Also an all-digital implementation of the block is possible. As shown in Figure 6.37, the receiver is continuously adjusted when the CIR changes during the timeslot. The gradient algorithm can be used to minimize the mean square error by varying the carrier phase, j; MF tap gains; and the Viterbi parameters.

6.6.1.3 Decision Feedback Equalizer

The equalizer shown in Figure 6.38 consists of a feed forward part that receives the input and a feedback part that is fed by the past detected symbols [1]. Both these filters consist of a *finite impulse response* (FIR) filter with their tap spacings equal to bit duration, and their respective outputs are added to form the equalizer output before threshold detection. The tap coefficients of both FIR filters can be computed by solving a set of linear equations or by iterative adjustments by processing the training sequence and subsequently the information symbols. For iterative tap, coefficient adjustment either by a slowly converging but less complex gradient algorithm or a more advanced and more complex but fast converging algorithm can be used. It has been pointed out in the literature that the equalization performance is highly dependent on the capabilities of the coefficient adaptation algorithm used, which is the main processing load of the equalizer [12,18–22].

The number of taps is generally between 2 and 5, which is dependent upon the maximum delay to be equalized. The adaptation of the taps must take place during the training as well as during the actual interval when the messages are received. A long preamble may be required during the training sequence, especially for low S/N ratio. After the initial preamble, the tap gains are adjusted to minimize the mean square error (LMS algorithm) during the message part of the timeslot. The error, as shown in Figure 6.38, is defined as the difference between the analog sample

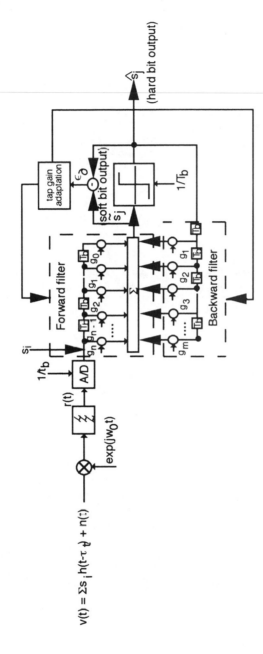

Figure 6.38 Block diagram of the DFE adaptive receiver.

at the input to the decision device; and the corresponding transmitted binary digit and can be expressed as

$$\epsilon_n = \tilde{S}_n - \hat{S}_n \tag{6.46}$$

In terms of the forward and backward filter coefficients, the error can be expressed as

$$\epsilon_n = \int \gamma(t) r(t + nT_b) \, dt - \sum_{k=1}^{\infty} g_k \hat{S}_{n-k} - S_n \tag{6.47}$$

where

$$g(t) = \sum_{i=1}^{\infty} g_i S(t - iT)$$

Here, the forward filter impulse response is $\gamma(-t)$ and is defined over the whole range $|t| < \infty$. Thus, the minimization problem is

$$\min_{\gamma(t), b_k} E[\epsilon_n^2] \tag{6.48}$$

As the number of taps increases, not only do these filters become more complex but the tap gain adjustment also becomes more involved.

6.7 DELAY REQUIREMENTS

The transmission delay in any network has to be controlled in order to avoid speaker annoyance, which is dependent on the magnitude of the returning signal and the amount of delay involved. On short connections the delay is small enough that the echo appears as a sidetone and the talker feels this as natural coupling to the ear [1]. As the round trip delay increases, attenuation to the echoes are desired to reduce annoyance to the talker. This is required because a long delay between a subscriber talking and receiving a reply disturbs the flow of conversation. For this reason a delay requirement has been imposed in the GSM network with the following contributory components [1].

- Speech coding algorithmic delay   < 20 ms;
- Interleaving delay   < 37 ms;
- Processing delay   < 8 ms;

- Other delay in the radio subsystem   < 15 ms;
- Overall delay   70 ms to 80 ms.

The delay requirement was the most difficult parameter to consider in the design specification of the GSM system due to complex design tradeoff decisions. Each of the components, when considered in isolation, will make good use of an increased delay budget, described as follows.

Speech coding algorithmic delay, otherwise known as the transcoder delay, is defined as a time interval between the instant a speech frame of 160 samples has been received at the encoder input and the instant the corresponding 160 reconstructed speech samples have been output by the encoder at an 8-kHz sampling rate. Ideally, one can assume that the longer the delay in the speech coding, the better the quality of the speech, or alternatively, the lower the necessary bit rate for a given speech performance.

Interleaving delay, as discussed in Section 6.4.1, the 2×456 bits of coded and error protected speech frame, are interleaved by a factor of eight. With the frame duration of 4.6 ms, the total interleaving delay amounts to 37 ms ($4.615 \times 8 \approx 37$ ms). This delay is the tradeoff between speech quality reduction due to burst noise caused by multipath and annoyance due to delay. However, 456 bits could fit exactly into four timeslots (114×4), which would have reduced the interleaving delay to about 20 ms. This would have deteriorated the burst error correction quality, resulting into lower speech quality and reduced spectrum efficiency of the system. On the other hand, an even deeper interleaving delay would cause speech quality improvement, but also added annoyance to the user.

Processing time: Low power consumption is the basic requirement for the mobile terminal, which imposes tough constraints on the implementation of the mobile terminal. Since mobile terminal implementation is based on ASIC design, which is based on a CMOS VLSI circuit where power consumption is a linear function of the clock rate, a relaxation of the delay requirement could make it possible to distribute the processing over a longer period with a lower clock rate, thus reducing the power consumption considerably. On the other hand, a higher clock rate does reduce the processing time but also increases the power requirement. A compromise is therefore required between the power consumption and the processing speed. The delay figure of 8 ms is merely to satisfy the power consumption requirement of a handheld mobile unit without excessive delay.

Other delay: Assume, the miscellaneous delay of 15 ms in the radio path; the overall delay of 80 ms represents the outcome of the various tradeoffs already described. The value of 80 ms is supposed to be short enough to avoid any conversational delay problems for the majority of calls. For a connection to PSTN, an 80-ms delay will come as an add-on to the existing delay. Consequently, there is a need for echo control in all calls originated or terminated by the GSM network. For this reason it is proposed to include an echo cancellor at the PSTN interface of the

GSM system to remove any echoes being returned from the PSTN to the mobile user.

6.8 DISCONTINUOUS TRANSMISSION

The capacity of a cellular radio system is determined by the level of cochannel interference. One effective way of reducing cochannel interference is to switch the transmitter off when no speech is present. This mode of operation is known as DTx. In a normal conversation, each person speaks, on average, for less than 40% of the time. It is easy to show that those cellular systems designed to take advantage of DTx that their system capacity could at least be doubled. As a result of switching off the transmitter, the total power requirement of the MS is reduced. In order to realize these advantages of DTx, the system designer faces some potential problems of system optimization. These include the following.

- Reduced radio transmission time;
- Insignificant speech quality reduction;
- No significant cost increase.

The process of simultaneously achieving all the three objectives starts with the detection of speech at the transmitter end and the natural-like substitution of noise during the silence period. The block diagram of the speech processing functions is shown in Figure 6.39.

In order to minimize the radio transmission, the noise frame must be distinguished from the speech frame, which can only be made by taking into consideration the spectral characteristics of the input signal. Provided the background noise is stationary over relatively long periods, compared to the speech the spectral characteristics of the noise will be very similar from frame to frame. In principle, then, it should be possible to detect the presence of speech by looking for deviations from the spectral characteristics of the background noise.

Speech detection is carried out at the transmitting end by a *voice activity detector* (VAD), which distinguishes between speech superimposed on environmental noise and noise without speech being present. The output of the VAD is used to control a transmitter switch. If the VAD fails to detect every speech event, then the transmitted speech will be degraded due to clipping. On the other hand, if the VAD identifies noise as speech often, then the effectiveness of DTx is diminished. Both of these result in poor performance [23].

At the receiving end, the background acoustic noise abruptly disappears whenever the radio transmitter is switched off. Since switching can take place rapidly, within words as well as between words, it has been found that this noise modulation can be very annoying to MS. In very bad cases the noise modulation greatly reduces

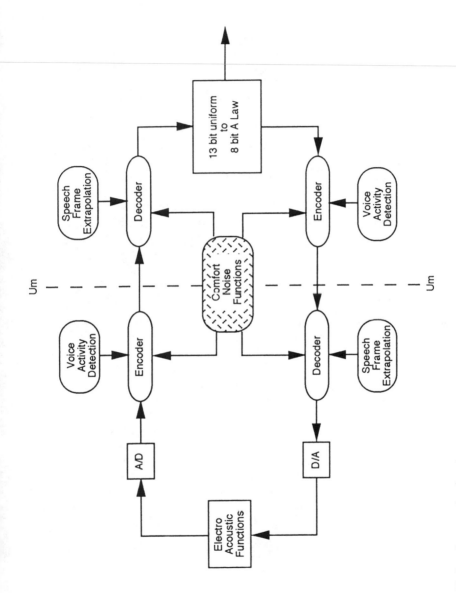

Figure 6.39 Speech processing functions.

the intelligibility of the speech. This problem can be overcome by generating at the receiver a synthetic signal known as "comfort noise" inserted at the receiver whenever the transmitter is switched off. If the comfort noise characteristics are well matched to those of the transmitted noise, the gaps between the talkspurts can be filled in such a way that the listener does not notice the switching during the conversation.

Since the noise at the transmitting end is constantly changing, a means must be found to keep the comfort noise generator updated to match with rapid changes. It is also necessary for the receiver to be able to switch between comfort noise and the received signal in synchronism with the switching of the transmitter. This DTx information is sent over the speech channels as it is not effectively possible to use signaling channel for DTx in GSM.

Another important aspect of the voice control system is known as *Speech Frame Substitution* (SFS). When speech data is received with a large error, the previously received speech frame is substituted in lieu of the present frame. This will enable the DTx system to maintain high speech quality. As stated, all these objectives must be met without significant cost penalty. We shall now describe details of each of the major subsystems associated with the DTx system. These subsystems are achieved by designing the following functional blocks: (1) VAD, (2) comfort noise functions, and (3) speech frame extrapolation. The reader is advised also refer to Section 6.3.

6.8.1 Voice Activity Detector

VAD operating in a mobile environment must be able to detect speech under a low S/N ratio condition. Under this condition it may be difficult at times to distinguish between speech and noise using simple level detection techniques when parts of the speech utterance can be well below the noise level [24]. The distinction between speech and noise under these conditions can only be made by measuring the spectral characteristics of the input signal. If the background noise is stationary over relatively long periods (compared to the speech), the spectral characteristics of the noise will be quite similar from frame to frame. In principle, it then should be possible to detect the presence of speech by looking for deviations from the spectral characteristics of the background noise. VAD incorporates an inverse filter, the coefficients of which are derived during noise-only periods. When speech is present, the noise is attenuated by the filter, leaving mostly deviations from the spectral characteristics of the noise. Deviations can now be due to speech. The energy of the inverse filtered signal is compared to a threshold that is updated during noise-only periods. If the energy is greater than the threshold, then the presence of speech is declared.

6.8.2 The Comfort Noise Functions

The main functions of the comfort noise subsystem are as follows.

- Noise measurement at the transmit end;
- Synthesis of the comfort noise;
- Update parameters that describe the background noise.

As stated, the information related to the spectrum is carried by the filter coefficients, and the level must be computed on the background noise at the transmit side and sent to the receiver in order to follow changes of the noise characteristics [23].

The noise parameters must be updated at the end of each speech-burst so that the comfort noise that is inserted during pauses of the speech matches the transmitted noise. They must also be updated at intervals during speech pauses so that when speech transmission begins after the pause there is no discontinuity between the comfort and the background noise that is transmitted.

It was experimentally verified that a good match of the background noise is possible by exciting an all-pole LPC synthesis filter with a random sequence scaled at a proper level. Since noise can rapidly change, requiring a high adaptation rate and demanding high activity on the channel, the tradeoff between these two conflicting aspects should aim for acceptable noise matching without a high overhead of activity. Based on these two conflicting requirements, an update interval of 480 ms has been chosen in the GSM system.

6.8.2.1 *Background Noise Parameters Evaluation*

The comfort noise parameters are computed at the transmit side whenever the end of a talkspurt occurs and at a regular rate during speech pauses.

When the end of a talkspurt is detected by the VAD algorithm, the radio channel is kept open for an additional hangover period (DTx hangover) lasting four frames. During these four frames there is a very high probability that no speech is present, so the background noise characteristics can be evaluated. The computation of the average level and spectrum parameters of the background noise is accomplished in a simple way by making use of the parameters computed by the RPE-LTP algorithm. The scale factors and the LAR parameters are averaged over the DTx hangover period, and the resulting values are quantized for transmission.

The information related to the background noise is arranged in a special *Silence Descriptor* (SID) frame that is sent over the channel. The encoding of the SID frame exploits the fact that only some parameters of the RPE-LTP frame are used to send the noise characteristics. In particular, the bit field in the frame devoted to transmitting the RPE pulses during speech transmission is partially filled with a special codeword that signals to the receiver that the received frame has to be considered as a SID-frame.

6.8.2.2 *Comfort Noise Generation Algorithm*

The comfort noise generation procedure has been tailored to the RPE-LTP receiver algorithm. When comfort noise is to be generated, the RPE excitation parameters (pulse amplitudes and the grid position parameter) normally input to the speech decoder are replaced by a locally generated random sequence. This random excitation signal is fed into the synthesis filter after scaling by the level parameter received in the special SID frame. The LPC synthesis filter uses coefficients received in the SID frame; these are representative of the average background noise spectrum. In addition, to minimize the effect of the long-term synthesis filter, the gain of this filter is set to the lowest available quantized value.

Updating the comfort noise parameters occurs each time a special SID frame is received; during speech pauses this adaptation occurs every 480 ms. Unfortunately, the background noise sometimes changes significantly during this period, giving rise to discontinuities in the comfort noise. For this reason the comfort noise parameters should be averaged over a few frames to obtain smooth transitions [24].

The DTx system will, of course, be working in an environment where high digital transmission error rates can be expected between the transmitter and receiver. For this reason it has been necessary to design the DTx strategy so that the effects of the loss and corruption of SID frames are minimized.

6.8.2.3 Speech Frame Extrapolation

The SFS is activated when the most vulnerable bits of the speech coder parameters are so heavily corrupted that they cannot be corrected by the channel decoder. When this situation is detected by the channel decoder, the SFS replaces the corrupted speech frame by the preceding (uncorrupted) speech frame and submits this to the speech decoder. The SFS method makes use of the fact that in very many cases the signal properties of consecutive speech frames are highly correlated. The application of SFS allows a higher channel error rate or the reduced value of C/I ratio.

6.9 TIMING ADVANCE MECHANISM

Using a TDMA system on the air presents a problem because each individual user can only transmit during the allocated timeslot and be silent during the remaining part of the frame. If the user transmits during an unspecified timeslot, it will interfere with other users. Assume that the mobile was allocated slot number 2 (TS2) and he is close to the BS and that mobile 1 has been allocated slot number 1 and is at the boundary of the cell site. Due to higher distance resulting in larger propagation delay, the transmitted signal of mobile 1 during TS1 reaches the BS later. If no precaution is taken, the mobile transmitting in TS1 will overlap with what the base receives from the other mobile during TS2. Thus, it is imperative that the base make measurements at the time of arrival of the specified user and give commands to adjust the user timing so that this overlap does not happen. From the BS point of view, the organization in the uplink direction is derived from the downlink by a nominal delay of three BPs. This delay of three BPs is a constant throughout GSM and is shown in Figure 6.40 along with the timing advance mechanism for this example. Thus the whole process of timing and its adjustments can be summarized as follows.

- Nominal delay between uplink and downlink three BPs;
- When at a large distance from BS, MS is instructed to delay its timing further;
- The mechanism avoids overlap of the adjacent BPs.

It should be noted that in a GSM system the MS is affected by considerations about propagation delays, which—even at the speed of light—are not negligible compared to the burst duration.

When the MS is far from the BTS, propagation delays cannot be neglected and an exact three-BP shift cannot be maintained both at the MS and at the BTS. It is imperative that the bursts received at the BTS fit correctly into the timeslots. Otherwise, the bursts from MSs using adjacent timeslots could overlap, resulting in a poor transmission quality or even in a loss of communication.

The only solution is that the MS advances its emission relative to its reception by a time corresponding to and from propagation delay. This value is called the timing advance. The exact shift between downlink and uplink as seen by the MS is then three BP minus the timing advance. The timing advance value can be computed only by the BTS and is then provided to the MS through signaling. The process of measurement is during the access burst with a long GP of 68.25 bits. During the access burst when the BS detects a 41-bit random access synchronization sequence, it measures the received signal delay relative to the expected signal from a MS of zero range. This delay, known as timing advance, is signaled using a 6-bit number to the MS, which advances its time base over the range from 0 bits to 63 bits, in units of 3.7 μs. A timing step of 3.7 μs corresponds to a time duration of a single bit with the data rate of 270 Kbps. The maximum timing advance of 233 μs is adequate for MS to be up to 35 km from the BS, which is the maximum allowable cell radius of the GSM system. In normal operation the BS continuously monitors the signal delay from the MS and will instruct the MS to update its time advance parameter about two times per second over the SACCH. The update period of twice per second assures that the MS never gets too far out of timing alignment. For large traffic cells there is an option to actively utilize every second timeslot only to cope with higher propagation delays, which is spectrally inefficient, but can be tolerated for a rural system. Using the alternate timeslot sets up the limit of a 120-km radius for the urban cell.

6.10 MOBILE-ASSISTED HANDOVER

Handoff is the process whereby the cellular system redirects the MS/BS radio path from one BS to another, thus affecting the transfer of the cellular subscriber from

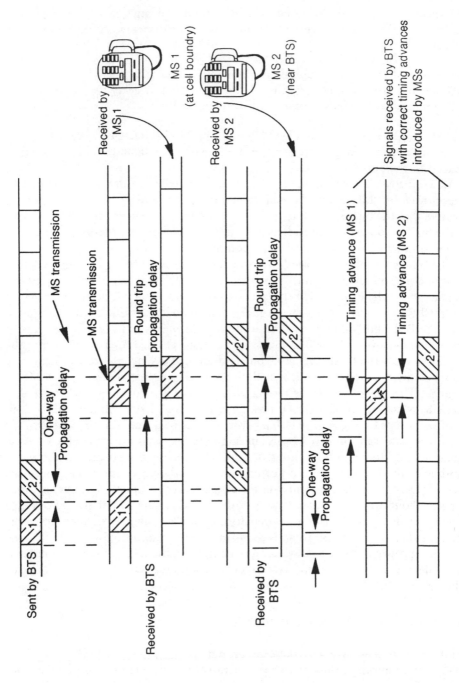

Figure 6.40   Timing adjustment of MS at the BS.

one cell site to another. Note that the mobile has to retune its channel frequency as it changes cells. The objectives are to achieve this process without the knowledge of the user and, once the handoff is made, to prevent the mobile from being immediately handed off again.

Thus, handover is defined as the process of automatically transferring a call in progress to a different cell in order to avoid adverse effects of user movements, such as loss of received signal strength or exceeding the timing advance adjustment capabilities.

There are four purposes for handover: (1) maintenance of high signal quality (rescue low-quality channel), (2) recovering cochannel interference, (3) traffic balancing among cells, and (4) recovering in the event of failure of a control channel. Rescue is the forced change of cell because the call will be lost if the cell is not changed; this is the most common purpose for handover. Another purpose is to reduce the interference that a call in one cell is causing to calls in another cell, and another is to ease traffic congestion by moving some calls in congested cells to cells with less congestion.

Signal quality refers to the minimum required carrier-to-interference ratio. For a high-quality handoff system one must take into account the following. Handoff must take place before a noticeable degradation of the signal quality occurs at the mobile. Before the handoff decision, the signal received at the mobile must be integrated for sufficient time to assure that the quality degradation is real and not due to multipath fading. The new channel to which the mobile is switched should be sufficiently higher in level (high S/N ratio) so that the new channel is not switched again immediately.

Unlike the signal quality directed handoff, which is done to accommodate the user, the traffic-balancing handoff accommodates the system. Its main purpose is to balance the load among the various cells so that an overload condition is not experienced in any one cell. Load balancing is most effective when there is a significant amount of overlap between adjacent cells and is achieved via a technique known as "directed handoff." A directed handoff is initiated when cell load reaches a preset percentage of the full load. Cells encountering this condition will endeavor to find an alternate cell and path with some of their load. The obvious underlying assumption for load balancing is that simultaneous overload condition, which does not occur in adjacent cells. With proper system design the load-balancing algorithm will provide help in handling peak load conditions, but it will not compensate for an inadequate quantity of voice channels. The algorithm for traffic balancing must consider that the new cell must be able to handle the call without the need for immediate handoff due to low S/N ratio and must not be itself in an overload state.

The other objective for handoff is due to a failure of the control channel when a voice channel is used as a backup control channel. Systems designed with this feature will require the mobile to be switched over to another channel if it is using the designated backup control channel for voice communication. The main objective for failure-driven handoff is to vacate the channel so that the channel is free of voice traffic and ready to be assigned as a control channel.

There are three cases of handover, all of which must be treated somewhat differently. First, there is handover from one radio channel to another of the same BS (same BSC, same or different BTS). Second, there is handover between channels of different BSCs under the control of the same MSC. Third, there is handover between channels under the control of different MSCs in the same PLMN. There are no procedures for handover between cells of neighboring PLMNS.

There are two modes of handover: synchronous and asynchronous. In synchronous handover, the old and new cells are synchronized so that their TDMA timeslots start at exactly the same time. This permits the MS to compute the timing advance for the new cell. The difference in timing advance between the old and new cells is then just twice the difference in arrival times of signals within the same slot to the MS from the two different BTSs.

In asynchronous handover, the old and new cells are unsynchronized, so the MS cannot independently correct the timing advance in this way. The timing advance must then be initialized at both the MS and the now BTS during handover. The MS sends the new BTS access bursts with a zero timing advance, and the BTS then determines the required timing advance from the round trip time delay of a given message. The BTS then sends the required timing advance to the MS. Asynchronous handover leads to a longer interruption of communication during handover than does synchronous handover (200 ms versus 100 ms).

By repetitively monitoring the SCHs of all the surrounding cells between transmission and reception of traffic bursts, the MS can compute beforehand the timing advances required for all surrounding synchronous cells. The MS is then presynchronized with any such cell to which it may be handed. This speeds up the handover process.

The received signal levels at the mobile are plotted as functions of its distance from the cell sites of two cells transmitting equal power shown in Figures 6.41(a, b). The intersection point is the equal level point on two curves. From the figure, as the intersection level moves higher the overlap increases. The two cases of extensive and minimal overlaps are illustrated. If the mobile is traveling from cell A toward cell B, the switchover threshold must be at the right of the intersection point. This is the only condition when the mobile is switched from a marginal signal level case to an improved signal quality that satisfies the handover criteria. Here 1 and 1' are appropriate switchover points for a mobile within cells A and B. The offset 11' can be regarded as a natural hysteresis. The second case is rather interesting where the threshold is at −120 dBm, on the left of the crossover point. In this case the switch over from cell A to cell B will be detrimental as the mobile will have a lower S/N ratio at cell B. In this case, switch over should be put on hold until point 2, where the S/N ratio improvement is greater than 20 dB. This holding of the switching point can be built into the software.

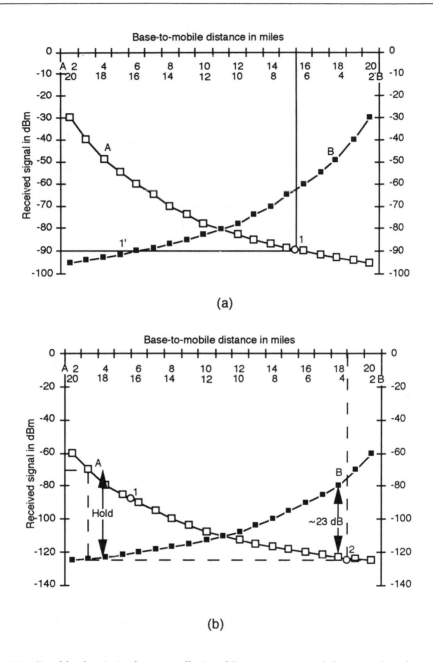

Figure 6.41 Signal level variation between cells A and B: (a) extensive and (b) minimal overlap.

6.10.1 GSM Application

Since the mobile continuously scans the six adjacent BCCHs, it is constantly aware of the RF quality of adjacent channels. The path loss is determined for each cell, including the currently serving cell, at least every 5 sec.

GSM will then check whether the path loss (with respect to preset path loss) for the current cell falls below 0 for a period of 5 sec (which indicates that the path loss has become too high). If the path loss of a new cell is below the path loss of the currently serving cell by at least as much as the parameter CELL RESELECT HYSTERESIS for a period of 5 sec, handover will happen. The objective is to operate each MS on the best quality RF carrier available at its present location. The *Received Signal Strength Indicator* (RSSI) should be used in conjunction with BER. Bad RSSI indicates a bad channel.

6.11 CONCLUSIONS

In this chapter we discussed the relevant technical characteristics for a GSM system, including speech and channel encoding, digital modulation, equalizer, delay requirement, discontinuous transmission, timing advance mechanism, and handover techniques. The spectral efficiency of a GSM is higher than their predecessor analog system. The coding scheme is based on the RPE-LTP technique with a raw speech rate of 13 Kbps. The coder provides an acceptable quality with a MOS of 3.54, which is highest among four coders in the final evaluation. Digital modulation is based on the constant envelope that is suitable for class "C" amplification, which is a must for a mobile subscriber. Both Viterbi and DFB equalizers are used by a GSM system. BS instruct users, located at the boundary of the cell, to advance their timing so that they do not overlap at the BS. Since mobile occupies one-eighth of the frame, during idle time it makes channel quality measurement and assists the BS in the handoff process.

Problems

6.1 Why is the number of carriers used in GSM 124 and not 125?

6.2 Compute for a 20-MHz system, the number of half rate channels assuming 200-MHz modulation bandwidth. Assume eight users per carrier for the full rate speech.

6.3 Find the speech data rate with 8,000 samples/s speech with 13-bit uniform coding.

6.4 What is the significance of constraint length K for convolutional coding? Will the system performance improve as K increases?

6.5   What is the significance of interleaving in GSM? Compute the interleaving depth for GSM assuming the frame duration to be 4.6 ms. Can we arbitrary increase the interleaving depth? Narrate advantages and disadvantages of higher and lower interleaving depths.

6.6   Why is speech processing delay a function of power consumption at the mobile terminal? Why are MOS and MOPS important for speech processing? What do they signify?

6.7   Provide answers to the following questions with respect to a RPE-LTP speech encoder/decoder used in GSM system.
   1.   Reasons of using Hamming Window during preprocessing of speech signal.
   2.   Why is LAR used instead of reflection coefficients?
   3.   Why is Short-Term Analysis and Filtering done in RELP? What is the purpose of Long-Term Prediction Analysis and Filtering?
   4.   Why are long-term predictor gain b' and delay D' required?

6.8   What type of errors does the interleaving mitigate? Why is interleaving so useful in mobile communication?

6.9   For a data rate of 280 Kbps, modulation efficiency of 1.2, and a doppler shift of 2 kHz, compute the required channel bandwidth.

6.10   Why do you think that a higher level of modulation may not be successfully used in cellular application.

6.11   Why is MSK a preferred modulation type for mobile application. Give reasons to justify?

6.12   Why is the baseband filtering used in GMSK modulation?

6.13   Discuss the following terms with respect to GSM system:
   1.   Compactness of modulation;
   2.   Good BER performance in fading environment.

6.14   Give reasons to justify that a higher level of modulation cannot be effectively used in cellular systems.

6.15   Why is discontinuous transmission required in GSM? What are the advantages of this technique?

6.16   Define the term "comfort noise." Will the spectra of noise change from frame to frame? How do you distinguish between noise frame and speech plus noise frame? Define the term speech frame extrapolation.

6.17   Show that the total timing adjustment range of 233 μs is adequate up to 35 Km of mobile separation from the BS.

6.18   Narrate all reasons of handoff. Define the term "Directed handoff."

6.19   Name the three classes of handover. What are the two modes of handover? Why does asynchronous handover take more time than synchronous handover? How is the "CELL RESELECT HYSTERESIS" used in the handover process?

6.20   The European GSM system carries eight users per carrier having the bandwidth of 200 kHz. The required C/I ratio with and without frequency hopping are 9 dB and 11 dB, respectively. Compare the capacity of this system with analog FM having the bandwidth of 25 kHz and the required C/I of 17 dB.

6.21 Calculate the spectral efficiency of the GSM system assuming multiframe duration of 120 ms, 200- kHz channel bandwidth, 124 carriers, and the data rate of 270.8 Kbps. Assume user information bit per slot to be 114 bits.

References

[1] Mehrotra, A., *Cellular Radio: Analog and Digital Systems*, Chap. 6, Norwood, MA: Artech House, 1994.

[2] ETSI/GSM Section 5.05, "Digital Cellular Telecommunication System (Phase 2); Radio Transmission and Reception," May 1996.

[3] Raith, Krister, "Capacity of Digital Cellular TDMA System," *IEEE Trans. on Vehicular Technology*, May 1991, pp 323–331.

[4] Hellwig, K., P. Vary, et al., "Speech Codec for European Mobile Radio System," *IEEE Vehicular Technology Conf.*, 1989, pp. 1065–1069.

[5] Southcott, C. B., D. Freeman, et al., "Voice Control of the Pan-European Mobile Radio System," *IEEE Vehicular Technology Conf.*, 1989, pp 1070–1073.

[6] ETSI/GSM Section 4.0.2, "European Digital Cellular Telecommunication System (Phase 2); Speech Processing Functions: General Description," April 1993.

[7] Coleman A., et al., "Subjective Performance Evaluation of the RAPE-LTP Codec for the Pan-European Cellular Digital Mobile Radio System," *IEEE Vehicular Technology Conf.*, 1989, pp 1075–1079.

[8] Natrig, John E., Stein Hansen, and Jorge de Brito, "Speech Processing in Pan- European Digital Mobile Radio System (GSM)-System Overview," *IEEE Vehicular Technology Conf.*, 1989, pp 1060–1064.

[9] ETSI/GSM Section 6.10, "European Digital Cellular Telecommunications System (Phase 2); Full Rate Speech Transcoding," September 1994.

[10] ETSI/GSM Section 5.02, "Digital Cellular Telecommunication System (Phase 2) Multiplexing and Multiple Access on the Radio Path," March 1996.

[11] Veterbi and Omura, *Principles of Digital Communication and Coding*, New York: McGraw Hill, 1979.

[12] Proakis, J. G., *Digital Communications*, Chap. 7, New York: McGraw-Hill, 1983.

[13] Stzernvall, Jan-Eric, and Jan Uddenfeldt, "Performance of a Cellular TDMA System in Service Time Dispersion," *IEEE Vehicular Technology Conf.*, 1987.

[14] Simon, Marvin K., and Charles C. Wang, "Differential Detection of Gaussian MSK in a Mobile Radio Environment," *IEEE Trans. on Vehicular Technology*, Vol. VT-33, No. 4, November 1984, pp 307–320.

[15] Hirade, Kenkichi, and Kazuaki Murota, "A Study of Modulation for Digital Mobile Telephony," *29th IEEE Vehicular Technology Conf.*, March 1979.

[16] Feher, K., "Modems for Emerging Digital Cellular Mobile Radio System," *IEEE Trans. on Vehicular Technology*, May 1991, pp. 355–365.

[17] Raith, K., B. Hedberg, et al., "Performance of a Digital Cellular Experimental Test Bed," *IEEE Vehicular Technology Conf.*, 1989, pp 175–177.

[18] Sevensson, L. "Channel Equalization for Digital Mobile Telephone Using Narrow-Band TDMA Transmission," *IEEE Vehicular Technology Conf.*, 1989.

[19] Steel, Raymond, *Mobile Radio Communications*, Piscataway, NJ: IEEE Press, 1992.

[20] Bune, P. A. M., "Effective Low-Effort Adaptive Equalizers for Digital Mobile Phone Systems," *IEEE Vehicular Technology Conf.*, 1989.

[21] Ziemer, Rodger E., and Roger L. Peterson, *Digital Communication and Spread Spectrum Systems*, New York: Macmillan, 1985.

[22] Pahalvan, K., and Allen H. Levesque, *Wireless Information Networks*, New York: John Wiley & Sons, 1995.

[23] ETSI/GSM Section 6.12, "Comfort Noise Aspect for Full Rate Speech Traffic Channels," January 1993.

[24] ETSI/GSM Section 6.32, "Voice Activity Detection (VAD)," April 1996. Goodman, David J., "Second Generation Wireless Information," *IEEE Trans. on Vehicular Technology*, May 1991, pp. 366–374. Bhargava, Vijay K., et al., *Digital Communications by Satellite*, New York: Wiley-Interscience, 1981.

CHAPTER 7
▼▼▼

SUBSCRIBER MANAGEMENT AND NETWORK MAINTENANCE

7.1 INTRODUCTION

A commercial structure that is becoming more common is based on the concept of service providers. Service providers are companies, usually distinct from the operator, who take the responsibility for commercial contacts with customers. This typically includes the establishment of subscription, the dispatch of bills to the subscribers, and the recovery of a bill. A subscriber who goes through a service provider has no direct contact with the operator. Usually, the operator bills globally to the service provider for all charges related to the subscribers, managed by the latter, and provides the service provider with toll ticket information necessary to produce the individual bills.

From the point of view of charging and accounting, two major features of GSM that differentiate it from other mobile telephone standards are that (1) GSM facilitates roaming subscribers between networks, and (2) GSM provides a number of supplementary services. At the early stages of GSM development it was recognized that international roaming not only requires a common technical standard but also a coordinated commercial approach for billing and accounting. For unified international roaming it was assured that a user who has service provided in one country by one of the network operators can gain access to the service of other network operators in other countries. Also, the scheme for charging to customers is unified from network to network. Thus, a compatible approach to tariff and accounting, including intersystem accounting, was agreed upon as a part of GSM MOU as discussed in Section 1.2 and Table 1.5 (MOU-Transfer Account Data Interchange [TADIG]). On the other hand, the MOU does not recommend levels of tariffs or any other matter that is regarded as anticompetitive in countries where different operators are in competition. Thus, the establishment of tariffs was left for determination on a national level.

Like any other system, GSM also has failures from time to time. Network maintenance will be required to minimize the loss of service due to failure. Failures tend to occur mostly in overload situations when the system is pushed to its limits, and such situations are difficult to predict on a testbed. Maintenance is faced with this large variety of failure sources. Thus, automation for fault detection and specific software is necessary to assist the maintenance teams in their task of fault recovery.

In view of this discussion, this chapter is segmented into three parts. In Section 7.2, we discuss the administrative aspects of subscription that are controlled by service providers. Section 7.3 deals with billing aspects of the service, and finally in Section 7.4 we bring out some important aspects of network management. The discussion of network management is rather general here; the specifics of the topic are treated in Chapter 9.

7.2 SUBSCRIPTION ADMINISTRATION

As the Plain Old Telephone Set (POTS) era draws to a close, GSM starts by explicitly separating the role of service providers, network operators, and the subscriber. The relationships between them are shown in Figure 7.1 [1]. This separation of roles promotes innovative new services without requiring additional network investment from service providers and allows an independent service provider to interface with the service users and network operators. Thus, service providers and operators are different companies in most cases. As shown by dotted lines in Figure 7.1, the relationship between network operator and subscriber can be direct. Thus, some operators may also play a dual role of operating the network as well as the interface with the customers. In some cases, the network operator chooses a mix of both. Usual functions of service providers include the following.

- Establishing initial subscriptions;
- Dispatching bills to subscribers;
- Receiving global bills for all subscribers from the system operator;
- Receiving toll ticket information from the operator;
- Storing subscribers information (such as name and address) in its data base.

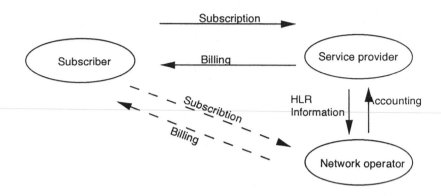

Figure 7.1 Role of network operators and service providers.

It is the responsibility of service providers to establish contact with new customers, provide marketing information in support of selling services, and bring new customers to the network operators. In those cases the role of service providers is distinctly different than that of the network operator. In other words, service providers fulfill the marketing end for the network operator. Network operators provide services, generate toll ticketing information, and provide collective bills for all subscribers served by individual service providers. It is the responsibility of service providers to extract information and supply bills to the subscribers individually. In this respect, service providers will have their own data base with an IMSI identity for the subscribers.

The specific functions associated with subscription include the following.

- Subscription materialized by SIM and the corresponding entries in HLR and AUC;
- Subscription identified by IMSI;
- Means required to create, upgrade, and cancel subscriptions;
- Subscriber's name and address stored in HLR or in some other date base;
- IMSI serves as a link if name and address are kept in a different data base.

From the traffic handling point of view, a subscription is materialized by a SIM and corresponding entries in databases (in the HLR in the AUC), where the subscriber is identified by an IMSI. Additional data specific to the subscriber, such as name and address, are needed for commercial reasons and, in particular, for recovering charges. The HLR may, in addition to its canonical functions, be the depository for this data. The data may also be kept in another data base when IMSI has to act as the link between different databases.

The following are general processes for the creation of a new subscription by service providers. In a typical scenario the operator prepares the AUC and the data in the SIMs before the actual contact with future subscribers and may even store some correspondence records in the HLR. The service provider then has at their disposal a number of SIMs that already contain stored IMSIs but that cannot be used at this stage for service (no initialization has been done in the HLR). When the subscription is created, the service provider enters data related to the customer in its own data base and provides an IMSI and a SIM to the new subscriber. In this scenario, the service provider need not perform any action on the "ready-to-use" SIM except possibly printing the subscriber name on it. In some cases the customer may choose his directory number (the MSISDN number) among a proposed list of numbers. Another scenario, although less attractive to the customer, consists of completing the SIM personalization process (including IMSI and secret key K_i) once the subscriber is identified by the service provider. In such a scenario the subscriber must wait a few days to receive a SIM by mail. The main advantage of this scenario is the removal of the need for an operator to have a real-time subscription management network. In both cases, some steps must be taken to properly initialize the HLR record. This includes the description of the subscribed-to services, the allocation of one or several MSISDNs, and the enabling of the IMSI or actual service. These steps required for the creation of a new subscription are as follows.

- Initial contact with subscribers occurs before they subscribe for the service.
- Scenario I: Prepare AUC and data in SIMs before actual contact with the future subscriber. Thus, service providers have at their disposal several SIMs that contain IMSI but cannot be used as the HLR is not yet activated. Upon subscription creation, the service provider enters the data related to the customer in its data base and provides an IMSI and SIM to the new subscriber. A minimal waiting period may be required in this case.
- Scenario II: In this scenario, SIM (including storage of IMSI and K_i) can be created after the subscriber is identified by the service provider. Some waiting on the part of the subscriber may be required in this case as operator puts this information in HLR before activating the service. Thus, this may require more waiting compared to the above scenario.
- Modification of service data after initial subscription may be needed.
- After sale's assistance to subscribers may be necessary.

Functions do not cease with the creation of a new subscription. From time to time modifications of the service provided or assistance to the users may be required, such as change of address, addition of new features, and deletion of some features that are no longer required by users. Another vital service in the interest of subscribers are stolen SIM cards. An operator in this situation has to deactivate the IMSI in the HLR as quickly as possible so that the user is not necessarily penalized.

Since a subscriber interface to the GSM is through a service provider, all trouble reports on behalf of the subscriber will be issued to the network operator. It will also act as a clearing house for all the customer inquiries and complaints.

Since the service provider interfaces with both subscribers and the network operator, it takes the charging information from operators for all its customers, prepares individual bills, and forwards them to subscribers on a monthly basis. This complete process of billing and accounting is the topic of discussion in the next section.

7.3 BILLING AND ACCOUNTING

In today's telecommunication services, a billing system is a large portion of the total investment. Billing is a critical process as it delivers all the revenue to the network operator [2]. Any problem in billing causes an instant crisis in a company's cash flow. Payments on services are the feedback mechanism though which one can gauge whether the financial forecast of the company is or is not being met.

7.3.1 GSM and Subscriber Mobility

In practical terms, the common technical standard means that even while roaming, a subscription can effectively be validated before each use of the MS. The subscriber, once in possession of a valid subscription to his home network, can roam to any other network without additional notification or credit checking and without suffering any connection or subscription fees from the visited network [2–4]. One bill from the home network can cover charges to the subscriber both for use of the home network and for roaming in other PLMNs within or outside the country. A visited network is obliged to supply service to a visitor from another network who possesses a valid subscription. This arrangement is bilateral between operators and, hence, balances out, assuming visiting roamers to the HPLM are about the same as its own subscribers visiting other PLMNs. The visited network will not receive any connection or subscription fee in relation to the visitor (it does not have the administrative overhead or collection costs associated with individual users). To be reimbursed by the home network, the visited network must transmit details of usage to the home network on a timely basis and in a standard format.

7.3.2 Subscriber Billing: National and International

There is a relationship between each PLMN and other networks to which they interconnect. These networks may be national or international carriers and may even be fixed or some other national mobile networks. Each PLMN has to negotiate with other networks with which they interconnect as to the basis of compensation for call delivery. In all cases an accounting relationship has to be established between PLMNs and other networks. In some cases this may be controlled by national regulation. There should be well laid out principles for how to charge and for what services to charge, but not how much to charge. The rules should be flexible enough not to restrict competition but rigid enough to permit the home network to process information supplied by other visited networks. These rules and regulations must be settled before a PLMN comes into operation. MOU does not restrict the PLMN for the choice of its tariff; for example, each PLMN can have a constant unit value for the traffic it provides or apply a fixed charge for the initial establishment of the call or even vary the unit of charge based on the nature of the call and the time of the day, for example.

7.3.3 Charging Principles

Subscriber charging is divided into two parts: (1) network access charge and (2) network utilization charge. Network access charge is the part of the collection charge intended to cover the cost of service that is independent of the actual use of the networks and services. It corresponds to what is generally known as the "network connection charges." The cost may depend upon the initial and subscription fees. In particular, it may depend upon the usage of PLMNs, which in turn depends upon usage of radio path, BS equipment, MSC, and other interconnecting networks between them. Thus, the network access charge is normally based on data registered in the subscription handling procedures. The access charge is not included in accounts between network operators and their establishment, which is strictly a national or an individual network function. The network access charges are collected by the HPLMN operator from his subscribers.

The network utilization charge is strictly based on a per call basis and thus depends upon network utilization. This is a function of the area and country from which the call is made, dialed number, time of the day when the call is established, location of the called number, and telecommunication and/or supplementary services used. The different components of network access charges include the following.

- Use of GSM PLMNs (Access to one or several PLMNs);
- Use of national/international fixed networks;
- Use of connections between different networks;
- Use of SS7;
- Use of a number of telecommunication and supplementary services.

The different components of network usage charges include the following.

- Network usage;
- Area and the country from which the call is originated;

- Called party number;
- Time of day a call is made;
- Types of services utilized (telecommunication or supplementary services).

The A-subscriber normally does not know exactly where the B-subscriber is located. Consequently, the A-subscriber is willing to pay the costs corresponding to the prefix dialed, but not necessarily the additional costs to any other place where the B-subscriber may be at the time of the call. As a consequence of the constant mobility of its subscribers, a PLMN needs a lot of flexibility. The general principle for charging is that the calling party pays. The tariff principle of all GSM PLMNs is such that the calling party pays the total charges for calls that they initiate, with the following three exceptions.

- Reverse or transfer charge calls;
- Call rerouting charges;
- Forwarded calls.

With respect to calls to a MS that has roamed away from its home PLMN, that is, the called subscriber is not at his nominal *Home Public Mobile Network* (HPLMN) address, the charges for rerouting the call from the nominal to the *Visitor Public Land Mobile Network* (VPLMN) should not be charged to the calling party. If appropriate, they may be recovered by the HPLMN from the called party. The forwarded leg of a call is treated as if it were initiated as a separate call by the forwarding subscriber. Thus, for those cases where a subscriber has roamed away from his HPLMN, the charges for the call rerouting of mobile-terminated calls from the subscriber's nominal address to his new forwarding address should be realized from the called party. These charges will be realized by the subscriber's home network operator, but the VPLMN will have to show in its record this forwarding leg of the call as a zero-priced call and forward the record to HPLMN. This zero-priced record forms the basis of duration of a forwarded call to the HPLMN by the visiting network and is based on which charges will be prepared for the terminating subscriber.

When a subscriber roams internationally and originates a call on a visited network, the VPLMN will bill the HPLMN through the Transferred Account procedure in respect to the use of the VPLMN's network. The charges raised will be at the VPLMN's network tariff computed as if the visiting subscriber were a home subscriber of the VPLMN. The HPLMN will use the charging record supplied through the Transferred Account procedure as the basis for subscriber billing.

Similar to the charging methods for mobile in HPLMN, or as a roamer in VPLMN, there is a charging principle for supplementary services. The party initiating the use of the supplementary service pays any charges for the use of the service. When a supplementary service, which is chargeable on a usage basis, is used by a roaming subscriber, the VPLMN will transmit a charging record to the HPLMN through the Transferred Account procedure. One of the supplementary services included in the GSM specifications is "Advice of Charge." This service is intended to allow a subscriber to have an indication of the level of charges he has incurred in a real-time basis. Roaming poses some specific problems in ensuring that there is the best possible level of correlation between the Advice of Charge and the bill.

Fairly complex rules have accordingly been proposed in relation to the Advice of Charge. Charges will be expressed not as an amount of money but as a number of units. However, each unit will have a relationship to a monetary value that must remain constant even when a subscription is being used in a visited network.

7.3.4 Billing and Accounting: Different Call Components

In relation to the individual possibilities of subscribers, the operator needs a flexible charging system. For this reason every call connection may be divided basically into five different call components.

- Originating call component;
- Terminating call component;
- Roaming call component;
- Call-forwarding call component (call-forwarding call);
- Transiting call component (land–land call).

The basic idea of these five components is discussed in the following subsections.

7.3.4.1 Originating Call Component

The operator can charge the A-subscriber for the originating part of the call until the geographical area where the mobile wants to reach (by dialing an area code) is effectively reached. Thus the charging analysis will be done by the originating network and based on the analysis of the called number. Figure 7.2 shows mobile-A charged for the call to a PSTN subscriber.

7.3.4.2 Terminating Call Component

The call component from the visited MSC/VLR to the mobile subscriber is recorded separately and can be charged either to the A-subscriber or the B-subscriber depending on the applications of the different operators. Here we take an example of mobile-terminated calls identified as B-subscriber. In this case the network operator may be charging the terminating part of a call to the B-subscriber when roaming. Some examples of the mobile terminating part of calls are shown in Figure 7.3. Here

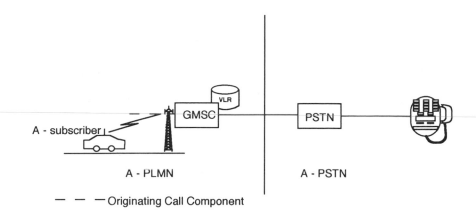

Figure 7.2 Mobile originating in A-PLMN a call to A-PSTN.

the called MS has roamed outside its HPLMN and GMSC routes the call to the VPLMN.

7.3.4.3 Roaming Call Component

If the B-subscriber (mobile) starts doing more than roaming, the routing leg of the call to his new location may cost more than the originating part. This routing leg cannot be charged to the A-subscriber because he is not responsible for the roaming of the B-subscriber. Therefore, the operator needs such a call component in order to charge it separately to the B-subscriber. Figure 7.4 shows the roaming part of the call where the originating subscriber is in country A and the destination subscriber is supposed to be in country B but at the time of the call is roaming in country C.

7.3.4.4 Call-Forwarding Call Component

The operator will charge the B-subscriber for rerouting a call to a C-PLMN due to roaming and call forwarding. The roaming part leads to HLR interrogation in the B-PLMN and subsequent switching to VMSC II VLR combination. The call-forwarding component is shown in Figure 7.5 between VMSC-II and MSC-III.

7.3.4.5 Transiting Call Component

In a situation where the call is set up over long distances, it might be of interest to have a separate charging data output just for the transiting part. The transiting part

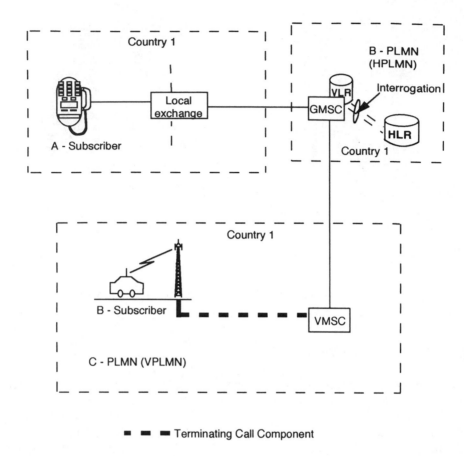

■ ■ ■ Terminating Call Component

Figure 7.3 Roaming mobile user is switched to VMSC in C-PLMN (VPLMN).

of the network is shown in Figure 7.6 where the call is set up between a mobile subscriber in country A to a mobile subscriber in country B. The transiting part is between GMSC I to MSC II.

7.3.5 Standard for Data Transfer

The standard method for exchange of data between PLMNs will be by *Electronic Data Interchange* (EDI). The exchange of data on magnetic tape will only be used as a fall back procedure, where EDI is temporarily unavailable. The frequency of data exchange is also important. Call data must be exchanged twice a month as a minimum. The generated bill is sent to a "CEPT clearing house." Each transfer should contain all the data available at that time and should not normally contain

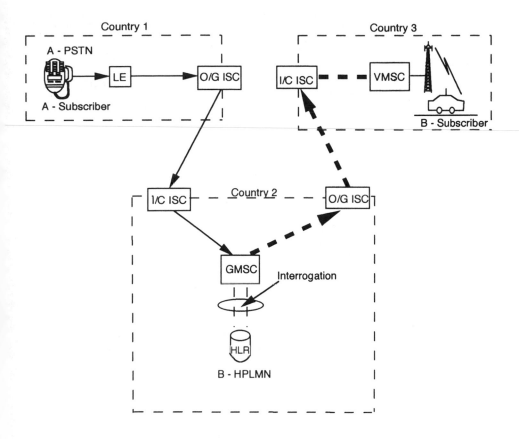

■ ■ ■ Roaming Component of the Call

O/G ISC = Outgoing International Switching Center (ISC)
I/C ISC = Incoming International Switching Center (ISC)

Figure 7.4 Subscriber belonging to B-PLMN in country 2 is being terminated in country 3 where one is roaming.

data relating to usage of more than seven days prior to the last transfer of data. The billing data shows the names of the HPLM and the VPLMN, total number of calls, and the charging period. Each network is responsible for setting its own standards for dealing with disputes/claims from its own subscribers. Claims between networks will be limited to principles only and should have substantial worth to pursue. It should not be concerned with an individual call. Disputes between networks, if not resolved bilaterally, can be referred to the MOU.

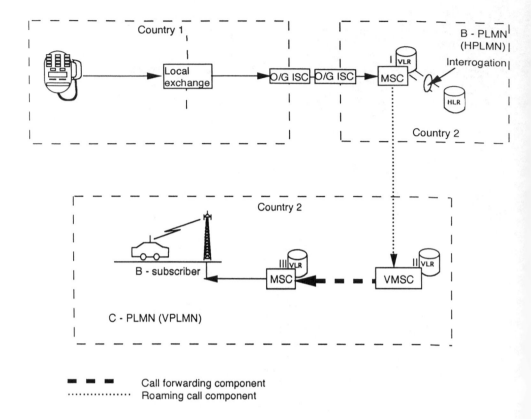

Call forwarding component

Roaming call component

Figure 7.5 Roaming mobile user is switched to VMSC in C-PLMN (VPLMN) and then forwarded to MSC III.

7.3.6 Toll Ticketing

Toll tickets are created by MSC/VLRs and by the GMSCS. Toll tickets are individual records enumerated for each call and contain all the information necessary to calculate call charges. Because of roaming, the MSC/VLR tickets pertain to subscribers that may belong to networks other than that corresponding to the issuing machine. Each ticket includes the IMSI to the billed subscriber. The IMSI is the basis for forwarding the ticket to the right place. In the GSM MOU, the rule is that the charges are collected from a subscriber by the operator with which he holds a subscription.

Tickets for home subscribers, whether internally generated or coming from other operators, must then be sorted on an individual subscriber basis. The operator is in charge of processing up to this point. In a scenario with a service provider, a

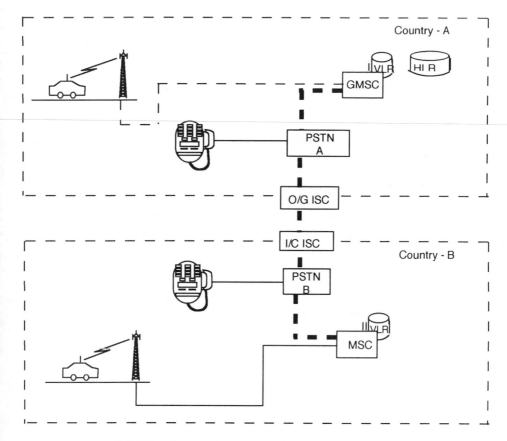

O/G ISC = Outgoing International Switching Center (ISC)
I/C ISC = Incoming International Switching Center (ISC)

Figure 7.6 Transiting call component GMSC-I to MSC-II.

global bill is sent to the service provider for all calls of the corresponding subscribers. The service provider in turn prepares individual bills. When a call is registered by *Toll Ticketing* (TT), the charging subsystem sends information of interest for charging of the call, for example, start time and A-subscriber number, which has been collected and temporarily stored in the different *Charging Data Record* (CDR) of individuals. When an output is requested, the information is transferred to the TT-function, which formats data into a TT-record. A typical call connection through the network is divided into basic call legs, and for each leg an individual CDR is created by the system. When the call is terminated, individual call components are sent to the TT output block. As was already explained, a call is divided into different call components. This means that the TT-records are produced in accordance with the principle of having one TT-record per leg. For a call between two mobiles without any call diversion, the following could be applied.

- One TT-record for the leg from the mobile subscriber to the GMSC (produced in the A-subscriber MSC/VLR);
- One TT-record for the roaming forwarding leg;
- One TT-record for the call leg from the VLR/MSC to the terminating subscriber.

The detailed example of a mobile-to-mobile call with all three components of the TT (TT1 through TT3) are shown in Figure 7.7. Here TT1 has records between mobile to MSC, TT2 has records for the usage of connection between two MSCs, and TT3 records the terminating end of the connection between MSC III and MS B. Similarly, there can be other components of a call associated with different calls. These components include the following.

- Originating leg of a call from mobile to MSC;
- Call leg terminating at a mobile subscribe (Land-to-mobile call);
- Roaming call forwarding leg of a call;
- Roaming leg of the call.

7.4 NETWORK MAINTENANCE [1]

As stated in the introduction, a system often fails in peak load conditions when the system is pushed to the limit. This type of failure is difficult to predict beforehand.

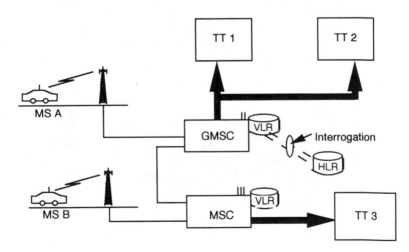

Figure 7.7 Components of a typical call.

The prime responsibility of the system operator is to minimize the loss of service quality during failure, for example, to provide enough extra resources or redundancy to minimize the effect of failure on customers. Failure can broadly be categorized into two types, namely, internal failure and external failure.

An example of internal failures of the system is that associated with components of the subsystem. Among external failures are those associated with power supply, disconnection of connection lines, and software failure. Software can be a major source of concern, especially in view of the fact that 100% working software for a system of this complexity is difficult to design and debug in laboratory surroundings. Complex situations can arise in the field that are difficult to formulate initially before the system is put into service. With all considered, the final objective from the operator's point of view is to limit the immediate effect of failure, which forces the operator to bring redundant equipment into service either online and supporting those failure conditions or offline and ready to be installed. In view of this, we need to focus in this section on three steps, namely, the minimization of failure occurrence, its effects, and fault detection.

7.4.1 Minimization of Failure Occurrence

For good-quality equipment an operator must consider different manufacturers during procurement and choose the best one according to specific criteria. One will perform a minimum set of tests on each piece of equipment during the installation phase before putting the equipment into commercial operation. Tests for a BS may cause some potential problem: Tests impact the neighboring cells; for example, it produces interference to other existing cells. Since handover parameters in neighboring cells may change during the test, modifications in the existing part of the network may be necessary. The tested cell must be barred to prevent MSs to camp on it, and handovers toward this cell must be forbidden. Also, modifications in the existing part of the network are necessary to test the new cell in a faulty operational configuration. Since handover parameters in neighboring cells are different, one must see whether or not the new cell has to be taken into account.

7.4.2 Minimizing the Effects of Failure

Failure of some components may be more serious than others. As an example, a single channel failure at the BS may not be as serious as a failure of a switching matrix at MSC or BSC. The network provider can take care of this problem by using, for example, additional transceivers at BTS or a standby switching matrix. Hot standby can be provided for critical subsystems, such as processors. For a less critical system, cold standby or even just a spare unit will suffice. The key method is to limit such catastrophic events by redundancy where interchangeable devices are used in parallel. For instance, if a cell needs four *Transmitters/Receivers* (TxRx) to accommodate the expected traffic, five will typically be installed, providing 25% redundancy.

Two adjacent points can consist of several links, and redundancy ensures that the required traffic load can be supported when one of the links is missing. In normal operation, the traffic is spread on all links. If acknowledgment and repetition protocols detect that one link is faulty, the corresponding load is transported to other links. By doing so, it is assured that there is no loss of traffic and capacity is maintained.

7.4.3 Fault Detection

In order to detect and isolate faults, the critical areas of hardware and software have to be monitored continuously. If a particular BS or MSC cannot be reached, then it should be indicated by a fault of a node or on the communication path to the node. To correctly alert the maintenance crew of the failure, an alarm could be sounded, a visual alarm could be viewed on the display unit (remotely or locally), an automatic email message can be generated, or a call could be made to a destination phone or a beeper. Remote detection of the fault is possible between the two units when the interface protocols are not consistent with each other. An alarm at the interface unit will signify that one end or the other of the system is not working according to the standard specified in the protocol. Similarly, loop sensing can be done for monitoring the transmission system. For example, an open loop condition can be sensed by the absence of a loop current or a voltage level.

Each PLMN network has to have the capabilities for automated corrective maintenance including automated fault detection, notification, and restoration without any aid from OMC facility. Also, each network element must have a manual corrective maintenance facility including fault verification, isolation, and repair without any direction from OMC. The corrective maintenance can be performed in response to network alarms and/or complaints by the customers.

Preventive maintenance activities are directed toward detecting and repairing potential faults before they surface in the system or are perceived by customers. There are two types of preventive maintenance: the scheduled types and the error analysis-driven. The scheduled type of maintenance is performed on a regular basis and mostly prescheduled. Examples can be hardware diagnostics and presence of consistence software throughout the GSM system. Error analysis-driven preventive maintenance includes analysis of error logs, maintenance logs, and performance data. The analysis of logs usually will require manual analysis and interpretation of results. The objective of error analysis is to identify system, subsystem, or components that are degrading with time. Once the system, subsystem, or the component is identified it may be scheduled for additional diagnostics or can be scheduled for repair.

The maintenance personnel can receive their work assignments either as work orders or trouble tickets. Customer can initiate service-related trouble reports via the service provider, which can subsequently be converted into work orders or trouble tickets. A trouble ticket can be assigned to a single responsible entity to resolve the associated trouble. Once the fault is fixed, a trouble ticket can be closed. A trouble ticket should contain the following information: customer trouble report, originator name and organization, trouble category, severity of fault, priority of problem, open date and time, and resolution description. If the customer is the source of the trouble report, it should contain information relating to customer name, customer identifier, customer location where the problem has occurred, and his ISDN number.

7.5 CONCLUSIONS

In this chapter we discussed the important role of the service provider, which acts as an intermediary between the network operator and subscribers. The GSM system limits payment by originating party except when the called party has roamed away from HPLMN or has transferred the call to somewhere else from the home PLMN. Different legs of a call generate different toll ticketing information. Periodic network maintenance minimizes the effects of failure. Fault detection and correction are steps for the proper resolution of system and network faults.

Problems

7.1 Narrate the functions of service providers and network operators.

7.2 Tabulate essential differences between two scenarios associated with a new subscription.

7.3 What are network access and utilization charges? Who pays for the forwarding part of a call (the calling or the called party)? Who pays for the supplementary services?

7.4 Draw the signal flow diagrams, providing details of various components of calls for the following cases.
1. Mobile originating a call in PLMN-1 and terminating in PLMN-2 with the same country and in a different country.
2. Mobile originating a call in PLMN-1 for a subscriber having its home in PLMN-2 in the same country but the call being received in country-2 where he is a roamer at the time.
3. Mobile originating a call in PLMN-1, country-1 for a subscriber located in PLMN-2 but the call being received in country-3, PLMN-3 at the time.

7.5 What is Toll Ticketing? Who generates TT, and what does it contain? What are different legs of TT-records?

7.6 Do you agree or disagree with the concept of a TT-record for each leg of the call? Draw a system block diagram similar to Figure 7.7 that includes originating and terminating legs, a call forwarding leg, and a roaming leg.

7.7 Define the network maintenance concept. What are two types of network failures? What does the operator do to minimize network failures?

7.8 Define the concept of fault detection. What does one mean by automated corrective maintenance? What is preventive maintenance?

7.9 What is a trouble ticket? Who generates these tickets? What does TT contain?

7.10 Who pays for the forwarding part of a call (the calling or the called party)? Who pays for the supplementary charges?

References

[1] Mouly, M., "The GSM System for Mobile Communications," 1992. 49, Rue Louise Bruneau, F-91120, Palaiseau, France.

[2] Maxwell, I. R., "Charging Principles and International Accounting," GSM Seminar, Budapest, October 1990, Session 2.3.

[3] ETSI/GSM specification Vol. 3.0.4, Version 4.0.0, October 1992.

[4] ETSI/GSM specification Vol. 2.20, Version 3.0.1, January 1990.

CHAPTER 8
▼▼▼

GSM PROTOCOLS

8.1 INTRODUCTION

Using the OSI model, the GSM system can be described by considering several functional layers arranged in hierarchical form. These consist of the physical layer, data link layer, and the so-called "layer 3." Layer 3 functions are designated as the application layer and should not be confused with the standard layer 3 functions of the OSI model. The application layer is composed of three sublayers: RR, MM, and CM. This RR, together with the data link layer and the physical layer, provide the means for point-to-point radio connections on which MM and CM messages are carried. The signaling protocols are shown in Figure 8.1.

The lowest layer of the radio interface, layer 1, provides functions necessary to transfer bit streams on the physical radio links. Digital signal processing techniques are used to perform equalization functions that recover transmitted bit patterns from signals distorted by the radio environment and channel coding functions (due to band limiting) that multiplex signaling and data channels onto the radio path, providing a level of immunity to errors. Speech coding functions also use complex digital signaling techniques to compress speech information into a manageable data rate and vice versa.

Layer 2 provides a reliable dedicated signaling link connection between the MS and the BS. The layer 2 protocol is based on the ISDN *link access procedure*

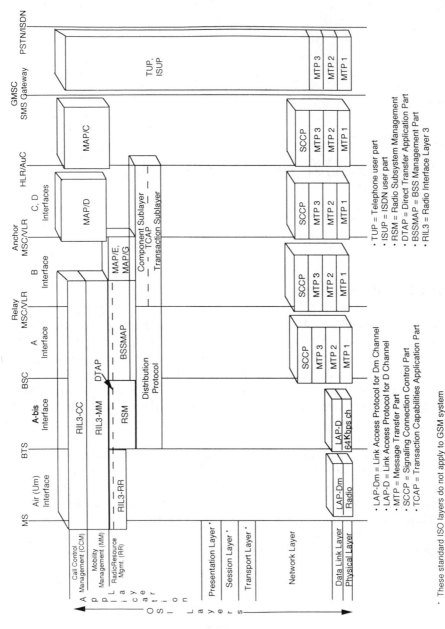

Figure 8.1 Signaling protocols of the GSM system.

(LAP-D) but adopted to take account of the limitations using a radio path. On the other hand, standard LAP-D protocol is used internally within BSS (between BTS and BSC). The *Message Transfer Part* (MTP) of SS7 is used on the BSC-to-MSC interface to provide a reliable data link service. The same protocol (MTP1) is kept between MSCs, between MSC to HLR/AUC, AUC to GMSC, as well as between GMSC and PSTN.

Layer 3 is split among three layers: RR, MM, and CM, as shown in Figure 8.1. The RR protocol entity provides control functions for the operation of common and dedicated channels. The RIL3 RR protocols establishes and releases radio connections between the MS and various BSCs for the duration of a call and, despite user movements, provides system information broadcasting, inter- and intracell change of channels, and ciphering mode setting, for example. The *Radio Subsystem Management* (RSM) protocol provides RR functions between the BTS and BSC. The *Direct Transfer Application Part* (DTAP) protocols provide RR messages between the MS and MSC. The *BSS Management Application Part* (BSSMAP) protocols provide RR messages between the BSC and MSC. The distinction between DTAP and BSSMAP is provided by a small "Distribution" protocol below them.

Mobility management, which best defines the dialog between MS and the network, deals with the management of MS location and the security functions (authentication and ciphering key management) necessary for mobile application. In addition to these functions, the MM sublayer also provides connection management services to the CC layer.

The higher layer that sits over MM is called the CM. The CM protocol controls the end-to-end call establishment (both mobile originating and terminating) and, in general, all functions related to call management. The CM sublayer also includes several independent entities, such as SMS and SS protocols. In essence, all messages and procedures required for call establishment, call release, and handling of SS and SMS are contained in this entity. Therefore, it is possible to provide several telecommunication services at one time and to change between different services if necessary. Thus, the overall objectives of layer 3 are to provide the means for the following.

- The establishment, operation, and release of a dedicated radio channel connection (RR);
- Location update, authentication, and TMSI reallocation (MM);
- The establishment, maintenance, and termination of a circuit-switched call (CCM);
- SS support;
- SMS support.

In addition to the aforementioned protocols, there are other protocols such as MTP3 and SCCP that are used above the data link layer between BSCs and MSCs and also between MSCs and different databases. TCAP protocol, which sits above SCCP, supports various transactions between two nodes of the network. TCAP manages the transaction on an end-to-end basis.

MAP protocol is used between MSC, VLR, HLR, and AUC in the form of query and response messages. These protocols are designated as MAP/B through MAP/H. For an example, the exact protocol used is MAP/C between GMSC and HLR/AUC.

In view of this discussion, the objectives of this chapter are to elaborate on all of the mentioned protocols. Sections 8.2 and 8.3 discuss layer 1 and layer 2 of GSM. Section 8.4 covers MTP3, SCCP, and TCAP protocols. In Section 8.5 we discuss application layer protocols, namely, RIL3-RR, MM, and CM protocols. DTAP and BSSMAP protocols are covered in Section 8.6. Voice and ISDN user protocols are covered in Section 8.8. Section 8.9 presents a summary of the protocols discussed in earlier sections and is arranged as U_m, A-bis, and A interfaces between mobile and BTS, BTS and BSC, and between BSC and MSC, respectively. This is where a serious reader will be able to conclude to which interface these protocols apply. The reader should note that this chapter is closely related to Chapter 4 on mobility management, where the protocols have been arranged according to different functions performed by mobile users and the network. This chapter is rather complex, and for complete understanding of the subject matter the reader is advised to go through the material slowly and in conjunction with Chapter 4.

8.2 PHYSICAL LAYER [1–3]

The physical layer is the lowest layer of the protocol architecture and represents the functions necessary to transfer bit streams on the physical medium. It includes functions such as channel coding and decoding, modulation and demodulation, equalization, TDMA access scheme, and radio transmission channel characteristics. Control functions such as power control, link monitoring, and diversity are applied at this level to enhance the performance of the system at level 1. Physical channels are shared among many users and are allocated to users on a temporary basis. One requirement for the physical channel is to have higher spectrum efficiency, which in turn requires that the channel capacity and the holding time be minimized.

Layer 1 is the physical layer, is responsible for the physical transmission of zeros and ones on the medium, and always contains specifications with respect to the size and shape of the pulses. Adapted rates are provided for the different functional channels that are lower than the ISDN bit rates. There are essentially two types of communication channels in GSM, namely, traffic and signaling. The individual subscriber traffic data over the channel (air interface) is at 16 Kbps. Between BTS and BSC both traffic and signaling may be carried at 64 Kbps or higher. One standard form is at the rate of 2 Mbps. Speech and data may be multiplexed to form a

2-Mbps PCM link as shown in Figure 8.2. The formation of a frame, consisting of 32 channels, is actually done at the level 2 (LAPD) protocol.

The physical layer interfaces with layer 3, RR through MPH-primitives (management physical). These primitives, exchanged within the RR layer, are related to functions such as the assignment of channels and measurement of channel level. This is a direct exchange of information from layer 1 to layer 3 bypassing completely layer 2. During the operation of a dedicated physical channel, the physical layer measures the signals of neighboring BSs and the signal quality of the physical channel in use. These measurements are transferred to RR layer 3 through MPH primitives. The physical layer also interfaces with data link layer 2 through pH-primitives (physical) for the transfer of layer 2 frames. These primitives are also responsible for the establishment of channels to the data link layer. The physical layer interfaces to other functional units in the MS and in the network for supporting TCHs. These interfaces are shown in Figure 8.3.

Service Access Points (SAP) of a layer are defined as gates through which services are offered to the adjacent higher layers. In the GSM system, SAP is defined between the physical layer and data link layers for BCCH, PCH + AGCH, RACH, SDCCH, SACCH, and FACCH as shown in Figure 8.4. Thus, a layer above requests services from the layer below by means of service primitives, which is a logical exchange of information and control between layers.

8.3 DATA LINK LAYER

The main purpose of a layer 2 protocol is to provide link connections to exchange signaling between different entities, namely, MS, BTS, BSC, MSC, VLR, HLR, and CCITT No. 7 Network. In GSM, three types of layer 2 protocol are used (Figure 8.1): $LAPD_m$ (link access protocol for signaling channels) on the air interface, LAPD on A-bis interface, and the MTP-2 (message transfer part—CCITT recommended). The protocols $LAPD_m$, between mobile user and BTS, and LAPD, used in the BSS, are similar to the ISDN protocol [2–4]. However, the $LAPD_m$ takes advantage of synchronized transaction to avoid use of flags and thereby increases the speed of operation and protection against errors. The MTP protocol uses standard ISDN functions [2–4]. On the terrestrial link, connecting the BSS to the MSC on the A interface, level 2 of the SS7 is used to provide OSI layer 2 functions. Layer 2 achieves the reliable transport of signaling data due to error detection and error recovery.

The detailed functions of the layer 2 $LAPD_m$ protocol are as follows.

- Establishment and release of signaling layer 2 connections;
- Multiplexing and demultiplexing of several signaling layer 2 connections on a dedicated control channel and discrimination between them by including different *Service Access Point Identifiers* (SAPI);

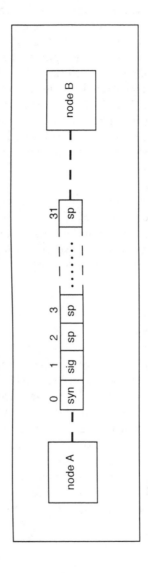

Figure 8.2 PCM link at 2 Mbps.

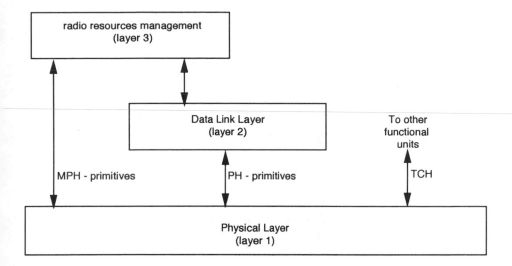

Figure 8.3 Interfaces with the physical layer.

- Mapping of signaling layer 2 service data units on protocol data units (in case of acknowledged operation service data units may be segmented and reassembled at destination);
- Thus, numbering of protocol data units modulo 8 to maintain sequential order;
- Detection and recovery of errors due to loss, duplication, and misorder;
- Flow control.

The establishment and release of a layer 2 connection coincides with the allocation, release, and change of dedicated radio channels. Signaling layer 2 connections are frequently established and released, and thus, the average lifetime of a connection is short. Multiplexing and demultiplexing deals with arranging different user data (eight channels per frame) in a frame format.

$LAPD_m$ uses one of the two modes of operation for the transmission of layer 3 messages: unacknowledged operation or multiple-frame operation.

On the DCCH both unacknowledged and acknowledged operations are used, whereas on the CCCHs only unacknowledged operation is applied. Thus, both modes are applicable for transmissions over one of the DCCHs (SDCCH, SACCH, and FACCH) in contrast to information transfer over CCCHs (BCCH, PCH, and AGCH) where only unacknowledged operation is possible.

For an unacknowledged information transfer, the use of layer 3 service implies that the information transfer is not acknowledged by the data link layer, and thus error check facilities are not provided. The transmission and reception of messages

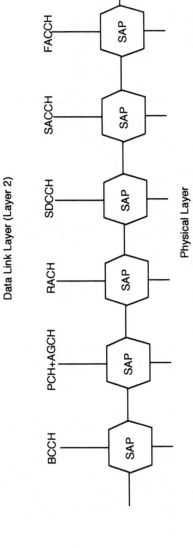

Figure 8.4 Interfaces between physical and data link layers.

here use data link service primitives, that is, DL-DATA-REQUEST and DL- DATA-INDICATION.

An acknowledged information service is provided to the network layer by a multiple-frame operation mode. Message units, received or to be transmitted, will once again be exchanged between layers 2 and 3 by means of DL-DATA- REQUEST and DL-DATA-INDICATION primitives.

In LAPD$_m$ each frame is fitted into a physical block that is 23-octets long. LAPD$_m$ uses no flags for frame delimitation; instead frame delimitation is done at layer 1, which defines the frame boundary. Thus, operation at this layer is fully synchronous. The format for LAPD$_m$ is shown in Figures 8.5(a,b). The two types of frame formats are type A and type B. Type A format is used on DCCHs (SDCCH, FACCH, and SACCH) for frames where there is no information field; while format B is used on DCCHs, which have an information field. The frame consists of an address, control, length, and information fields. The address field identifies the correct level 3 process. The control field is used for control information, for example, type of frame, sequencing, and acknowledgments. The length field identifies the frame length, which contains the actual information as opposed to fill bits. On the other hand, fill bits ("11111111" or "00101011") are used when the frame does not contain any information (idle frame).

Details of the address field are shown in Figure 8.5(c). Bit 1 is an extension bit that is presently set to "1". In the future it may be allowed to set to "0" for multioctet addressing capabilities. Bit 2 is the Command/Response bit that identifies whether the frame is for command or response. Mobile will send a command with C/R set to "0" and response with C/R bit set to "1". The reverse is true for BTS. Bits 3 to 5 identify the level 3 process where "000" is used. The bit configuration "011" is defined and represents transmission of SMS. At the present time all other combinations are reserved. The two-bit *Link Protocol Discriminator* (LPD) is used to specify a particular recommendation of the use of LAPD$_m$. Presently "00" and "01" are used. The LAPD$_m$ value of "01" is used for the data link protocol for *Short Message Service Cell Broadcast* (SMSCB). The number eight bit is reserved for future use.

Details of the control field are shown in Figure 8.5(d). It carries the sequence number and specifies the type of frame, either command or response. Three types of control field formats are used: numbered information transfer (I format), supervisory functions (S format), and unnumbered information transfer and control functions (U format). The I format is used to perform an information transfer between layer 3 entities. Each I frame has a send sequence number $N(R)$. The S format is for data link supervisory control functions such as acknowledge I frames, request retransmission of I frames, and request a temporary suspension of I frames. The U format is used to provide data link control functions and unacknowledged information transfer. For further details of I, S, and U formats the reader is referred to GSM/ETSI specification 4.01 version 4.1.0.

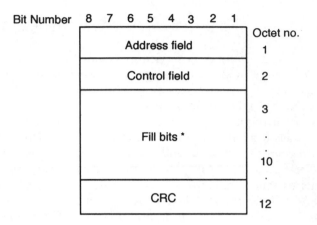

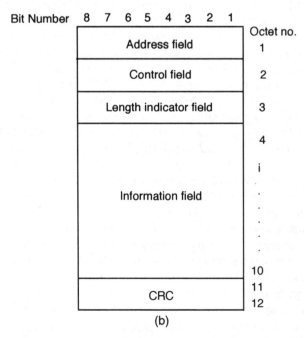

Figure 8.5 (a) Format type A on LAPD$_m$, (b) format type B on LAPD$_m$, (c) address field of LAPD$_m$, (d) control field, and (e) length indicator field.

8	7	6	5	4	3	2	1
0	LPD		S A P I			C/R	EA

EA: Extension bit (more bits follow)
C/R: Command/Response bit
SAPI: Service Access Point Identifier
LPD: Link Protocol Discriminator

00	I frame with information field < L1 octets
01	S frame
10	I frame with information field = < L1 octets
11	U frame

(c)

Figure 8.5 (continued).

Details of the length indicator field are shown in Figure 8.5(e). Similar to bit 1 of the address field, this bit is always set to "1". Future evaluations may use value "0" to indicate a multioctet length. Bit 2 is used to indicate when additional data follows (Binary 1). When this value is set to "0", no other frame follows; in other words, the present frame is the last frame of data. Bits 3 to 8 indicate the remaining length of the message in number of octets.

Eight-, sixteen-, and twenty-four-bit CRC can be used. For 16-bit CRC, the polynomial is given by $X^{16} + X^{12} + X^5 + 1$.

Details of LAPD frame format on A-bis interface are shown in Figures 8.6(a,b). The main distinction between LAPD and LAPD$_m$ is the absence of address and control fields. Thus, the protocol is only used for the unacknowledged mode of operation, which applies to BCCHs and CCCHs only. Both FCCH and SCH under BCCH do not require acknowledgment. Similarly, no acknowledgment is needed for PCH and AGCH. The LAPD frame, as seen in Figure 8.1, is used internal to BSS, namely, between BTS and BSC.

On the A interface, data link protocol is supported by MTP layer 2, which provides the means for protected message exchange between BSC and MSC, between MSCs, between MSC and HLR/AUC, and between HLR/AUC and GMSC.

8.4 MTP3, SCCP, AND TCAP PROTOCOLS

All network components are connected by CCITT No. 7 signaling links using MTP and SCCP functions and part of the *Transaction Capabilities Application Part*

Control field bits	8	7	6	5	4	3	2	1
I-Format		N(R)		P		N(S)		0
S-Format		N(R)		P/F	S	S	X	X
U-Format		L		P/F	U	U	U	U

N(s): Transmitter send sequence number
N(R): Transmitter receive sequence number
S   : Supervisory function bit
U   : Unnumbered function bit
P/F : Poll bit, when issued as a command
      Final bit, when issued as a response

(d)

Figure 8.5 (continued).

8	7	6	5	4	3	2	1
LENGTH INDICATOR FIELD						M	EL = 1

BIT 1: Extension bit. Currently defined to be 1.

BIT 2: More data bit (represented as binary "1").

BITS 3-8: Length indicator. **(e)**

Figure 8.5 (continued).

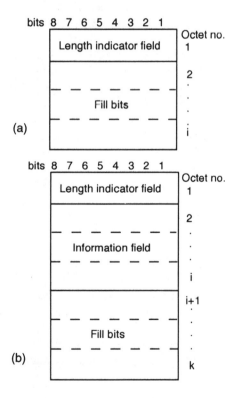

(a)

(b)

Figure 8.6 Format types: (a) A-bis, and (b) B-bis on LAPD.

(TCAP) functions. The link layer function is provided through MTP, and connection multiplexing is achieved using SCCP in connection-oriented mode.

The MTP of SS7 is used on the BS-to-MSC interface to provide a reliable data link service. The MTP is the basic layer information exchange consisting of three sublayers (levels): MTP levels 1 and 2 (MTP 1, MTP 2) provide the means for protected message exchange between adjacent nodes of a signaling network; whereas level 3 (MTP 3), which is of interest here, is built by functions for message routing as well as for operation and maintenance of the network.

The SCCP provides addressing and routing information for the transfer of data between software applications operating within the capabilities of the signaling network. SCCP is added to layer 3 of MTP to keep CCITT No. 7 in line with the OSI model. The SCCP enhances MTP to provide service according to layer 3 service of the ISO OSI model. It allows addressing messages on virtual or connectionless services between arbitrary nodes of the signaling network. This signaling network can span several PLMNS. The SCCP messages can either be connection-oriented or connectionless. This is contrary to MTP, which only uses the connections service.

MTP and SCCP in the past have worked quite well for communication with messages and telephony signals between two nodes. Each and every message was acknowledged, thus adding higher reliability. Thus, component handling was satisfactory for the operation. But these days computers are making more and more use of the signaling network. Computers use dialogues to communicate with each other, and they invoke operations at the distant end of the chain.

Signaling in a GSM network could mean that operators in different and widely separated countries may use the CCS signaling. This means that packets of signaling frame(s) would be traveling between countries and may be passing through a large number of switches. This demands a very comprehensive and reliable system for routing of signals. These functions for routing and reliability are placed in the SCCP layer. As discussed, the SCCP supports the connection-oriented and connectionless methods of signal communication.

In the connection-oriented method, the originating point first sends a pilot packet before sending true data. This initial packet travels across the network and arrives at its destination. While finding its way through the network, this pilot leaves behind a trace of the path it has taken from start to end. When a path is traced successfully through the network, the rest of the packets follows the same path; hence, the name virtual circuit is sometimes used.

In the connectionless method, each packet is furnished with the address and has to find its own way to the destination. As one would expect, they do not follow the same path and do not always arrive in the same order as transmitted. But as they are all provided with a serial number, they can easily be sorted back into the original order at the destination point. The other name designation for this service is Datagram. There are some special functions in the SCCP to ensure that an entire frame has arrived intact. In the event of an error, repeat transmission takes place.

This signaling network is used not only for data transmission in connection with speech connection but also for the transfer of short messages.

The TCAP provides a set of query/response procedures that are used to obtain a service that is requested. Dialogs are structured by the TCAP, a particular application service element to support various transactional applications in SS7. The purpose of TCAP is to provide a common and general system for the transfer of information between two nodes. It supports a wide variety of applications and is useful over exchanges in a telecommunication network.

TCAP is structured into two sublayers, namely, component and transaction sublayers. The component layer supports the exchange of *protocol data units* (PDUs or TCAP components), invoking remote operations and reporting their results. The transaction sublayer supports related PDU exchange. These transactions can terminate prearranged, that is, using timers, where no indication is given to the user.

Transation Capability (TC) offers some general standardized protocol functions and thus reduces the need for developing new protocols every time new features are introduced. In order to satisfy this argument, TCAP offers in general component handling (a component is a protocol data unit exchanged between two telecommunication users) and dialogue handling, which is used when two TC users exchange a number of components (especially suitable for computer exchanging messages).

With the help of a component, the user can invoke an operation remotely and receive a reply. Two users can exchange much information in dialogue mode. But TCAP also offers the facility of several dialogues running concurrently between two users or between two computers. In this case, each dialog will have several operations running simultaneously. Figure 8.7(a,b) shows the component and dialog handling between user one and user two.

8.5 APPLICATION LAYER

As stated in the introduction, layer 3 of the GSM protocol is composed of three sublayers: RR, MM, and CM. The RR, together with the data link layer and the physical layer, provide the means for point-to-point radio connections on which MM and CM messages are carried. In this section we provide details of these protocols [2,3,5]. It should be noted that RR protocol applies to air (U_m) interface, while MM and CC interfaces apply between MS and MSC.

8.5.1 RR Layer

Functions and protocols of the RR layer are related to a common concept that is best explained using an object-oriented approach. The management of RR connections deals with means of establishing, maintaining, modifying, and releasing the

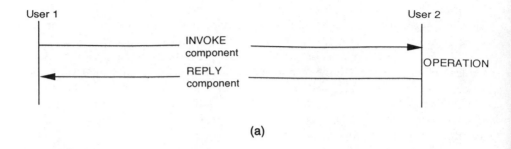

(a)

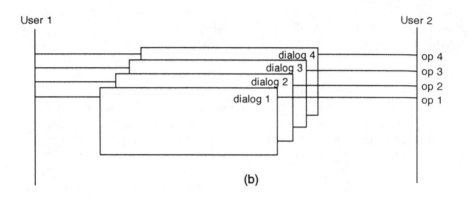

(b)

Figure 8.7 (a) Component handling, and (b) dialog handling.

basic means of communication on the radio interface and through the radio subsystem. The RR layer can be regarded as consisting of three main protocols.

- BSC protocol, allowing the BSC to control MS actions such as channel allocation, channel release, and handover;
- The BTS-BSC protocol, allowing the BSC to control the BTS;
- The BSC-MSC protocol, allowing the MSC and the BSC to exchange the necessary data for setting up, controlling, and releasing the A-interface connections.

In particular, the RR sublayer provides procedures for the following phases: (1) during system information broadcasting; (2) during connection establishment; (3) during connected phase; and (4) during connection release phase. We shall briefly describe all these phases. For more elaborate details the reader is referred to Section 4.08 of ETSI/GSM specifications.

8.5.1.1 Procedures for System Information Broadcasting

During this phase MS is not allocated any dedicated channel. It simply monitors CCCH and BCCH. In idle mode, the MS listens to the BCCH and to the paging subchannel for the paging subgroup where he belongs. It should be noted that since mobile belongs to different groups, they only monitor the paging during the designated period. This is a power-saving approach for mobile. MS also measures the radio propagation on BCCH for connection with other cells. The information broadcast is grouped into the following classes.

- Information for unique identification of current network, location area, and cells;
- Information used for cell selection and adjoining cell signal level measurements for handover procedure;
- Information describing the current control channel structure;
- Information defining different options supported within the cell.

BCCH broadcasts network-related information such as the ID of the network, LA, and cells. The threshold level at which the mobile can operate satisfactorily is also a part of the BCCH broadcast. The control channel structure of the logical control channel (CCCH) and the various available options within the cell are also parts of the broadcast message over the BCCH.

8.5.1.2 Procedures During Connection Establishment

This group of procedures consists of immediate assignment and paging messages. The purpose of the immediate assignment procedure is to establish an RR connection between the MS and the network. The immediate assignment procedure can be initiated by a request from the MM sublayer to establish an RR connection or by the RR entity in response to a paging request from the network. The assignment message is initiated by the MS and by sending channel request messages, a maximum of $M + 1$ times (several repeats due to contention possibility) over the RACH. The channel request message on RACH consists of eight bits. Among those eight bits, five are randomly allocated by the MS and serve as a discrimination between potentially colliding random accesses. The three remaining bits define the kind of channel required. The network in turn allocates a dedicated channel to the MS by sending an IMMEDIATE ASSIGNMENT message or an IMMEDIATE ASSIGNMENT EXTENDED message. The IMMEDIATE ASSIGNMENT message contains channel assignment information for one MS only, while the IMMEDIATE ASSIGNMENT EXTENDED message contains assignment information for two MSs at the same time. If the channel is not available for the assignment at the time the MS requests it, the network may send to MS an IMMEDIATE ASSIGNMENT REJECT message. The message will contain the request reference from the MS (random number, which the mobile included in its original channel request message over RACH). An immediate assignment procedure is terminated on the network side when the main signaling link is established. For details of messages during this phase the reader is referred to Section 4.4.1.

The network can initiate the establishment of an RR connection by a paging procedure. Network initiation is achieved by broadcasting a paging request message on the appropriate paging subchannel. There are three types of paging messages: Paging Request types 1 to 3, as discussed in Section 4.4.2. After the mobile receives the page, he initiates the channel request as described. The information received on the Channel Request message can be used by the network to assign a suitable channel to MS.

8.5.1.3 Procedures During Connected Phase

Procedures in this phase are (1) MS level measurement report; (2) handover procedure, that is, intra- and intercell handover; (3) frequency redefinition; (4) transmission mode change; (5) ciphering mode setting; (6) additional channel assignment; and (7) partial channel release.

When in RR-connected mode MS sends the level Measurement Report messages to the network over SACCH in unacknowledged mode. These measurements include measurement results about reception characteristics from the current cell and from the neighboring cells. If neighboring cell information is not available, the MS indicates this in the Measurement Report message to the system. MS measurements are regularly transmitted to the BTS. These measurements are grouped with the measurements performed directly by the BTS and are sent to the BSC through the A-bis interface. The BSC may thereby have a global idea of the uplink and downlink situation of the MS, both in the current cell and for neighboring cells on which the MS could potentially camp. These measurements are extensive and include the reception level and the quality of transmission. The rough quality is judged by the frame error rate of the SACCH, while the final quality is determined through demodulation and decoding processes.

The handover execution procedure has been designed with the aim of minimizing the speech interruption time. Procedures are provided on the radio path for the case of totally asynchronous cells, finely synchronized cells, and pseudosynchronized cells. The latter case allows for the shortest speech interruption time and can be estimated to be around 150 ms or less.

The A interface specification allows handover to be performed regardless of the relative position of the current and target cells in the equipment hierarchy. A handover between two BTSs managed by the same BSC may be autonomously handled by the BSC, details of which are covered in Section 4.9. The network initiates the handover procedure by sending a Handover Command message to the MS on the main SDCCH. For handover between cells belonging to two different BSCs (and even two different MSCS), the A interface provides all the procedures needed for the BSCs and MSCs to cooperate for the execution of the handover. After lower layer connections are successfully established, the MS returns a Handover Complete message on the SDCCH. The reader is also advised to consult Section 4.9 for further details of the handover process.

The frequency redefinition procedure is used by the network to change the frequencies and the hopping sequences of the allocated channels. This only applies to the frequency-hopping case. For this procedure, the network sends to MS a Frequency Redefinition message containing the new parameters together with the start time for the new frequency pattern.

The transmission mode change procedure allows the network to request the MS to change the channel mode. The channel mode will include channel coding, decoding, and transcoding modes used in the selected channel by mobile. The network shall initiate a procedure by sending a Channel Mode Modify command to MS. The message will include the reference of the channel and the new mode the mobile must adopt.

The network initiates the ciphering mode setting procedure by sending a Cipher Mode Command message to the MS on the SDCCH, indicating whether ciphering will be used or not and, if yes, which algorithm will be used. After taking the action on the Ciphering Mode Command message, the MS sends back a Ciphering Mode Complete (BSSMAP message between BSC and MSC) message to the network. Details of the ciphering process have already been covered in Section 4.4.5. The complete ciphering process makes use of RIL3-RR and BSSMAP messages.

The purpose of the additional assignment procedure is to allocate an additional dedicated channel to the MS while keeping the previously allocated channel. The network initiates the process by sending an Additional Assignment message to the MS on the main SDCCH. Upon receipt of the message, the mobile activates the new channel. After activating the channel, the MS sends an Assignment Complete message to the network.

The purpose of partial channel release is to deactivate part of the dedicated channels in use. The network initiates this procedure by sending a Partial Release message to the MS on the main SDCCH.

8.5.1.4 Procedure During Connection Release Phase

The purpose of this procedure is to deactivate the dedicated channel in use. The procedure is applied for the release of TCH as well as for SDCCH signaling channel. When the TCH is released, MS returns to the CCCH and starts monitoring the PCH. At that time the MS is idle again. The channel release process has already been discussed in Section 4.4.6. The process makes use of RR, BSSMAP, and DTAP messages, as shown in Figure 4.9(b).

8.5.2 MM Layer

The mobility management layer is mainly concerned with three families of functions: location, registration, and security. Another function of the MM-sublayer is to provide connection management services to the upper CM layer. Based on how the procedure is activated, one can categorize the MM layer into three groups.

- MM common procedure;
- MM specific procedure;
- MM connection management procedure.

8.5.2.1 MM Common Procedure

The MM common procedure can always be initiated, while the RR connection is already in existence. The procedures belonging to this group and initiated by the network are as follows.

- TMSI reallocation procedure;
- Authentication procedure;
- Identification procedure;
- Abort procedure.

The procedure initiated by the MS is the IMSI detach procedure. We first discuss procedures initiated by the network followed by the IMSI detach procedure, which is initiated by the MS. The reader is advised to refer to Figures 4.10(a, b).

The purpose of the TMSI reallocation is to provide user identity confidentiality, that is, to protect the user from being identified and located by others. Location confidentiality allow the user ID (IMSI) not to be transmitted in Plaintext over the radio interface by using an alias, the *Temporary MS Identity* (TMSI), which is allocated by the network and sent to MS in a ciphered mode. This allocation is typically performed as part of the normal location updating procedure. The TMSI is then used instead of the IMSI in paging messages and for the MS to identify itself in the first message on a dedicated channel. If a TMSI, provided by the MS, is unknown in the network, which may happen due to data base failure or other similar reasons, the network may require the MS to provide its IMSI. The network initiates the TMSI reallocation procedure by sending a "TMSI Reallocation Command" message to MS. This message contains a new combination of TMSI and LAI. Upon receipt of the "TMSI Reallocation Command" message the MS stores the LAI and new TMSI in its SIM. If the received identity is IMSI, the MS will delete its old TMSI. In either case, MS will send a "TMSI Reallocation Complete" message to the network.

The IMSI detach procedure is invoked by MS if the MS is deactivated or if the SIM card is detached from the ME. A flag (ATT) broadcasted in the "System

Information Type3" message on the BCCH is used by the network to indicate whether the detach procedure is required.

The IMSI detach procedure may not be started if a MM specific procedure is active. In the case that the MM specific procedure is active, this procedure may be delayed or omitted.

GSM provides for several security features such as authentication and data and location confidentiality. The authentication feature ensures to a very high level of probability that the user is the who one claims to be. This feature protects against fraud and consists of a challenge-response mechanism as discussed in Section 5.3, where a random text is transmitted to the MS and the MS answers with a response obtained by ciphering the random text with the user secret key. This secret key is stored in the SIM, where the authentication algorithm is also computed. User data confidentiality is achieved by ciphering data and signaling bits transmitted on the audio path. The algorithm is synchronized with the TDMA clock (frame number) and adds little complexity to the MS. The ciphering key is obtained as a side product of the authentication procedure and differs from call to call. An authentication request by the network is initiated by an "Authentication Request" message at any time while a RR connection is in existence. MS in turn sends the "Authentication Response" message back to the network. The security parameters for authentication and ciphering are tied together; that is, from a challenge parameter RAND both the authentication response SRES and the ciphering key can be computed simultaneously. The reader is advised to refer to Figure 4.7, where details of MM messages between MS and MSC are displayed.

The identification procedure is used by the network to request the MS to provide specific identification parameters to the network, such as the user's IMSI or IMEI. The MS should also be ready to respond to an "Identity Request" message at any time as long as the RR connection exits between the mobile and the network. The network initiates the process by sending an "Identity Request" message, and the mobile responds by sending an "Identity Response" message back to the network. These MM messages flow directly between MS and MSC, as shown in Figure 8.1. Complete details of the associated protocols between MS and MSC are shown in Figure 4.6.

The abort procedure is invoked by the network to abort any ongoing MM connection establishment or previously established connection. The procedure is activated when the network sends an "Abort" message to the MS. Before sending the "Abort" message the network shall release locally any ongoing MM connection. At the receipt of the "Abort" message, the MS shall release any MM connection.

8.5.2.2 MM Specific Procedure

A MM specific procedure can only be started if no other MM specific procedure is running or no MM connection exists between the network and MS. If this procedure is active, CM request for MM connection will either be rejected or delayed until the running MM specific procedure is terminated. The procedures belonging to this class are as follows.

- Normal location updating procedure;
- Periodic updating procedure;
- IMSI attach/detach procedure.

Normal location updating happens when the MS selects a new cell based on the measurement and finds out that the location area where this new cell belongs is different from where it was registered earlier. This procedure is also activated if the network indicates that the MS is unknown in the VLR as a response to an MM connection establishment request.

Periodic location updating is performed at regular intervals; the duration is controlled by the network. Its main purpose is to allow the consistency of the location data bases to be maintained and to restore them, if need be, for example, after a failure. The Location Updating Request message will contain the information that the message is for periodic location updating. The application of MM messages are shown in Figure 4.12.

IMSI attach is performed at switch-on time. It is the counterpart of the IMSI detach procedure that is performed at switch-off time. The IMSI attach procedure is invoked if the detach/attach procedures are required by the network and an IMSI is activated in a MS within the coverage area from the network or a MS with an IMSI activated outside the coverage area enters the coverage area. IMSI attach is performed by using the location update procedure.

8.5.2.3 MM Connection Management Procedures

These procedures are used to establish, maintain, and release a MM connection between the MS and the network over which an entity of the upper CM layer can exchange information with its peer. A MM connection establishment can only be performed if no MM specific procedure is running.

Mobile-terminating MM connection establishment is considered to be in progress when a RR connection has been established as a result of a Paging Request message received by the MS up to the point where a CM message is received that initiates a mobile terminating CM connection. Mobile-originating MM connection establishment is considered to be in progress when a CM Service Request is sent up to the point when a response is received from it. The connection request processes has been discussed in Section 4.4.1.

After the MM connection has been established, it can be used by the CM sublayer entity for information transfer. A CM sublayer entity, after having been advised that a MM connection has been established, can request the transfer of CM messages.

An established MM connection can be released by the local CM entity. After the release of MM entities, MS will enter the wait state and expect the release of the RR connection. In the network, if the last MM connection is released by its user, the MM sublayer may decide to release the RR connection. In order to establish another MM connection, an RR connection may be maintained.

8.5.3 CC Layer

The call control sublayer includes several protocol entities, for example, CC itself, SS, and SMS. The CC entity includes the control functions defined for the ISDN network signaling. All messages and procedures required for call establishment, call clearing, call information, and some miscellaneous procedures are covered in this entity. In particular, it covers the following procedures.

- Mobile-originating call establishment;
- Mobile-terminating call establishment;
- Signaling procedure during active state, which includes user notification, call rearrangement, DTMF protocol procedure, and in-call modification;
- Call clearing initiated by the network;
- Call clearing initiated by the mobile;
- Miscellaneous procedures, including in-band tones and announcement and status inquiry procedures.

Mobile-originating call establishment can only be initiated after the establishment of a peer-to-peer connection between the originating mobile and the network, since call establishment initiation is done by the CC layer by requesting a sublayer to establish a mobile-originating MM connection and entry "MM Connection Pending State." There are two kinds of a mobile-originating call: basic call and emergency call. The request to establish an MM connection will contain a parameter to specify whether the call is a basic or an emergency call. The procedure essentially consists of call set-up message from the mobile, call proceeding, alert, and call connect/call reject messages from the network. After receiving the connect acknowledgment from the mobile, a TCH can be assigned. As to when the TCH is assigned is a decision dependent upon the network. A comprehensive detail of mobile originating a call is covered in Section 4.6 and Figure 4.14. TCH assignment is based on whether the procedure is Off-Air Call Setup and is covered in Figures 4.13(a,b).

Similar to the mobile-originating case, the mobile-terminating call establishment can only be initiated after the establishment of the MM connection by the network. The procedure starts when the mobile call setup message from the network is followed by a call confirm or a release confirm message from the mobile, based on the result of its compatibility checking. This is shown in Section 4.7 and Figure 4.15. Upon receipt of the call confirm message by the network channel assignment, a command is initiated by the network followed by channel assignment complete message by the mobile. Conversation starts after the mobile answers the phone. It is once again a network-dependent decision when to initiate the assignment of a TCH during the mobile-terminating call establishment phase. This is dependent upon whether or not the network operator uses OACSU or not.

The GSM system allows signaling during the active state of a call. The user notification procedure allows the network to notify a MS of any appropriate call-related event during the active state of a call (for example, RIL3 -CC Call Proceeding). The MS can also notify the remote user of any appropriate call-related event during the active state of the call by sending a notify message containing a notification indicator to the network. Upon receipt of the message, the network will forward the notify message with the same indicator to the destination party. A change of user terminal equipment at the mobile, or for non PLMN user, can be initiated by call rearrangement procedure. A notify message is sent either by the network to mobile or by mobile to network to inform the destination party. The use of DTMF is only allowed when the speech teleservice is being used or during the speech phase of alternate speech/data and alternate speech/facsimile teleservices and will be disallowed in all other phases. DTMF messages are only supported in the direction from mobile to the network. When the mobile depresses a single key in its DTMF key pad, MS sends a "Start DTMF" message containing the value of the digit pressed by the user. Upon receipt of the "Start DTMF" message, the network sends a "Start DTMF Acknowledgment" message back to the mobile user. The release of the DTMF key by the mobile initiates the "Stop DTMF" message. Each pressing and releasing of the DTMF key constitutes a single DTMF transmission, as shown in Figure 8.8(a). Multiple DTMF transmission is possible by successively pressing DTMF keys. This is shown in Figure 8.8(b). In call modification this is used to alternate between speech and data or group 3 fax, and speech followed by data. To alternate between speech and data the mobile must identify this mode of operation. Also, the type of change between the modes must be identified by means of the repeat indicator, for example, Mode 1 "alternate" Mode 2, or Mode 1 "and then" Mode 2. In some cases the system may only change the configuration parameters while keeping the original channel, while in other cases a new channel may be assigned.

Call clearing can either be initiated by the network or by the mobile. The network initiates call clearing by sending a DISCONNECT message to mobile. Upon receipt of the disconnect message, mobile sends a RELEASE message to the network. Upon receipt of the RELEASE message the network sends a RELEASE COMPLETE message, releases the MM-connection, and returns back to the null state. Section 4.4.6 and Figure 4.9(a) describe the network-initiated call-clearing process. Call clearing by mobile is initiated when one sends a DISCONNECT message to the

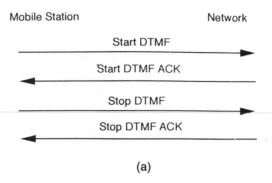

(a)

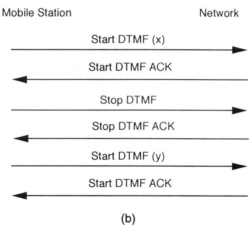

(b)

Figure 8.8 (a) Single and (b) multiple DTMF transmission.

network followed by a RELEASE message from the network. Upon receipt of the RELEASE message from the network, mobile will initiate a RELEASE COMPLETE message and also release the MM connection and return back to the null state. Figure 4.9(b) covers the call release initiated by mobile. When tones and announcements are provided together with a call state change, the appropriate messages, such as ALERTING and DISCONNECT, are sent to the mobile. When a call control entity wishes to check the correctness of a call state at a peer entity, a STATUS INQUIRY may be sent requesting the call state. Upon receipt of the STATUS INQUIRY message, the receiver will respond with a STATUS message. In the next section we describe the RR, MM, and CC messages that are originated by the network or the mobile.

8.5.4 Message Formatting [5]

Standard messages in the application layer protocol consists of protocol discriminator, transaction identifier or skip indicator, message type, and other information elements as required. The organization of a typical message is shown in Figure 8.9(a). The protocol discriminator field is half a byte long and routes the messages to the appropriate sublayer of the application layer (distinguishes between RR, MM, and CCM messages). The coding of the protocol discriminator is shown in Table 8.1. Bits five to eight of the first octet of every message belonging to the protocol "Call control" and "Special services" contains a *Transaction Identifier* (TI). The skip indicator, once again bits five to eight of the first octet of every message, belongs to RR and MM. Some examples of the coding for message types for RR, MM, and CM protocols are shown in Table 8.2.

Example 1

We now provide an example of an IMSI detach indication message content that belongs to the MM sublayer. The message is from mobile to network. Protocol discriminator, skip indicator, and the message types are shown in Tables 8.1 and 8.2. The message itself is 1-octet long for MS classmark and the mobile identity, which can be 1- to 9-octets long. Figure 8.9(b) shows the layout of the actual message. SI is '0000' as this field is only active for RR messages.

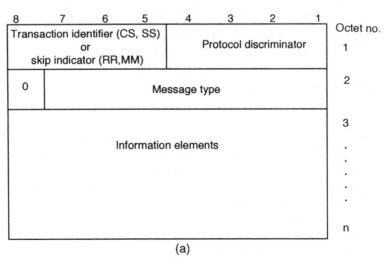

(a)

Figure 8.9 (a) General message organization, (b) IMSI detach indication message content, and (c) location updating reject.

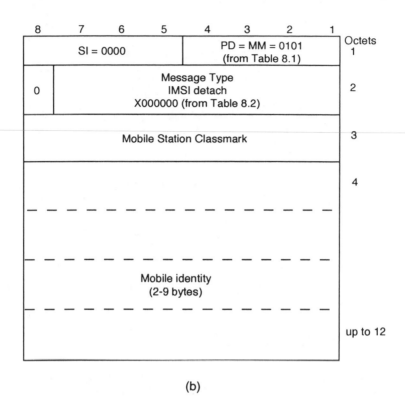

(b)

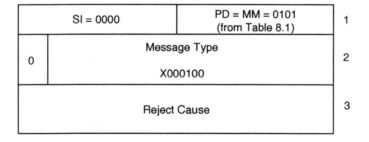

(c)

Figure 8.9 (continued).

Table 8.1
Coding for Protocol Discriminator

Bit Number				
4	3	2	1	Call for sublayer
0	0	1	1	Call control; call related to SS messages
0	1	0	1	MM messages
0	1	1	0	RR messages
1	0	0	1	SMS
1	0	1	1	SS
1	1	1	1	Reserved for test procedures

Example 2

As a second example of an MM message we code a "Location Updating Reject" message. The complete message is only three-bytes long and is shown in Figure 8.9(c).

8.6 BSS APPLICATION PART

The MTP and the SCCP are used to support signaling messages between MSC and BSS. One user function of the SCCP is the BSSAP, which uses one signaling connection per active MS, which has one or multiple transactions for the transfer of layer 3 messages [2,3,5–7]. Figure 8.10 provides the hierarchical relation of BSSAP in the GSM system. BSSAP is divided into two functions: *Direct Transfer Application Part* (DTAP) and BSSMAP.

DTAP protocol is used by BSS to transfer layer 3 messages between MS and MSC without interpreting the message at BSS. In this sense, BSS is transparent to the DTAP process, for example, no signal analysis is done at BSS.

BSSMAP is the process within BSS (A-bis interface) that controls RR in response to instructions from MSC. BSSMAP is used in the assignment and switching of RR at both call setup and handover. Both connectionless and connection-oriented procedures are used to support the BSSMAP.

The structure of the user data fields for DTAP and BSSMAP are shown in Figures 8.11(a,b). Both messages contain a discrimination parameter, a length field, and the actual layer 3 message. A DTAP message also contains a *Data Link Connection Identification* (DLCI). The discrimination parameter is 1-octet long and is set to a transparent value of "1". For the nontransparent case, the value is set to "0". The length indicator is coded in one octet and represents the number of octets of the layer 3 message. The DLCI parameter represents the type of connection such as originating data link over the radio interface.

Table 8.2

Some Examples of Message Type and Their Coding for RR, MM, and CM Sublayers

8	7	6	5	4	3	2	1	
Bit Number								
Some examples of messages in MM								
0	0	1	1	1	–	–	–	Channel Establishment Messages
					0	1	1	Additional assignment
					1	1	1	Immediate assignment
					0	0	1	Immediate assignment extended
					0	1	0	Immediate assignment reject
0	0	0	0	1	–	–	–	Channel Release Messages
					1	0	1	Channel release
					0	1	0	Partial release
					1	1	1	Partial release complete

(Similar message coding for Ciphering, Handover, Paging, System information, and Miscellaneous messages)

8	7	6	5	4	3	2	1	
Some examples of messages in the MM sublayer								
0	x	0	0	–	–	–	–	Registration Messages
				0	0	0	0	IMSI detach indication
				0	0	1	0	Location updating accept
				0	1	0	0	Location updating reject
				1	0	0	0	Location updating request
0	x	1	1	–	–	–	–	Miscellaneous Messages
				0	0	0	0	MM status

(Similar coding for Security and CM messages)

8	7	6	5	4	3	2	1	
Some examples of messages in the CM sublayer								
0	x	0	0	–	–	–	–	Call Establishment Messages
				0	0	0	1	Alerting
				1	0	0	0	Call confirmed
				0	0	1	0	Call proceeding
				0	1	1	1	Connect
				1	1	1	1	Connect acknowledge
				1	1	1	0	Emergency setup
				0	0	1	1	Progress
				0	1	0	1	Setup

(Similar coding for call clearing, call information, and miscellaneous messages)

8.7 MAP PROTOCOL [2,3]

The MAP protocol is seen as a remote data base access performed by the exchange of messages that are grouped into simple dialogues, mostly in the form of query and response. In the OSI model MAP resides above the TCAP protocol as shown in Figure 8.1. MAP only uses the connectionless class of SCCP protocol. The protocol

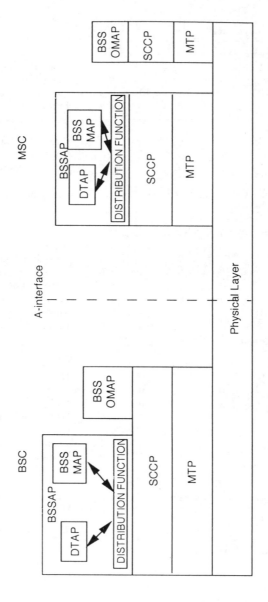

Figure 8.10 Hierarchical relation of BSSAP in the GSM system.

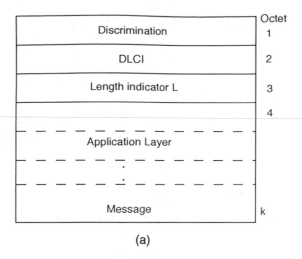

(a)

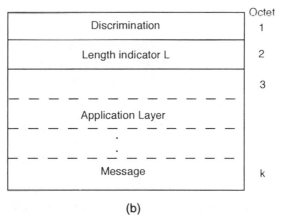

(b)

Figure 8.11 Structure of (a) DTAP protocol and (b) BSSMAP protocol.

is designed to interact with MSC, VLR/HLR, AUC, and GMSC so that they can communicate with each other. MAP initiates different operations and accepts the return of results or the return of application specific errors. MAP contains a number of functional blocks known as *Application System Elements* (ASE), which are used for communication between two peer-to-peer nodes as shown in Figure 8.12. All messages contain either mandatory or optional application parameters. The functions

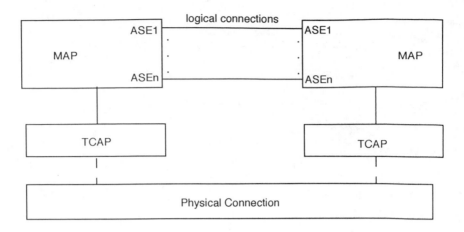

Figure 8.12 MAP communication between peer-to-peer nodes.

can be grouped into two categories, namely, mobility management and basic service support. Both of these are discussed in subsequent subsections.

8.7.1 MAP Protocols for MM

The mobility management consists of four different procedures.

Location registration: This procedure updates the information of the LA where the mobile is presently roaming. When the subscriber arrives in a new VLR area, the HLR is informed about the address of its new VLR where he can subsequently be paged. The HLR updates the subscriber data with respect to its new location. This service is initiated by MAP_UPDATE_LOCATION, which consists of service primitives of mandatory and some optional parameters as outlined in Section 6.1.2.1 of GSM/ETSI specifications, and uses the MAP/I protocol, which is shown in Figure 8.13.

Location cancellation: This is once again connected with location registration. When a subscriber roams into a new location VLR area, HLR is updated with respect to its location as described, and correspondingly the old VLR is informed by the location cancellation process and deletion of data with respect to the subscriber. This service can be invoked automatically by MAP_CANCEL_LOCATION and consists of various mandatory and optional primitives. The MAP/I protocol, as shown in Figure 8.13, is used for this service.

Update location area: This service is used between MSC and VLR to update location information in the network. It is initiated by a MS when changing its LA or when registering for the first time. This service is initiated by MAP_UPDATE_LOCATION_AREA and consists of several mandatory and

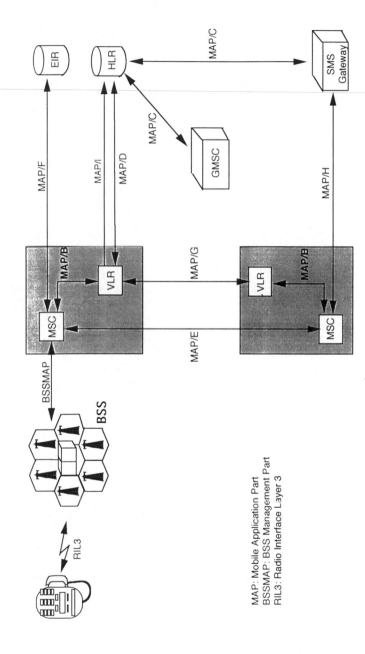

MAP: Mobile Application Part
BSSMAP: BSS Management Part
RIL3: Radio Interface Layer 3

Figure 8.13 MAP protocol connections.

optional parameters as described in Section 6.1.1.1 of GSM/ETSI 09.02. It uses the MAP/B protocol as shown in Figure 8.13.

Deregistration of MS: An MS may optionally deregister in the VLR when being switched off. This information is stored in the VLR and is used during mobile-terminated call setup to reject the call without paging.

Identification service: The MAP_SEND_IDENTIFICATION service is used between a VLR and a previous VLR to retrieve IMSI and authentication sets for a subscriber registering fresh in the VLR. The MAP/G protocol, shown in Figure 8.13, is used for this service.

8.7.2 MAP Protocols for Basic Services Support

Basic services support comprises retrieval of subscriber data during call set-up. The data connected with the service or terminal capabilities, paging of the subscriber for mobile terminating calls, initiation of security procedures at MS access, and handing over an existing communication to a different MSCs.

For *retrieval of subscriber data during call setup,* two different cases need to be discussed: data associated with mobile-originating call setup and data associated with mobile-terminating call setup. For mobile-originating call setup, MS can only successfully access the system when the authentication is positively checked. Thereafter, the service request is passed to the VLR to check subscription and supplementary service status as discussed in the next section. An indication is sent back to the MSC either to proceed with the call further or to block the call. For mobile-terminating call setup, first the inquiry of the current location of the subscriber is needed to deliver the call at the correct destination. Subscription and termination parameters may have to be checked and certain connection parameters may be negotiated with the MS before terminating a call.

Page and search: This service is used between VLR and MSC to initiate paging of a MS for mobile terminated call setup and mobile-terminated SMS. The MAP_PAGE service, which contains some mandatory and optional primitives, are used for paging purpose while the MAP_SEARCH_FOR_MS service is used when MS has to be paged in all LAs of the particular VLR where the mobile is presently located. For a mobile-terminating call, the subscriber must be paged by sending broadcast messages over all cells served by the VLR where the mobile is currently located. If the VLR fails to get the mobile response, the entire MSC area may be paged. The whole activity is supported by the search procedure. MAP/B protocol is used for this.

Access management: While accessing the network, whether this is after successful paging or for some mobile-originated action, the subscriber has to establish authenticity toward the network. Therefore, the VLR is informed about subscriber identity and type of access and initiates authentication, encryption on the user link, and reallocation of TMSI. Only if all these processes are done successfully is the service provided to the mobile.

Handover to a new cell: To maintain an existing communication in progress, even when roaming into some area served by a different MSC, MAP supports handover between MSCs. Here, when the call has to be handed over from MSC-A to MSC-B, MSC will receive MAP_PERFORM_HANDOVER service message. The MAP/E protocol is used for this purpose.

Purge service: The MAP_PURGE_MS service is used between VLR and HLR to cause HLR to mark its data for an MS so that any request for routing information for a mobile-terminated call or a mobile-terminated short message will be treated as if the MS is not reachable. The MAP/D protocol is used for this.

Authentication service: The MAP_SEND_AUTHENTICATION_INFO service is used between the VLR and the HLR to retrieve authentication information from the HLR. The VLR requests HLR to send RAND, SRES, and K_i parameters as discussed in Section 5.3. If the HLR cannot provide these, an empty response is returned. The VLR in this case may reuse old authentication triplets.

Security management service: The MAP_SET_CIPHERING_MODE service is used between VLR and MSC to set the ciphering mode and to start ciphering if offered by the system provider. The MAP/B protocol is used for this.

IMEI management service: The MAP_CHECK_IMEI service is used between the VLR and the MSC and between the MSC and the EIR to request check of IMEI. If the IMEI is not available in the MSC, it is requested from the MS and transferred to the EIR in the service request. The MAP/F protocol is used for this. The service primitives consist of some mandatory and some optional parameters as described in section 6.7.1 of GSM 09.02.

Fault recovery service is used by the HLR to indicate to a list of VLRs that a failure has occurred. Service is initiated by a MAP_RESET command. The MAP/D protocol is used for this purpose.

8.8 COMMON CHANNEL SIGNALING BETWEEN MSC AND FIXED NETWORK

Common Channel Signaling (SS7) is used as a means of transmission of data for different types of users, such as telephony and ISDN users. Hence the system can be divided into two parts: *User Parts* (UP: TUP and ISUP) and a common *Message Transfer Part* (MTP: MTP1, MTP2, and MTP3), as shown in Figure 8.1 [2,3].

Only UPs of the same type are compatible for communication with each other. Messages are transported between nodes 1 and 2 using MTP as a common transport medium as shown in Figure 8.14. Using the OSI model it could be said that the MTP represents OSI layers 1, 2, and 3 and UP represent OSI layers 4 and above, as shown in Figure 8.1.

User generates the data message at layer 4 and sends it to the MTP layer 3 where the *Service Information Octet* (SIO) is added to indicate whether the message

node 1   node 2

Figure 8.14 User data transport protocol between nodes using MTP layer.

is a signaling network management message or a message from different users such as TUP or ISUP. MTP 2 is responsible for reliable transmission, and therefore it adds a Flag for delimiting the messages, a *Check Sum* (CK) for error detection, a *Length Indicator* (LI) for the message (length of layer 3 message SIF + SIO combined), and the *Error Correction* bits (CORR). The message formed in MTP2 is forwarded to MTP1 where bits are converted into the appropriate level and shape for transmission over the medium. The user frame is shown in Figure 8.15.

8.9 STANDARDIZED INTERFACES IN GSM

In this section we summarize three standard interfaces in GSM system, namely, air or U_m interface between mobile and the BTS, A-bis interface between BTS and BSC,

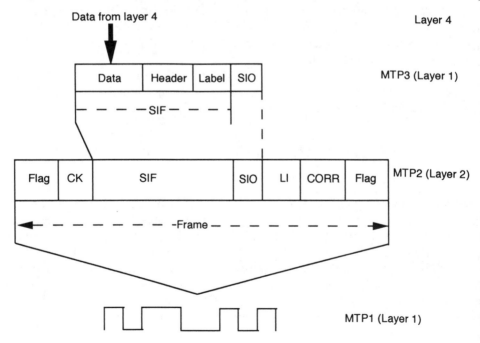

Figure 8.15 User frame formation and transmission over SS7 network.

and the A interface between BTC and the network subsystem MSC. In effect, this section summarizes the discussion of earlier sections. Via this section an interested reader should be able to realize protocols that apply to different interfaces.

8.9.1 U$_m$ Interface

In this section we briefly describe the interface between mobile and the BTS, namely, the U$_m$ interface. The physical layer of the U$_m$ interface uses the TDMA frame structure defined by the GSM, while the data link layer is based on the LAPD$_m$ protocol derived from LAPD. Above the LAPD$_m$ level is the application layer which is divided into three sublayers. The RR sublayer, for the protocol between the MS and the BS subsystem, controls the radio connection, details of which have been provided in Section 8.5. The MM sublayer between the MS and network subsystem manages subscriber mobility in the network, in particular due to the location procedures that let the mobile locate itself each time it changes from one area of coverage to another. The sublayer CC, SMS, and SS handle procedures needed to set up services such as an incoming or outgoing calls, SMS, or SS.

8.9.2 A-bis Interface

This is an interface between BTS and BSC. There are three internal configurations as shown in Figure 8.16: (1) *Base Control Function* (BCF) and a single *Transceiver* (TRX) combination, (2) several TRXs with one physical connection to BSC, and (3) several TRXs each served by its own physical connection. Here, the TRX is a functional entity (not physical) that supports eight channels of the same TDMA frame. The BCF is the functional entity that handles common control functions within a BTS such as frequency hopping, external alarms, power supply, and time base.

As discussed earlier there are three layers of interfaces between BTS and BSC—namely, the physical layer (layer 1), signaling links (layer 2), and an upper layer (layer 3) of signaling. The physical layer either transmits at 2,048 Kbps or at 64 Kbps. Four coded speech at 16 Kbps may be multiplexed to form a 64-Kbps data channel. The transcoder/rate adapter circuit is used in this case on the BSC-to-MSC interface to convert 16-Kbps speech data to a 64-Kbps rate. Thus, the A-bis interface in this case carries both speech and signaling channel data at 16 Kbps or 64 Kbps.

Layer 2 uses the standard LAPD protocol defined by CCITT. The addressing of TRXs and BCF is made using separate *Terminal Endpoint Identifiers* (TEIs) for each TRX and BCF. There are three logical links (as shown in Figure 8.17) between BSC and BTS having different SAPIs.

- *Radio signaling link* (RSL), SAPI = 0: Used for traffic management procedure. One line per TRX.

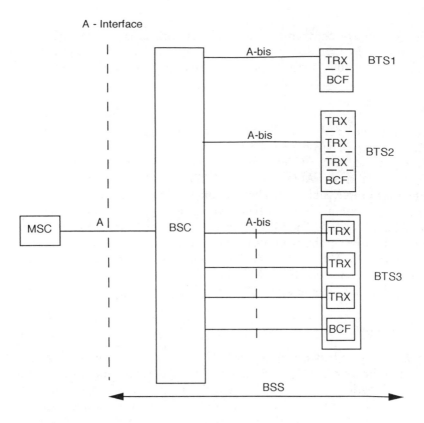

Figure 8.16 BSS subdivision and interfaces.

- *Operations and management link* (OML), SAPI = 62: Used for supporting network management procedures. One link per TRX and BCF.
- *Layer 2 Management link* (L2ML), SAPI = 63: Used for layer 2 management messages to TRX and BCF. One link per TRX and BCF.

The messages over the RSL for traffic management can be broken down into four categories: (1) radio link management; (2) dedicated channel management; (3) common channel management; and (4) TRX management, as shown in Figure 8.18.

The group of messages for radio link management support the establishment, use, and release of radio interface signaling links. Messages that are transparent to BTS also belong to this group. Messages belonging to the dedicated channel management group deals with the messages used to manage radio channels assigned to a particular user, for example, channel activation, authentication, and encryption. Common channel management messages are those used by common control channels,

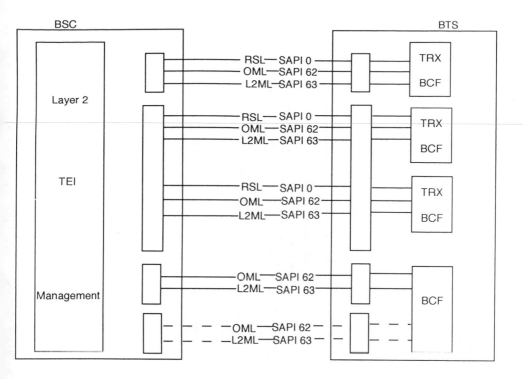

Figure 8.17 Logical L2 links of A-bis interface.

such as, random access and paging from and to mobile subscriber. TRX management messages are those used to carry information common to a single TRX, for example, idle channel information and flow control.

There is no direct link between BTS and OMC. All messages from OMC go first to BSC and then are routed to BTS. O&M messages for layer three at A-bis interface can be divided into three groups, namely, O&M formatted, MMI interface messages, and TRU O&M messages, as shown in Figure 8.19.

O&M-formatted messages convey commands, responses, reports, and files, for example. These messages arrive into many frames in layer 2 and then are sent to layer 3. MMI interface messages set up the transparent transport mechanism across A-bis interface. TRU O&M messages are those messages that allow the BSC to control remote TRU via A-bis interface.

8.9.3 A Interface

An interface between MSC and BSC are connected through the physical layer (layer 1) and transmits data by using one or more 2,048-Kbps transmission systems. The

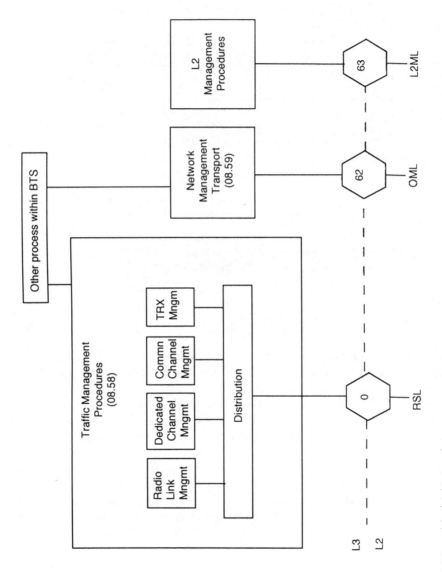

Figure 8.18 L3 model of A-bis interface.

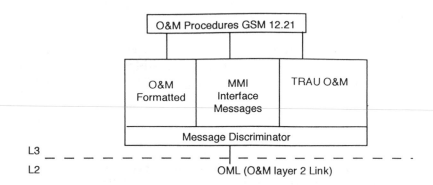

Figure 8.19 Layer 3 structure for O&M message transport over the A-bis interface.

configuration is shown in Figure 8.2, where 31 channels of speech and signaling data are used along with one synchronization channel. A single channel of signaling data is used for call setup, handover, and release of call. The complete protocol structure is shown in Figure 8.2, which consists of MTP protocol for layer 2, SCCP, and BSSAP protocols. We already discussed MTP and SCCP protocols in Section 8.4.

BSS application layer protocols actually pass only the RR and O&M messages. Call control and mobility management messages pass directly from MS to MSC. Thus, layer messages are classified under the following three categories.

- DTAP messages   Direct Transfer Application Part;
- BSSMAP messages   BS System Management Application Part;
- BSS O&M messages   BS System Operation and Maintenance.

The BSSAP handles two groups of signals, as shown in Figure 8.20.

DTAP messages pass directly between MS and MSC without going through BSC/BTS combination. These messages deal with call control and mobility management. BSSMAP messages pass through BSC/BTS combination and perform tasks such as resource management and handover control.

Finally, BSS O&M messages are used to transfer operations and management information over the A interface.

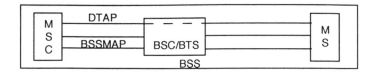

Figure 8.20 DTAP and BSSMAP message flow between MSC and MS.

8.10 CONCLUSIONS

The design of a GSM system has been done with a view of providing compatible interfaces between mobile-BTS, BTSs-BSCs, and BSCs-MSCs. By standardizing A and A-bis interfaces, the flexible, efficient, and cost-effective design of BS and switching equipment have been achieved.

Problems

8.1 Define the terms: primitives, entity, SAP, and procedure.

8.2 What is the basic difference between $LAPD_m$ and LAPD?

8.3 What is the difference between acknowledged and unacknowledged modes of operation in $LAPD_m$ messages?

8.4 What is the reason for using fill bits in $LAPD_m$ messages?

8.5 What is the difference between connectionless and connection oriented signaling?

8.6 Enumerate the basic functions of MM, CC, and RR layers. Illustrate with some examples.

8.7 What is the main purpose of TMSI reallocation?

8.8 What do you need to abort a message?

8.9 Show the complete message coding for: call confirmed, call proceeding, location updating request, channel release, immediate assignment, and partial release.

References

[1] ETSI/GSM Section 04.04, Version 4.0.0, "Layer 1 General Requirements," October 1992.

[2] Lycksell, Edgar, "The A and A_{bis} Interfaces of the GSM System," October 1990, Session 3.3.

[3] Pautet, Marie-Bernadette, and Michel Mouly, "GSM Protocol Architecture: Radio Sub-System Signaling," *IEEE VT Conf*, 1991.

[4] ETSI/GSM Section 04.06, Version 4.1.0, "MS-BSS Interface Data Link Layer Specification," January 1993.

[5] ETSI/GSM Section 04.08, Version 4.4.0, "Mobile Radio Interface Layer 3 Specifications," April 1993.

[6] ETSI/GSM Section 08.01, Version 4.0.0, "European Digital Cellular Telecommunications System (Phase 2) BS System to Mobile Switching Center Interface General Aspects," January 1993.

[7] ETSI/GSM Section 08.06, Version 4.1.0, "European Digital Cellular Telecommunications System (Phase 2) Signaling Transport Mechanism Specification for the BSS-MSC Interface," January 1993.

CHAPTER 9
▼▼▼

DETAILS OF MS, BSS, MSC, AND NMC

9.1 INTRODUCTION

Due to the integration of different technologies such as RF, switching, and transmission, modern telecommunication systems are increasing in complexity and need an improved network management system for its proper operation. One example of such a system, which includes these concepts, is the GSM system. These technologies have been extensively used in the developments of mobile, BSS, MSC, and network management subsystems.

The development of mobile terminals, for GSM, has represented a formidable challenge to the industry, in requiring the implementation of a complex system with a very large amount of processing power, to meet benchmarks of size and cost set by first-generation analog cellular terminals. This challenge has essentially been met by *Application Specific Integrated Circuits* (ASICs) development in radio and signal processing systems. Due to advanced digital signal processing, subscribers can obtain a substantially improved quality of service, in comparison with existing systems, such as *Nordic Mobile Telephone* (NMT) or *Total Access Communication System* (TACS). This naturally results in greater complexity of terminals that must, nevertheless, observe the constraints of physical dimensions and power.

BSS is a combination of digital and RF equipment that communicates with the MSC, OMC, and MS. The BS subsystem (BSS) includes transmission and control equipment needed to ensure connectivity with mobile subscribers distributed into various cells in the area of mobile coverage. Since there is no fixed correlation between terrestrial circuits connecting to MSC and the air interface between BTS and mobile, it also includes some switching capabilities.

MSC architecture generally encompasses microprocessor-based distributed processing and control. Each microprocessor is dedicated into specialized switching system tasks and includes call processing, switch matrix control, trunk control-to-fixed network, and radio channel control.

Powerful systems for network management are indispensable for network operators because they allow operation at a reasonable cost to the subscriber. In a situation of intensive competition, a rapid reaction to new market requirements (for example, the introduction of new features) provides a competitive edge and can be achieved by a carefully designed network management system.

In view of this, the objective of this chapter is to elaborate on mobile, BSS, MSC, and NMS subsystems and their architecture. In some sense this is a further elaboration of Chapter 2 on GSM architecture.

9.2 MOBILE STATION SUBSYSTEM

A GSM MS is equipment that is intended to access a set of GSM or DCS 1800 PLMN telecommunication services. Services are accessed while the equipment, capable of surface movement within the GSM or DCS 1800 system area, is in motion or during halts at unspecified points.

As stated in the introduction, the mobile terminal has large digital signal processing capabilities that enable it to get a substantially improved quality of service compared to the first-generation analog cellular system. Radiotelephone terminal design calls on many different skills. Serving both as a subscriber terminal and a means of establishing radio communications, the radiotelephone terminal provides a prime example of a combination of electronic and information techniques. Due to unique multipath interference surroundings, terminals have to work out the development of essential countermeasure algorithms against multipath. GSM terminals must also take account of antenna radiation characteristics, electromagnetic compatibility, and human body effects of RF propagation [1,2]. The design demands mastery of electronic circuit LSI techniques and digital signal processing combined with optimum utilization of processors. Constraints with respect to spectral purity demands an extremely rigorous approach to the design of circuits operating at transmission and reception frequencies. The terminal must have appropriate acoustic quality and must offer a user-friendly man-machine interface. Power supply design, with corresponding elements such as batteries and chargers, must be of light weight and volume. Industrialization of these terminals, which are intended for public use globally, resembles that of a mass-market product. In view of this, we will discuss the attributes of the MS including access configuration and capabilities, modes of its usage, features of mobile, and typical terminal architecture.

9.2.1 Mobile Attributes

MS service access configuration, MS access capability, and modes of use are among the list of MS attributes [2].

9.2.1.1 MS Access Configuration

The service access configuration must satisfy requirements of the customer and will depend on the combination of tele-, bearer, and supplementary services of one's subscription. The actual configuration will depend on the manufactures' implementation and can comprise of a single unit or a mobile termination unit with additional terminal equipment and/or terminal adapters. A comprehensive list of the different type of services is provided in Chapter 1, while the mobile terminal configuration is discussed in Section 9.2.3.

9.2.1.2 MS Access Capability

The MS access capability is defined in GSM Specification 04.03 and includes the following frequency bands of operation and the types of speech coding [2].

Frequency Bands of Operation

Three frequency bands are as follows.

- Standard GSM band;
- Extended GSM band (includes standard band);
- DCS1800 Band.

MSs may support one or more of these bands. If the MS is capable of both GSM and DCS-1800 operation, then it shall provide internally either a manual or automatic procedure to select the type of network. Simultaneous use of GSM and DCS-1800 modes are not supported.

Full Rate/Half Rate Services

Both full rate and half rate services are specified. Each basic service may either be supported by a MS on full rate, full and half rate, or not supported at all. As discussed in Section 3.2.5, the full rate speech is at a gross rate of 22.8 Kbps while the half rate is at the gross rate of 11.4 Kbps.

9.2.1.3 Modes of Use

Three modes of operation are as follows.

- Vehicle-mounted stations;
- Portable stations;
- Handheld stations.

Vehicle-Mounted Stations

This equipment is mounted in a vehicle, and the antenna is physically mounted on the outside of the vehicle. Vehicles include cars, motorcycles, trucks, buses, trains, and ships on internal or coastal waterways.

Portable Stations

This equipment may be hand-carried, and the antenna is not physically attached to the portion of the equipment containing the mobile termination. Portable stations may support all power levels required in the system. Portable stations can be vehicle-mounted and are usually composed of a portable plug-in unit and vehicle-mounted adapter.

Handheld Stations

This equipment is hand-carried or worn on the person, and the antenna may be physically attached to the portion of the equipment containing the mobile termination. Handheld stations are intended to be easily carried by a person and should have the following characteristics.

- Total weight is less than 0.8 kg;
- Volume is less than 900 cm^3;
- Power source to provide at least one hour of call duration or 10 hours in the state of being able to set up or receive calls.

The equipment can be vehicle-mounted and usually is a standard handheld station plugged into an interface in the vehicle that provides battery charging and externally mounted antenna connections.

9.2.2 Features of Mobile: Basic, Supplementary, and Additional

In Table 9.1 basic, supplementary, and additional MS features are listed as recommended by GSM 02.07. A basic MS feature is directly related to the operation of basic telecommunication services (for example, keypad function). A supplementary MS feature is directly related to the operation of SS (such as display of calling line number). An additional MS feature is a feature that is neither a basic nor a supplementary feature (for example, abbreviated dialing) [1,2].

MS features are qualified as mandatory or optional. Mandatory features are marked by "M". Optional features are marked by "O". Mandatory features must be implemented as long as they are relevant to the MS type. The implementation of optional features is left to the manufacturers' discretion. For all present and future MS features, manufacturers are responsible to ensure that MS features neither conflict

Table 9.1
Basic, Supplementary, and Additional Features of Mobile

Name of the Features	*Mandatory/Optional*
Basic MS Features	
1. Display of called number	M
2. Indication of call progress signals	M
3. Country/PLMN indication	M
4. Country/PLMN selection	M
5. Keypad	O
6. IMEI	M
7. Short message	M
8. Short message overflow indication	M
9. DTE/DCE interface	O
10. ISDN 'S' interface	O
11. International access function ("+" key)	O
12. Service indicator	M
13. Autocalling restriction capabilities	O
14. Emergency calls capabilities	M
15. Dual tone multifrequency function (DTMF)	M
16. Subscription identity management	M
17. On/Off switch	O
18. Subaddress	O
Supplementary MS Features	
1. Call charge units meters	O
2. Control of supplementary services	M
Additional MS Features	
1. Abbreviated dialing	O
2. Fixed number dialing	O
3. Barring of outgoing calls	O
4. DTMF control digits separator	O

with the air interface nor cause any interference to the network or to any other MS or to itself. The following paragraphs provide a brief explanation of these features.

Display of called number: This feature enables the caller to check the dialed number before the call is setup.

Indication of call progress signals: Indications, such as tones, recorded messages, or visual display are given based on signaling information returned from the PLMN. For data calls, this information may be signaled to the DTE.

Country/PLMN indication: The country/PLMN indicator shows in which GSM PLMN the MS is currently registered. This indicator is necessary so that the user knows when "roaming" is taking place and that the choice of PLMN is correct.

Country/PLMN selection: When more than one GSM PLMN is available in a given area, this procedures provides selection of a PLMN.

Keypad: A physical means of entering numbers, generally, though not necessarily, through a standard keypad.

IMEI: Each MS has a unique identity and shall transmit this on request from the PLMN. For details see GSM 02.16 and 03.03. The IMEI is incorporated in a module that is built within the MS and is physically secured.

Short message indication and acknowledgment: This feature allows the delivery of short messages to a MS from a service center. Such messages are submitted to the service center by a telecommunication network user who can also request information of the status of the message by further interrogation of the service center. The service center then transmits the message to an active MS user.

Short message overflow indication: An indication shall be given to the SM user when an incoming message cannot be received due to insufficient available memory.

DTE/DCE interface: A standard connector for attachment of a DTE to the MS and use in conjunction with data services.

ISDN 'S' terminal interface: A standard connector for attachment of equipment to the ISDN standard.

International access function: Provision is made for a direct, standard method of gaining international access. For this purpose, the MS may have a key whose primary or secondary function is marked "+". This is signaled over the air interface and would have the effect of generating the international access code in a network.

Service indicator: An indication is given to the user that there is adequate signal strength to allow a call to be made and that the MS has successfully registered on the selected PLMN.

Dual Tone Multifrequency: The MS shall be able to initiate DTMF in accordance with specifications GSM 02.03 and 03.14. Optionally, the MS may provide a suppress function that allows the user to switch off the DTMF function.

Subscription Identity Management: The IMSI is contained in a SIM. If the SIM is removable by the user, its removal detaches the MS, causing a call in progress to be terminated and preventing the initiation of further calls. As discussed in Section 2.3.3, the SIM may be a Plug-in SIM or a IC SIM card.

On/Off switch: The MS may be provided with a means of switching its power supply on and off. Switch-off shall be "soft," so that on activation the MS completes the following housekeeping functions: termination of a current call, detach (where applicable), and storing required data in the SIM before actually switching off.

Subaddress: This feature allows the mobile to append and/or receive a subaddress to a directory number for use in call setups and in those supplementary services that use a directory number.

Call charge units meter: The MS may incorporate a call charge units indicator. This call charge indicator will give information about the actual call charge units consumed. The call charge indicator will have a last conversation counter and cumulative counters for each PLMN.

Control of supplementary services: It is mandatory that the supplementary services are controlled by mobile.

Abbreviated dialing: The directory number or part of it, stored in the MS together with the abbreviated address. After retrieval the directory number may appear on the display. An incomplete directory number must be supplemented by means of the keypad function or a second stored number.

Fixed number dialing: This feature provides a mechanism so that, by the use of an electronic lock, it is possible to place a bar on calling any number other than those pre-programmed in the SIM.

Barring of outgoing calls: This feature, distinct from the supplementary service of the same name, allows outgoing calls to be blocked. The barring condition may be activated/deactivated by using a key or keyword, for example (exception: transmission of emergency calls). The barring may be selective; that is, applied to individual services (such as telephony and data transmission) or individual call types (such as long-distance, international calls) or supplementary services.

DTMF control digits separator: Provision has been made to enter DTMF digits with a telephone number; and upon response from the called party, the ME shall send the DTMF digits automatically to the network after a delay of 3 sec (±20%). The digits shall be sent according to the procedures and timing specified on GSM 04.08.

9.2.3 Mobile Configuration

The terminal equipment access configuration is shown in Figure 9.1. Functional groups TE1, TE2, and TA are conceptually the same functional groups as those in the ISDN. *Mobile termination* (MT) performs the following functions.

- Radio transmission termination;
- Radio transmission channel management;
- Speech encoding/decoding;
- Man-machine interface capabilities;

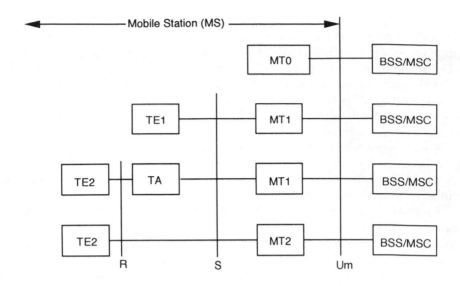

Figure 9.1 Mobile access configuration.

- Error protection across the radio path;
- Flow control of user data and mapping of flow control for asynchronous transparent data services;
- Data rate adaptation of user data to channel rate;
- Mobility management;
- Multiple terminal support.

As shown in Figure 9.1 there are three types of MT. MT0 includes functions belonging to mobile subscriber and has no terminal interfaces. MT1 includes functions belonging to mobile subscriber, and with an interface that compiles with the GSM recommended subset of the ISDN user-network interface recommendations. MT2 includes functions belonging to mobile subscriber, and with an interface that compiles with the GSM recommended subset of the CCITT X or V series interface recommendations.

9.2.4 Terminal Architecture

Figure 9.2 shows the architecture of a GSM mobile terminal developed by Alcatel. The terminal mainly consists of the following three sections.

- Radio subsystem;
- Baseband processing subsystem;
- Control subsystem.

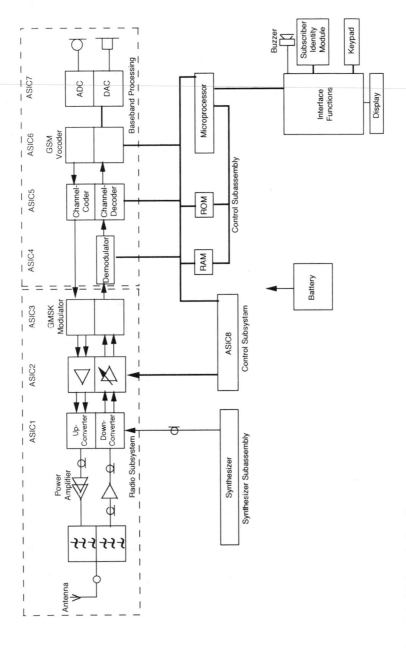

Figure 9.2 GSM terminal architecture.

The radio subsystem filters and amplifies the signal picked up by the antenna. For transmission, it generates, modulates, and amplifies the signal to be transmitted. This subsystem includes the following.

- ASIC1, incorporating the transmission and reception frequency converters;
- ASIC2, providing various intermediate reception functions and controlling the power amplifier;
- ASIC3, incorporating the GMSK modulator and filters on the receive side.

The baseband processing subsystem comprises the acoustic interface (that is, microphone, earpiece, and speaker), digital signal processing (that is, A/D converter, voice coding, channel coding/decoding), and demodulation of the signal received from the antenna. These functions are achieved in the following.

- ASIC4 demodulator;
- ASIC5 channel coder/decoder;
- ASIC6 GSM voice coder;
- ASIC7 analog-to-digital and digital-to-analog converters.

The control subsystem controls the radio and baseband processing subsystems and manages the external terminal interfaces (such as keypad, SIM, and buzzer), thus fully controlling operation of the terminal. ASIC8 executes real-time tasks (such as digit calculation, GSM clocks, and communication with terminal blocks) for the microprocessor.

As stated, ASIC4 realizes demodulation of the received GMSK signal and, in particular, performs the following functions.

- Characterization of the radio environment;
- Synchronization with the BSs;
- Demodulation of the binary data packets transmitted during each pulse;
- Radio control function assisted by the controller.

The functions provided by ASIC4 have the following constraints.

- GSM processing execution times (low execution times implies less circuit delay);
- Reduced consumption;
- Flexibility in terms of adaptation for terminal debugging tasks (easy to maintain).

The GSM terminal architecture, shown in Figure 9.2, is based on optimization of the following criteria.

- Manufacturing costs;
- Operating autonomy;
- Equipment volume and weight.

The strategy is to achieve simultaneous optimization of the three criterion and is based on adapting low power-consumption technologies. This led to the choice of a direct-conversion type radio architecture for both transmission and reception. For the transmission circuit, advantages of this approach include elimination of IF converter functions (reference signal, converter, and filter circuits), using the same digital components for generation of the GMSK modulation signals, and correction of transmission frequency. Likewise the reception circuit no longer includes customary conversion—that is, IF amplification and IF filtering functions—but instead a direct conversion to baseband is used. The utilization of ASIC technology has made it possible to develop an optimized LSI architecture leading to low-volume and low-consumption terminals despite the greater complexity of the GSM system.

The terminal design takes advantage of operational sequencing of each circuit; therefore, three states are defined for each mode.

- Active;
- Standby (digital circuit clocks are stopped, and analog circuit biased toward cutoff);
- Idle (power supply off).

Activation of the circuits is organized to optimize consumption, irrespective of the operating mode, according to the processing operations to be executed. For example, only one very low consumption circuit remains active in standby mode with others in an idle state. This circuit switches the main processor to the active state at the instant it is preprogrammed for the reception of a call, and the main processor then switches the circuits required for reception—first to standby and then to the active state. If the call concerns the terminal, other circuits are activated and subsequently returned to the standby or idle state as soon as possible.

9.3 BASE STATION SYSTEM

As stated earlier, the BSS is a combination of digital and RF equipment that communicates with the MSC, OMC, and the MSs. As shown in Figure 9.3, the BSS can be divided into a BSC and one or more BTSs. The BSC performs call processing, interfaces with OMC, and interfaces over the A link with MSC. A single BSC can support many BTSs. The BTS can be colocated with the BSC or remotely located. The external and internal interfaces of the GSM system are shown in Figure 9.4.

The BSS includes four different types of equipment, which are shown in Figure 9.5 [3]. The BTS includes radio and signaling resources needed to handle

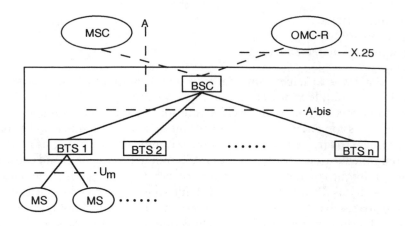

Figure 9.3 Block diagram of BSS.

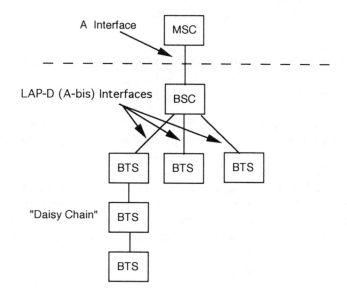

Figure 9.4 External and internal interfaces of BSS.

traffic within a given cell. The BSC controls all the BTSs and their radio resources needed to cover the BSS area. The speech *transcoders* (TC) are generally located on the site of the MSC. Transcoders convert the GSM speech coding (13 Kbps) format into PCM (64 Kbps) speech for the PSTN interface and converts 64-Kbps PCM speech to 13-Kbps data for GSM mobile application functions. Transmission elements (T)

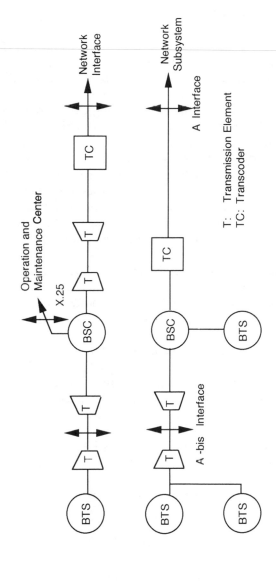

Figure 9.5 BSS subsystem architecture.

can be installed on the interface between the BTS and the BSC (A-bis interface) and/or on the interface between the BSC and the remote transcoders (A interface).

9.3.1 BS Controller Architecture

As stated, one or several BTSs can be connected to a BS Controller. For omnidirectional rural configurations, BSC and BTS are generally on the same site. Star configurations are also used in rural areas when several BTSs are chained or looped and controlled by a single remote BSC as shown in Figure 9.4. A sectorized urban configuration uses three BTSs at the same site. A single BSC architecture developed by Alcatel is shown in Figure 9.6. It consists of a switching matrix, terminal and trunk control units, and processors to interface with O&M through X.25 [3].

The interface toward the MSC is via A interface. The interface toward the BTS is called the A-bis interface. The signaling protocol used at the A interface is common channel signaling, No. 7. The BSC performs RR through the terminal control unit, switching matrix, and digital trunk controller. Here the main functions are to establish and to release radio resources in response to MS and MSC requests, and also the handover of MS either within the same cell or between cells belonging to the same MSC.

Through a Central Processor and X.25, interface BSC downloads new software releases through OMC. Software codes and other parameters from the BTSs are backed up in the mass memory of the BSC. On the other hand, all data of interest to the OMC is buffered at the BSC and transferred to the OMC when being asked or transmitted periodically.

The digital trunk controller terminates the 2-Mbps digital trunk to the MSC. The trunk controller is a VLSI implementation and has a SS7 protocol resident in its mass memory that is part of the controller. Both the designs of the terminal controller and trunk controller are based on a 16-bit microprocessor, 1 Mbyte of memory, and dedicated VLSI components for interface to the network, switch matrix, and BTSs.

A digital switch is a two-stage matrix and has 64 access ports. Each port can switch individual channels of 64 Kbps from the PCM timeslot of one port to the timeslots of an another port. The switch is used for switching 64-Kbps channels and interprocess communication data between the intelligent controller boards.

9.3.2 BTS Architecture

The BTS contains radio components of the BS and the digital modules that handle the interface to the BSC. The functional architecture of the Alcatel 900 BS is shown in Figure 9.7 [3]. The unit essentially consists of the following subassemblies.

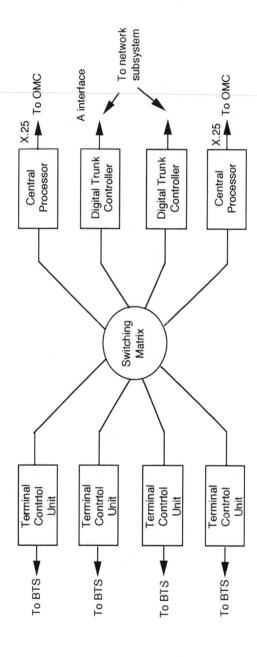

Figure 9.6 BSC architecture.

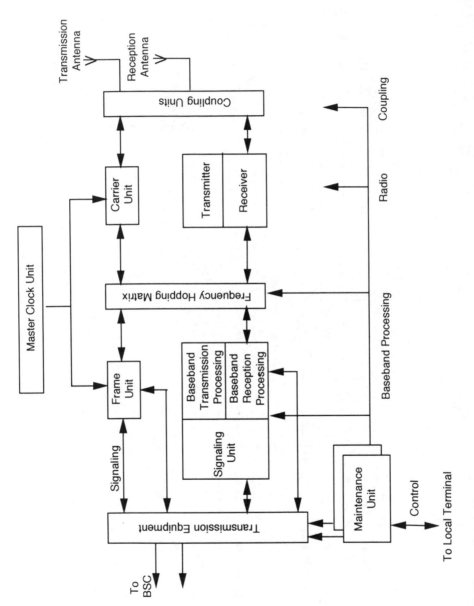

Figure 9.7 BS transceiver architecture.

- Coupling network;
- Radio section;
- Baseband processing section;
- Maintenance unit.

The antenna coupling unit mainly consists of a receiver multicoupler, transmitter filter, and a combiner. The receive antenna provides a means to connect any one of the receivers to any of the inputs to the antenna in a multiple receiver configuration. Both omnidirectional antenna, which covers the entire GSM frequency band, can be utilized. The maximum number of channels that can be supported by a single antenna is dependent upon peak power requirements of various channels. Antenna heights are influenced by coverage considerations. Multiple antennas for diversity are used to combat adverse effects of radio path. The matrix provides an RF connection path for any one of the available antenna inputs to any of the receivers. Control is provided to the matrix from the corresponding receiver. Antenna selection can be changed from slot to slot. Single antenna operation (one antenna for both transmit and receive bands) is also possible but requires a duplexor switch.

The coupling equipment includes a receiver multicoupler along with transmit filters using cavities, which can be tuned remotely from the OMC so that carrier frequencies at the BS can be changed without the need of someone at the BS. The number of cavities installed will actually depend upon the number of receivers at the BS. The cavity combiners are especially useful when a large number of carriers are to be mixed; as it provides a lower loss combining solution than hybrid combiners. Thus, when few carriers need to be combined and the power loss is acceptable, hybrid combining is used. For multiple carrier's combining, cavity combining is preferred as the power output is attenuated by more than 3 dB for each additional stage of hybrid (two carriers will require one hybrid, four carriers will require two stages of hybrid, eight carriers will require three stages of hybrid, and so on). To eliminate spurious signals and noise from the transmitter, which falls outside the transmit frequency band from reaching the antenna, a bandpass filter is required. One filter will be needed for each transmit antenna. A receiver multicoupler (bandpass filter followed by a low-noise amplifier) splits the signal received at the antenna and feeds it to different receivers. A typical receiver multicoupler configuration is shown in Figure 9.8. A receive filter at the input eliminates out-of-band noise.

The transceiver is the central element of the RF system. It generates all the RF frequencies required to perform transmit and receive functions, contains the digital circuits required for eight timeslots, as well as provides channel equalization and control logic. A channel equalizer is essential for the proper operation of the GSM system in multipath surroundings. The transceiver RF output drives the power amplifier to a maximum output level consistent with the rating of the transmit antenna. The receiver portion of the transceiver accepts the amplified and filtered receive signal from the receiver front end, performs down conversion, analog-to-

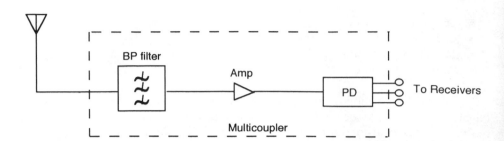

Figure 9.8 Receiver multicoupler configuration.

digital conversion, calculation of the receive signal strength, and provides the signal for baseband processing.

The high dynamic range receiver is required to allow for the varying distances (and therefore signal strengths) of mobiles from the BS. Until power control information is passed to the MS by the BSS, the received signal strength from the mobiles will obviously vary widely until their output power has been set in accordance with power control data passed over by BCCH.

The master frequency generator and *master clock unit* (MCLU) provide complete timing and frequency reference. Timing information is distributed in the frame units, carrier units, and frequency hopping units. To fulfill all timing supervisory functions, the MCLU has to perform complete timing defined by GSM as: starting at units of quarter bits (with a time period of 0.92 ms), extending to timeslots, TDMA frames, multiframes, superframes, and finally hyperframe with a duration of approximately 3.5 hr.

The frequency hopping unit performs switching according to the frequency hopping rule between the frame units and the carrier units in order to allow frequency hopping on a timeslot-to-timeslot basis. GSM recommendations provide for frequency hopping as an optional technique to counter multipath effects to enhance air interface confidentiality. Frequency hopping can either be done at the baseband or by fast tuning burst by burst at RF level. For burst-by-burst tuning, data is passed through different transmitters at every frame. When only a few carriers are present at the BS, frequency hopping by changing the synthesizer frequency is used, which requires a broadband combiner, which results in choosing a hybrid combiner with the associated higher loss. If the BS contains a large number of channels, hybrid loss will be excessive; thus, one chooses to use baseband hopping, which results in a fixed-cavity combining structure having lower loss.

The baseband functions of the frame unit in the transmit direction include data and speech rate adaptation, channel encoding, interleaving, encryption, and frame formation. Baseband operation, in the receive direction, consists of demodulation, de-interleaving, and equalization. Maintenance units monitor the health of different components in the transceiver unit.

All major subsystems (such as clock circuits, RF amplifiers and excitor) are continuously monitored and alarmed when necessary. The transmission unit adjusts the signal level and transmit and receive signals from the BSC.

9.3.3 Remote Transcoders

The transcoder converts standard 64-Kbps PCM speech into 13 Kbps (full rate speech) carried by the GSM air interface. The process is reversed in the opposite direction where 64-Kbps PCM speech is converted back into 13-Kbps full rate speech.

When transcoding is implemented at BTS, each 13-Kbps speech channel from the GSM air interface is converted by the transcoder at the BTS to a 64-Kbps PCM timeslot to BSC. This configuration is shown in Figure 9.9(a). When transcoding is not done at the BTS, the 13-Kbps digitized speech produced by the transcoder is padded with extra bits and can be carried on 16-Kbps channels. Four of these can now be supported by a single standard 64-Kbps channel and transported to the BSC. Similarly, four 16-Kbps channels can be carried over a single 64-Kbps channel between the BSC and MSC when the transcoder is not used at the BSC. Figure 9.9(b,c) show these cases. If we locate the transcoder at the BTS or at the BSC, only a single channel of voice can be carried over a 64-Kbps PCM channel. Similar savings can be achieved when a 2-Mbps link is used for transporting the data from the BTS to the BSC and subsequently to MSC. Here instead of transporting 30 digitized 64-Kbps voice, we can carry 120 speech channels over 2-Mbps link when the transcoder is not located at the BTS or at the BSC. Thus, the most effective placement of the transcoder is at the MSC. The channel efficiency, therefore, is maximized if the transcoder is located at the MSC. Although the most desirable location for the transcoder is at the MSC and the least desirable is at the BTS, it is technically possible to locate the transcoder at the BTS as well as at the BSC. The various transcoder configurations are shown in Figures 9.9(a–c).

9.4 MSC AND GMSC ARCHITECTURE

Similar to BSS architecture, the MSC, or its modified version for the Gateway, is based on the digital microprocessor based technology requiring LSI circuits. It is a stored program control, distributed microprocessor based, radio-telephone exchange that includes digital switching techniques.

In this section we will define the conceptual architecture of this subsystem based on large-scale integration technology and having multiple dedicated microprocessors for each major function. The various subelements of this subsystem are as follows.

- Switch matrix;
- *Automatic message accounting* (AMA);

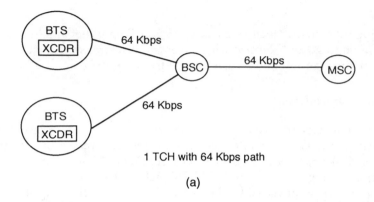

1 TCH with 64 Kbps path

(a)

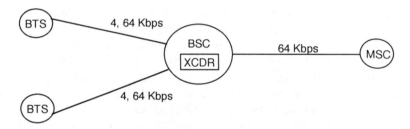

4 TCH with 64 Kbps path between BTS - BSC

(b)

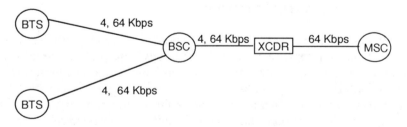

4 TCH with 64 Kbps path between BTS - BSC and between BSC - MSC

(c)

Figure 9.9 Transcoder at (a) BTS, (b) BSC, and (c) MSC.

- Timing system;
- X.25 interface subsystems;
- Land call management;
- Mobile call management;
- Mobility management;
- Service management subsystem;
- SS7 interface;
- VLR subsystem;
- Interface to AUC.

The architecture, based on distributed processors, as discussed here will have expansion capabilities as the need arises. Thus, processing power can be added at a future date to accommodate new features, capabilities, and growth. For example, as the GSM system grows in the number of BSs, PSTN trunks, or C7 signaling links, a new processor can simply be added or the capabilities of the existing power can be expanded to cope with this growth. Figure 9.10 is the conceptual architecture of MSC.

9.4.1 Switch Matrix

Due to advancements in LSI technology, a single-stage, nonblocking matrix based on the Time Speech Interpolation concept can be chosen. The matrix itself should be modularized and expandable as required by including additional modules. Each matrix module can be of 8,000 ports; thus, for the total matrix size of 256,000 ports, 32 such matrix modules can be used. Due to the nonblocking nature of the matrix, no labor-intensive load monitoring is necessary. Also, a single-stage switch provides a constant delay through the switch.

The basic architecture of the switch is shown in Figure 9.11(a). Basically, a time switch operates by writing and reading data out of a single memory. In this process the information on the selected timeslots is interchanged. The basic functional operation of a memory switch is shown in Figure 9.11(b). Here multiplexing and demultiplexing occur at the input and output of the matrix itself. In some cases this can be done separately before the data is applied to memory.

The exchange of information between two different timeslots is accomplished by a *timeslot interchange* (TSI) circuit. The incoming data is written at sequential locations in memory, and the data at the output is read according to locations stored in the control storage. As shown in the control storage memory address, j and i are stored at memory locations i and j; thus, the input data of channel i are put at slot j at the output, and input data of channel j is put at slot i at the output. There are two basic ways in which time stage memories can be controlled, that is, written sequentially and read randomly or written randomly and read sequentially.

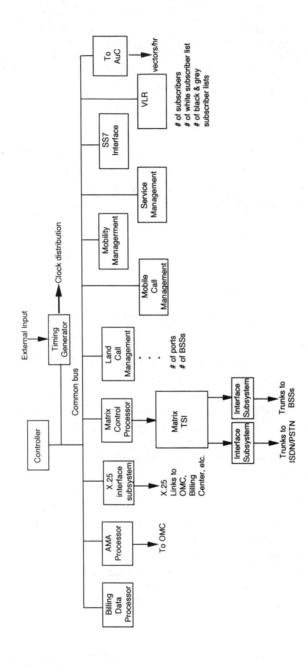

Figure 9.10 Conceptual architecture of MSC.

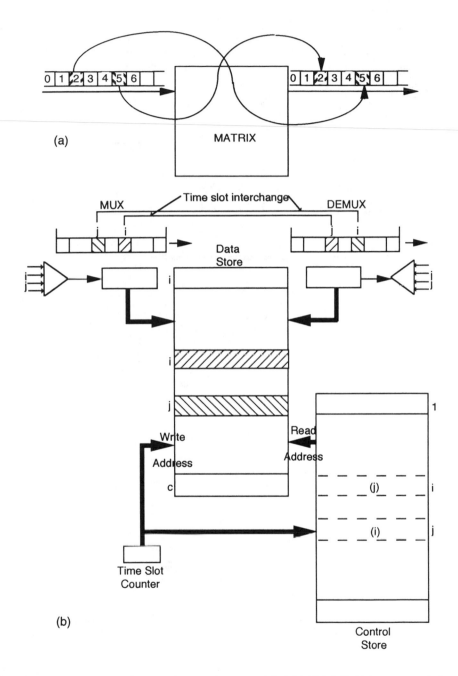

Figure 9.11 Basic operation of (a) timeslot interchange, and (b) MUX/TSI/DEMUX memory switch; and examples of (c) sequential writes and random reads, and (d) random writes and sequential reads.

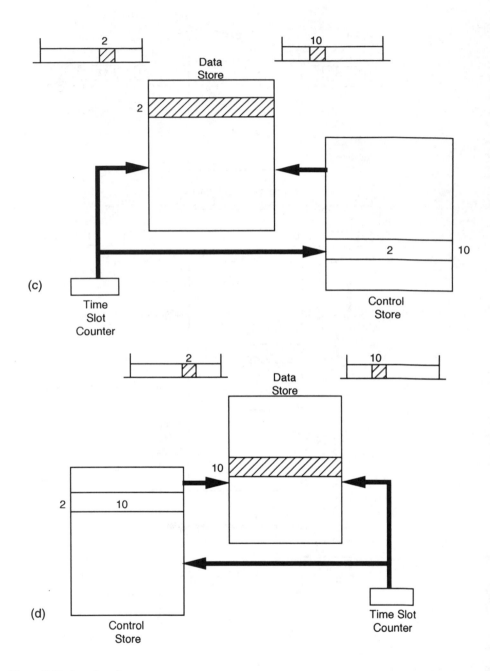

Figure 9.11 (continued).

Figures 9.11(c,d) show both modes of operation and indicates how the memories are accessed to translate information from timeslot 2 to timeslot 10. The first mode of operation, shown in Figure 9.11(c), has data for each incoming timeslot stored in sequential locations within the memory. As indicated, the data received during timeslot 2 is automatically stored in a second location within memory. Output, information retrieved from the control store specifies which address is to be accessed for that particular timeslot. As indicated, the tenth word of the control store contains the number 2, implying that the contents of data stored at address 2 are transferred to the output link during outgoing timeslot 10. The second mode of operation, depicted in Figure 9.11(d), is exactly the opposite of the first mode. Here data at the input are written into memory locations as specified by the control store, but outgoing data are retrieved sequentially under control of an outgoing timeslot counter. As shown in the example, information received during timeslot 2 is written directly into data store address 10, where it is automatically retrieved during outgoing TDM channel number 10.

9.4.2 Automatic Message Accounting

The dedicated AMA processor, as shown in Figure 9.10, collects billing data for different calls, formats the data appropriately, and sends them to the billing center through X.25 links. The billing records can optionally be copied to a magnetic tape. From these records, collected data usage statistics can be generated and sent to the OMC either on a periodic basis or whenever demanded by the OMC. Traffic statistics files can also be copied onto magnetic tape.

9.4.3 Timing System

Timing is extremely important for the proper operation of digital networks. Incorrect synchronization, producing frequency differences between digital switches, will cause loss of frame synchronizing. A network often uses a "Master Slave" configuration for synchronization in a plesiochronous manner, which allows the slave to recover the frequency from the master but does not require it to be phase locked to the master. The network normally has an internal high-precision oscillator or can be externally connected to a stable source such as a cesium beam. Normally there are a primary and a secondary timing sources. In this configuration the secondary timing source is at hot standby and switched upon failure of the primary source. Clock stability of the order of 1×10^{-9} may be required to ensure about 1 cycle slip/day and is illustrated in the following example.

Example

Assume a frame rate of 8,000 frames/s and clock stability of 1×10^{-9}. Cycle slip is $10^9/(8,000 \times 60 \times 60)$ or 34.7 hr $\approx$ 1.5 day. Thus, on average, a single clock cycle is slipped every 36 hr.

9.4.4 X.25 Interface System

X.25 interfaces are required to send data to the OMC and billing center. Thus, a cluster of X.25 interface cards are required at the MSC. The X.25 network is built using two basic building blocks, namely, signaling points and packet switches. The signaling point, in this case, are the OMC, the billing center, and the MSC. The network is shown in Figure 9.12. Packet switches are required as the data is packetized before transmission.

9.4.5 Land Call Management

A land call management subsystem performs functions associated with the monitoring of land trunk and trunks to the BSS, the translation of the dialed digits, processing incoming and outgoing calls, for example. A list of relevant functions may include the following.

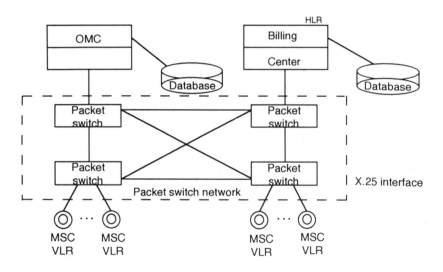

Figure 9.12 X.25 network architecture.

- Maintaining idle trunk list to PSTN;
- Maintaining transit trunk idle list to other MSCs;
- Performing traffic metering on trunks to PSTN;
- Routing database translation;
- Maintaining the idle trunk list to BSS;
- Performing traffic metering on busy trunks to BSS;
- Distributing mobile page to BSC and subsequently to BTSs;
- Processing land-originated and terminated call setup;
- Monitoring alarms on 2-Mbps trunks to PSTN and notifying fault-to-fault isolation subsystem.

9.4.6 Mobile Call Management

The mobile call management subsystem performs functions associated with the origination and termination of calls from the mobile. In particular, the subsystem is responsible for setting up calls originated by the mobile and terminated at the mobile. It is also responsible for the handover process, which involves two different BSCs.

9.4.7 Mobility Management

The mobility management subsystem is responsible for identifying the location of the mobile by processing location updates, IMSI attach and detach processes, and the associated traffic.

9.4.8 Service Management Subsystem

The main functions of the service manager are to provide circuit optimization, test interfaces, control of bridges for multiparty conference, receive channel test, receiver usage statistics collection, sender usage statistics collection, and recording and play-back of various announcements. The system also performs tests on tones and announcements that the MSC puts on trunks to BSS and PSTN. A maximum of 10 to 15 different announcements are generated in a typical MSC, and the duration of the announcements is generally arranged in steps of 4 sec, 6 sec, 8 sec up to a total of 120 sec.

9.4.9 VLR Subsystem

The VLR subsystem is mainly responsible for roamers in the area. In this process it performs the following functions.

- Allocates and manages TMSIs;
- Processes location updates;
- Processes IMSI detaches;
- Generates location event records;
- Provides implicit deregistration;
- Performs deregistration procedures;
- Allocates and releases the MSRNs on demand;
- Allocates and releases the handovers numbers on demand.

9.4.10 Interface to Authentication Center

The HLR Database/Authentication subsystem performs the basic database and authentication center functions for the HLR. It translates the MSISDN to get the IMSI and routes these to the appropriate HLR where the mobile is presently located. It also performs IMSI detach, location update, and database interrogation functions and manages authentication in the database. Several messages are exchanged between MSC and AUC simultaneously for these functions.

9.5 NETWORK MANAGEMENT SYSTEM

As stated in the introduction, powerful systems for network management are essential for network operators in order to be able to perform operation at a justifiable expense. The use of a suitable network management system has become a decisive factor in the technical and commercial success of the network operator. Well-organized network management allows a rapid reaction to changed requirements on the communications network. Under certain conditions it may be necessary, for instance, to quickly adapt the tariff structures to a new market situation or to introduce new or modified service features in the network. One such example is the expected increase in traffic during the Olympic Games in Atlanta or a trade show in a certain metropolitan city.

Applications of the network management system routinely include capturing billing data from MSC and handling some intricate special situations such as network expansion and fault management of the whole system. Modern network management integrates the operational functions of the separate network elements at a central point. Thus, it is possible to control the interplay of individual network elements from a central point. As an example, the MSC can provide a great deal of information regarding their own status (traffic parameters, alarms, and warning) and permit the operator at the central facility to have controlling intervention if necessary. The function of processing many individual information items from a network having hundreds of network elements is to provide a meaningful status overview, and the provision of control functions to permit timely intervention can only be practically achieved through a suitable computer-aided network management system. The network management system can be partitioned into two parts: first, a regional management of the network; and second, a national level network management.

The OMC may be considered a "regional manager" for the network hardware and software. Only parameters of the regional network are controlled through the OMC. Since there are several network operators in a PLMN, there will be several independent OMCs. Similar to the OMC, the NMC may be regarded as an "entire network manager" for the system hardware and software. The controlling parameters in this case will be associated with the overall network. Thus, a partitioned authority exit, one for the regional network and one at the national level, is shown in Figure 9.13. Here OMC1 and OMC2 control their own networks and the necessary information is fed to the NMC for their control, decision, and intervention if necessary. In this section we will discuss both of these network management centers separately and highlight their differences.

9.5.1 Functions of NMC

The NMC has a view of the entire PLMN and is responsible for the management of the network as a whole [4]. The NMC resides at the top of the hierarchy and provides global network management as shown in Figure 9.14. The NMC receives its information from the network equipment via the OMCs and filters the information. The NMC can, thus, focus on issues requiring national coordination. The NMC also coordinates issues regarding interconnects to other networks, such as the PSTN. In summary, network management for the PLMN consists of the following.

- Equipment for the operation of network elements in the field (for example, BTS, BSC, MSC);
- OMC as the central device for the operation of individual network within PLMN;
- NMC for the additional global technical management of the PLMN and further administrative and commercial control of functions at the total network.

OMC monitors and controls various elements of its network and feeds the summary view to the NMC. Since the NMC knows the status of the entire network, it can direct OMC operators to alter their approach for correcting a local regional problem because they know to what extent the present problem will be felt in other regional networks. OMC operators are simply ignorant about this situation.

In view of this role of NMC, the OMC need not be permanently staffed outside the normal working hours. It will be possible to control the regional network by remote access from the NMC after the regular working hours, reducing the overall cost of supervision and maintenance of the network. A summary of all relevant functions carried out by NMC includes the following.

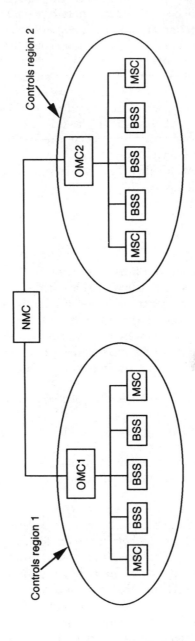

Figure 9.13 Hierarchical arrangement of NMC and OMC.

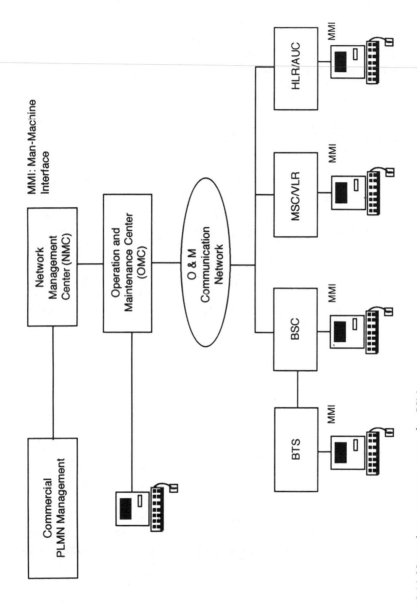

Figure 9.14 Network management system for GSM.

- Provides integrated operation of the network;
- Monitors the network for high-level alarms;
- Presents status of all regional networks;
- Provides network wide traffic management;
- Monitors the status of automatic controls applied to the network equipment in response to overload condition;
- Can take regional responsibility when an OMC is not manned;
- Supports network planning.

9.5.2 Functions of OMC

The OMC is a centralized facility that supports the day-to-day operations and management of a cellular network as well as provides a database for long-run network engineering and planning tools. An OMC manages a certain area of the PLMN, thus providing regionalized network management. From the network management point of view, an OMC operator can look into any part of the system closely. In case of a failure of some part of the system, the operator can decide what has really gone wrong with that part of the system. The operator can also reconfigure parts of the system by placing the troubled elements of the system out of service.

Important fields of application of an OMC include the areas of fault and alarm management, configuration and operations management, performance management, and security management.

9.5.2.1 Fault and Alarm Management

In case of a fault, the operator can execute tests and diagnostics from a central place and force the change of states of the network element based on test results and his individual judgment. The operator can also initiate traffic control or change the subscriber channel (forced handover), for example, if necessary. The operator can detect an error on the basis of alarm signals from the network elements and carry out recordings of these signals at a central point so that the pattern of errors over time can be analyzed for future corrections in a similar or parallel case. Since recorded data is in a centralized database, a quick search can be made based on a specified field such as event source, fault timing, or the effected system.

OMC selective testing can be carried out from a central point, and the status of individual modules in the network can be observed and modified. Thus, errors can be analyzed and countermeasures can be applied as necessary. Thus, the OMC has the facility to produce the network view and solve problems on a network basis.

9.5.2.2 *Configuration and Operations Management*

Configuration management permits an overview of the logical and physical structure of network elements. For example, the hierarchical internal structure of a BSS, MSC, and assemblies in the field may be displayed at any time. This is also true for the network element software and hardware. Thus, for example, a new software version supplied can be loaded and run at the right network element by the OMC. Configuration may be modified from the central location. Additionally, OMC also supports widescale changes and modifications to the network, as, for example, installation of new BTS or BSC. In addition, OMC can also be responsible for the introduction of new features and perform audits to check the integrity of hardware and software.

9.5.2.3 *Performance Management*

OMC provides capabilities for collection, analysis, and presentation of performance-related data from all network elements. The operator can schedule the collection of specific data from a certain network element for a certain interval starting at a fixed time. He can force the start of data collection or inhibit data collection if the congestion is high. The gathered data can be analyzed locally.

9.5.2.4 *Security Management*

Because of its comprehensive capabilities for intervention in the network, the OMC should also be looked at from the point of view of security aspects. Security management, the handling of access authorizations, is of real significance at OMC. A series of measures are generally used to prevent unauthorized access. This is achieved both using user- and device-related access authorization mechanisms incorporated in the software and smart cards with coded user identifiers. In particular, some of the measures employed are as follows.

- *Password authentication of OMC users:* All authorized users are allocated a password that must be used before gaining proper access to the system.
- *Logging of OMC access attempts:* Every time an attempt is made to access the system, a log is kept of the user identity, terminal identity, and time of attempt. Should a certain number of failed log in attempts occur, the user is disallowed for any further attempts for log in. At that time the system administration has to reset the system for the individual to log in again.
- *Automatic Log Off:* With this facility a user who has successfully accessed the system will be automatically logged off should the terminal kept idle for a predetermined length of time.

9.5.3 Architecture of OMC

As discussed previously, the OMC is used as a central system for the operation and maintenance of network elements like BSS, MSC, and HLR. The structure of the OMC is illustrated in Figure 9.15. Here the central computer is a UNIX server. The central system interfaces with BSS, MSC, HLR/AUC, and NMC. The user, through work station (through X.25) and X-terminal via LAN, has access to all operational and maintenance functions of the mobile network [4].

For the case of small regional networks with a few dozen BTSs, small SUN Sparc servers with one or two local X-Terminals will be sufficient, while high-performance UNIX servers with remote user workstations are required for a network with a few hundred BTSs.

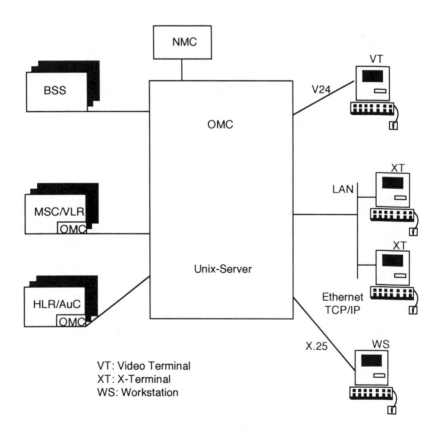

VT: Video Terminal
XT: X-Terminal
WS: Workstation

Figure 9.15 Architecture of OMC.

9.6 CONCLUSIONS

In this chapter we discussed the architecture of mobile terminal, BSS, MSC, and NMC. The enormous complexity of these subsystems are met by the development of ASICs. In particular, the mobile terminal has large digital signal processing capabilities that enables getting substantially improved quality of service compared to its analog cellular predecessor, that is, the first-generation system. BSS is a complex part of the GSM system, and it mainly consists of BSC, BTS, and transcoders. The optimum placing of the transcoder is really not at BSS. However, it can be a part of BSS under certain circumstances. A single stage of digital switching based on TSI is a desirable approach for MSC switch. A very large role of MSC can be performed based on the distributed processor concepts where each individual processor satisfies a dedicated function. Due to the enormous complexity of the GSM system, an integrated form of network management system, based on NMC and OMC, hierarchy is necessary.

Problems

9.1 What are various attributes of mobile? Why are they essential for our discussion here?

9.2 What are basic, supplementary, and additional MS features? Draw a clear distinction between them.

9.3 Name the three basic subsections of mobile terminal. Describe clearly the functions of radio, baseband processing, and control subsystems.

9.4 What are the various criteria used for the optimization of user terminals? State the advantages and disadvantages of direct conversion transmitter and receiver.

9.5 What are the main functions of terminal controller, switching matrix, and digital trunk controller for BSC?

9.6 What are the main subsections of BTS? Describe their functions clearly.

9.7 Why is transcoder required? What is the optimum location of a transcoder?

9.8 For the following configuration, find the net gain with and without power divider.

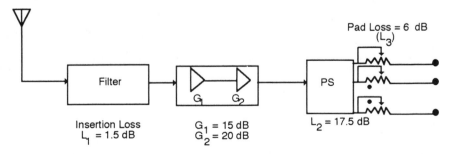

9.9   What are the advantages of distributed processing concepts with respect to MSC and GMSC?

9.10 Assuming clock stability of 5×10^{-9} and a frame rate of 8,000 frames/s, compute the clock slip rate.

9.11 Explain the basic concept of TSI. What is meant by sequential write random read and random write sequential read?

9.12 Why is a timing system required in MSC? What is the Master Slave concept? What is plesiochronous synchronization?

9.13 With respect to MSC, describe the functions of VLR, AMA processor, land and mobile call management, and mobility management.

9.14 Narrate the functions of NMC and OMC. Why has the hierarchical structure been adopted? Can NMC play the role of OMC in case of catastrophic failure of OMC?

9.15 What are the main functional areas of OMC? Describe the role of (1) fault and alarm management, (2) configuration and operations management, and (3) security management.

References

[1] Bursztejn, J., "GSM Terminals," *Electrical Communication,* 1993, pp. 128–140.

[2] ETSI/GSM Section 02.07, Version 4.4.0, pp. 2–15.

[3] Varin, J., et al., "GSM BS System," *Electrical Communication,* 1993, pp. 155–163.

[4] Schmid, E. H., et al., "GSM Operation and Maintenance," *Electrical Communication,* 1993, pp. 164–171.

CHAPTER 10
▼▼▼

MICROCELLULAR SYSTEMS

10.1 INTRODUCTION

The expansion of economic and commercial relations and improvements in transport have inevitably resulted in increased employee journeys. The search for improved efficiency, which is implicit in all modern economies, entails rapid circulation of information. Also, companies seeking to improve corporate and individual efficiency want to be able to contact their employees faster and more often. Putting these two elements together, communication requirements are bound to increase. Meeting these requirements involves multiplying and expanding the facilities for accessing telecommunication networks. The mobile communication systems that have prompted this issue is a solution to these needs. Since users of these systems are still users of fixed networks, they are accustomed to having excellent quality of services. It is, therefore, natural for them to expect no less from new mobile systems. Faced with these requirements, hardware manufacturers and network operators have designed and deployed radio systems that solve the difficulties involved in this type of transmission and problems associated with user mobility.

The mobile communications development plan is ambitious. The aim is to provide millions with the means to communicate on the move and, in so doing, cut the cord that has traditionally linked the terminal to the telephone network. It also aims to diversify the services available to provide a better response to the extremely varied needs of users.

Europe already has a proven track record of successfully managing and exploiting the latest mobile communications technology through GSM. Technical advances and significant changes in regulatory environments, backed by a strong political European vision and commitment from the telecommunications industry, led to the commercial emergence of second-generation digital systems and a much wider telecommunications market. This market encompasses cellular radio (GSM900/DCS1800), radio-based local-loop systems (DECT), low-power short-range cordless telephones and cordless PABX (CT2- CAI, and DECT), paging systems (ERMES), and wireless computer data links (HiperLan trunked radio systems for public service operations [TETRA]). All these digital systems are the subject of European-wide standardization by the *European Telecommunications Standards Institute* (ETSI).

In view of this, the objective of this chapter is to summarize those telecommunication projects that can be directly linked to present microcellular systems. In that regard we touch on mobile demand, technology, and trend in Section 10.2. Section 10.3 highlights technical requirements of microcellular system. Section 10.4 deals with the present operational CT2, DECT, and PCS 1800 systems.

10.2 FIRST-, SECOND-, AND THIRD-GENERATION SYSTEMS, TECHNOLOGY, AND FUTURE TREND

The first-generation standards available in networks since 1981, provided service to approximately 49 million customers worldwide at the end of 1994. The analog cellular deployment has been characterized by the available frequency (450, 800 or 900 MHz) and various standards leading to poor economics of scale and limited national and international roaming capabilities. Many of these systems were single source from manufacturers with proprietary interfaces, thus inhibiting cost reduction and economics of scale. However, it should be acknowledged that most of these standards have been successful in satisfying the early demand for mobility. Leading among them are four families: AMPS, TACS, NMT, and NTT, which have in themselves generated significant economics of scale and enhancement in terms of quality, service features, and handset choices.

The drivers for digital or second-generation standards were to achieve more capability from the available spectrum and introduce new (nonvoice) features and more spectrally efficient systems. Other regional drivers include the need for roaming and greater choice through open (nonproprietary) interfaces. Here the goal is to achieve the economics of scale. There are four families of digital cellular standards: GSM, D-AMPS, PDC, and CDMA. They are summarized in descending order of market demand in Table 10.1.

By the end of 1994, five million customers were using these digital systems, more than 90% of which were using the GSM family. The forecast for the end of

Table 10.1
First- and Second-Generation Cellular Systems

First Generation (Analog, 49 Million by 1/1/95)	Second Generation (Digital, 5 Million by 1/1/95)
AMPS 800	GSM 900
	1800 (DCS), 1900 (PCS)
TACS 900	Satellite
NMT 900 and 450	D-AMPS
NTT	PDC 1500
C-NET	CDMA 800
Radiocomm	
N-AMPS 800	

1995 is that approximately 13 million cellphones will be in use, again with over 90% using GSM. It is also estimated that by June 1995 over 40,000 GSM sites will be installed worldwide, reaching approximately 50,000 by the end of 1995. At the end of 1994 the worldwide customer total for cellular was 54 million with a prediction of over 200 million by the year 2000. Worldwide growth of cellular from 1989 is shown in Figure 10.1. From the figure it is clear that an explosive growth of cellular is predicted until the end of century.

While the assessment is in terms of customer numbers, there is a significant underlying variance in terms of usage per customer. In some of the newer markets where cellular has recently been introduced, business levels of around 400 min/mo per subscriber is not unusual, whereas in older established markets consumer levels of less than 30 min/mo/subscriber is not ruled out. The service revenues are predicted to increase significantly with the growth in both customer numbers and usage/

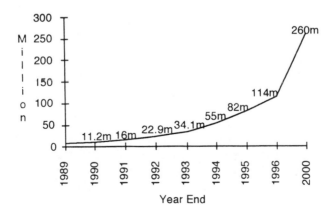

Figure 10.1 World cellular growth til the year 2000.

customer. However, forecasters should note that these calculations may also have significant variances depending on the degree of downward price adjustments. Service revenues for 1994 were approximately $50 billion and are forecast to be around $100 billion by the year 2000. This 100% increase in revenue, with the projected fivefold increase in customer numbers, is due to lower per customer usage with increased penetration and also reduced tariffing to attract new (consumer) customers.

From a cost/coverage perspective, the customer will choose from what is available in the market. However, it should be stressed that the level of coverage required may not need full cellular, and the budget constraints may also determine one's choice.

However, the "cordless" alternative (often called fixed cellular, wireless in the local loop) can be served by some cellular operators. Choices of systems will be more blurring especially at the boundaries of Cordless/Cellular and Cellular/Satellite. Naturally, the ongoing evolution of GSM will be a critical success factor as it moves along the path toward third-generation systems.

The scope to add further radio access techniques to the GSM platform is also being considered. Already there is significant interest in dual-mode operation between GSM and DECT. In the near term there are opportunities for convergence between GSM and DECT (cordless). In the long run scope exists to consider the alternative radio access techniques, such as CDMA linked to the current GSM platform.

Thus, when we combine all these issues of demand, supply (based on standards), and the respective barriers to growth, the trends in the cellular market can be characterized by the three Cs.

- Competition;
- Consumerization;
- Convergence.

The extent of liberalization and privatization is being complemented all over the world by an increase in competition in telecommunications. Digital mobile has typically been licensed such that the cellular operators do compete, although naturally some are aligned to PTTs in addition to the newcomers. The choice of standard technology has often been determined by both the available frequency and governments. In the future it would appear that market factors will play a stronger part in this decision as clearly governments may not wish to be seen backing significantly one technology against another (as illustrated by the recent technology-independent U.S. PCS auctions). In some parts of the third world countries cellular is trying to become the first phone. Clearly, the cost of mobility has to fall out, but where there is no fixed line available it has become the natural choice to choose the mobile phone.

Penetration levels in the Nordic countries are over 15% of the population and should reach around 25% before mid-1996. In North America, Australia, and Hong

Kong the customer base is over 10%, and the United Kingdom (the largest market in Europe today) is heading for 10% of the population by the end of 1995. In these higher penetration markets the following are highly visible.

- Competition, and high profile advertising/branding;
- Innovative distribution channels, for consumer and business;
- Quick credit check arrangements;
- Relatively low-cost colored handsets, and accessories.

These signs support the forecast that a cellular consumer age is coming. The penetration levels may not be as high for cameras, TVs, and VCRs, but the key enablers are more and more being put together in place.

There is a lot of attention associated with convergence between mobile and transportation, financial services, and the information and entertainment industries as shown in Table 10.2.

10.3 TECHNICAL REQUIREMENTS OF THE MICROCELLULAR SYSTEM

In this section we briefly describe the salient characteristics of the microcellular system. These are channel segregation, multiple-BS registration, diversity transmission, proper choice of modulation, performance criterion, propagation model, and multipath model.

10.3.1 Channel Segregation

For highly deformed cells, fixed channel assignment to cells are not possible. A decentralized channel assignment, where each cell finds and uses interference-free channels itself, is the preferred choice [1]. The basic channel segregation scheme is an improvement over a dynamic channel assignment scheme. In this scheme each cell acquires its favorite channels by learning through past experience on channel

Table 10.2
Cellular Radio Convergence With Other Industries

Industries	*Reason for Convergence*
Transportation	Mobility, value added, and differentiation
Financial services	Market research, credit checking, and enhanced mobility; in some cases this may be linked to more efficient billing or other back office processes
Information and Entertainment	Mobile data, messaging, and mobile multimedia services

usage. Each BS has a table where a priority function is defined for each channel. A priority function to the channel is defined as the probability, $P(i)$, of a successful transmission over the channel. The algorithm consists of the following four steps.

Step 1: When a new call arrives at the BS, the BS selects the channel that has the highest priority.
Step 2: The BS senses the channel, and if found idle, the BS starts communication on the channel and increases the priority function of the channel for next time use.
Step 3: If the channel is found busy, the BS looks for the next priority channel and decreases the priority function of the busy channel.
Step 4: If all channels are sensed busy, the call is blocked.

10.3.2 Multiple BSs Registration

The scheme allows the subscriber and the BS to perform location registration procedure without aid of a central station. The conventional technique is as follows. A subscriber station monitors received signal level from surrounding BSs [1]. The subscriber selects a BS with the highest signal level. The location registration request signal is sent to the BS with mobile ID. The scheme allows the location registration procedure to be decentralized. The location registration process is carried out by having multiple-cell registration. Since cells are overlapped, multiple registration is possible. The subscriber gives priority to the BSs based on signal level measurements from different cells.

10.3.3 Diversity Transmission

Antenna diversity as shown in Figure 10.2(a) is a very simple scheme since it requires only one receiver with a simple RF switch. It provides increased capacity, extends the range, and minimizes the time dispersion effects. Unfortunately, the scheme works well when the S/N ratio is high and performs poorly when the level is low. Supposedly, a diversity receiver should improve the signal quality while the signal is below a certain level, but unfortunately the scheme generates more noise than a receiver with no diversity. The switching noise generated is due to an abrupt change in amplitude and phase of the signal.

There are several methods of overcoming this difficulty. (1) The scheme shown in Figure 10.2(b) switches the antenna during preamble such that the switching transients dies down during an active period of data transmission. (2) The two schemes shown in Figure 10.3(a,b) use two antennas at the BS. The use of diversity at the mobile is essentially done at the BS, which also removes the complexity of the mobile transceiver, which is desirable. In single-frequency time division duplex operation where a single frequency is used for both up and down transmissions, the

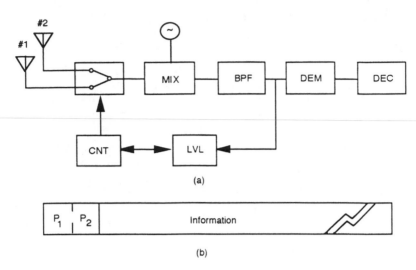

Figure 10.2 Antenna selection diversity: (a) block diagram, and (b) signal frame [1].

antenna used for downlink transmission is the one that was selected in the previous reception at the base. Here the base selects the best antenna based on the received signal strength and uses it for the next frame for transmission [1].

Correlation between up and downlinks are assumed to be high as both transmissions are at the same frequency. For the system with dual-frequency operation, coherence between the two frequencies is very low as the uplink and downlink channels are purposely widely separated in frequency. In this scheme, test signals are sent from the BS via the two antenna successively at the end of a framed signal. The subscriber measures, compares the two received signal levels, and sends back the result to the BS to select the antenna for the next transmission of the signal.

10.3.4 Modulation

Two different choices of modulations are obvious: constant envelope modulation belonging to MSK family, like GMSK, or a linear modulation belonging to QPSK family, like $\pi/4$-DQPSK. For higher throughput $\pi/4$-DQPSK may be better when compared to GMSK. Since the power requirements of the mobile is very low for handportable, the choice of a linear modulation is quite practical. Since the channel for microcellular system is Recian, fading is relatively infrequent; thus, an inexpensive low-speed AGC can be adequate for the mobile [1].

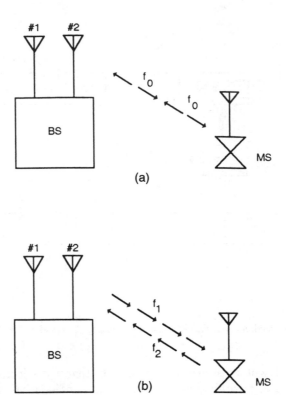

Figure 10.3 BS antenna selection diversity: (a) single-frequency time division duplex system and (b) dual-frequency full-duplex system [1].

10.3.5 Performance Criterion

The performance of a microcellular system can be judged in the same way as that of the cellular radio. The ratio, SIR, of the desired signal power (S) to the power of the active interfering cochannel signals (I_i) for the duration of the call is given by

$$SIR = S/\sum_{i} I_i \qquad (10.1)$$

For a TDMA system the uplink interfering power is due to users assigned to the same frequency at the concurrent times at other ports. The summation in (10.1) is over all the cochannel/coslot active users while the desired user is in communication. The definition of S/I for the downlink is analogous to that for the uplink, except it is calculated for the portable transceiver. Figure 10.4 shows the interference model.

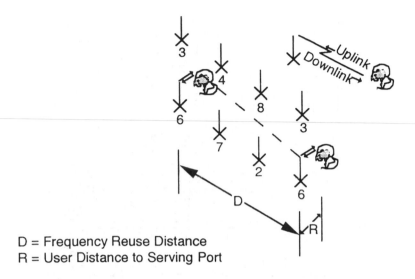

D = Frequency Reuse Distance
R = User Distance to Serving Port

Figure 10.4 Desired signal model [1].

10.3.6 Propagation Model

The model that matches well with measurements on microcellular layout on metropolitan cities is a one-dimensional model differentiating between *line-of-sight* (LOS) and *non-line-of-sight* (NLOS) paths. The model is represented by

$$L = -20 \log\left(d^\alpha \left[1 + \frac{d}{G}\right]^\beta\right) + K \qquad (10.2)$$

where d is the distance from the BS, G is the specific distance from BS where the exponent increases to β, K is some fixed loss, and α is the attenuation exponent for line-of-sight propagation.

Measurements conducted on microcells using low omnidirectional antennas in Stockholm show that the exponent changes after some distance from the BS. Layout uses twenty BSs, whose transmitted powers were measured at different streets. Transmitting and receiving antennas were situated at 5m and 2m, respectively. The frequency of measurements is 870 MHz. Figure 10.5 shows the measurement layout.

Figure 10.6 shows how the signal varies when turning from a LOS street onto a NLOS street. The LOS street is Kommendorsgatan, and the NLOS street is Grevgatan for BS designated as two. Mobile turns from Kommendorsgatan to Grevgatan for BS designated as two. The signal around the corner decreases by 20 dB to 25 dB in a short transition distance of 20m.

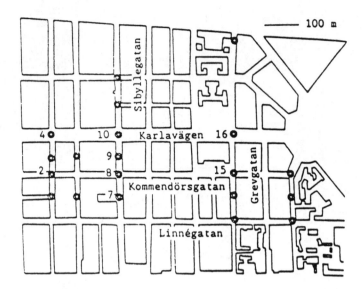

Figure 10.5 Antenna placement on Ostermalm. Signal strength levels from 20 antennas were recorded on all of the streets shown [1].

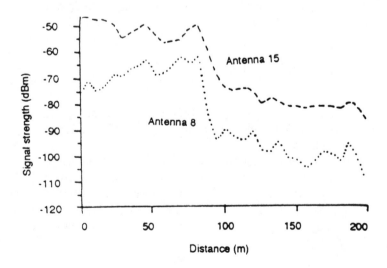

Figure 10.6 Mobile going east on Kommendorsgatan and turning right at Grevgatan [1].

10.3.7 Multipath Model

A multipath model of the microcellular is not as severe as in cellular radio. A model that fits well is the Recian fading model, represented by the density function, $p(x)$, as

$$p(x) = \frac{x}{\sigma_i^2} e^{-(x^2+A^2)/2\sigma_i^2} I_0\left[\frac{Ax}{\sigma_i^2}\right] \qquad 0 \le x < \infty, A \ge 0 \qquad (10.3)$$

$$K = \frac{A^2}{2\sigma_i^2}$$

where $I_0(l)$ is the modified Bessel function of the first kind and zero for higher orders. A represents the peak-to-zero value of the specular radio signal, which consists of a LOS signal and scattered time invariant signals received from various paths. K specifies the Rician distribution completely. As K approaches infinity, the Rician density becomes Raleigh. σ_i^2 denotes the average power that is received over different paths that vary with time due to moving objects. Having discussed the salient characteristics of microcellular system, we take up the studies of cordless telephones (CT2) and DECT systems.

10.4 CORDLESS TELEPHONES

In the telecommunications world, an approach comparable to wireless LANs is the development of cordless systems that offers spatially limited terminal mobility. Cordless telephone systems can be defined against the broader background of radio communications services. When compared with cellular radio systems, they are characterized by a shorter radio range (50m to 300m depending on the application and the standard implemented) and a narrower scope of services. On the other hand, service access and terminal prices are significantly lower than in cellular systems (by a factor of four to five). The most widespread application today is residential cordless telephony. Cordless handsets supporting multiple standards (CTO, CT1, CT1 +) are used. However, they have their disadvantages due to analog transmission and the availability of a limited number of channels and dialing security. New technical characteristics were defined to solve some of these problems and, eventually, resulted in the implementation of the CT2-Common Air standard for business applications.

In 1988, CEPT decided to introduce a new European system in a different frequency range. In 1992, this resulted in the *Digital European Cordless Telecommunication* (DECT) standard, which focuses on applications other than telepoint and provides wireless PABX services with high speech quality and local loop replacements as well as cordless data services for wireless LAN applications.

The main characteristic of a BS is the number of radio channels, which determines the number of simultaneous communications. CT2-CAI BSs provide one radio channel per transceiver. DECT BSs contain a single transceiver providing twelve radio channels. We elaborate on the technical aspects of both these systems in the following subsections.

10.4.1   Cordless Telephones (CT2-CAI)

Many systems already provide mobility, but only in part. Until now, no overall solution to mobility or system flexibility requirements has been achieved [1]. Business requires mobility so that it can provide identical services to the called and calling parties, no matter where they are in the company or what kind of terminal they use. The most basic need is to ensure that the caller be capable of contacting a party by dialing a single telephone number without needing to know where the party is located in a building or any other building of a company. Such needs are especially urgent for enterprises whose activities involve a high degree of mobility, such as stores, hospitals, and repair shops. Those needs are also felt strongly by certain categories of users in companies, such as maintenance personnel. Such users, although they can remain inside their office during working hours, often move from place to place. They would, therefore, like to have a single, portable telephone that they can carry around. Cordless systems are able to meet such needs. By supplementing wired connections or even replacing them, companies have the benefit of total flexibility, installation, configuration, and management of their communication systems.

The cordless telephone, CT2-CAI, mobile terminals can now provide the same facilities as fixed terminals for voice and data transmission with enough confidentiality for a business communication system. The characteristics of mobility are the same as in a cellular system. They are related to the availability of our basic functions: terminal identification, automatic location of terminals in a multicell system, intercell call handovers, and dynamic call allocation.

CT2-CAI terminal identification is based on a single 27-bit number assigned to each terminal, regardless of the service. In CT2-CAI subscriber location, the BSs call the terminals in the various cells. CT2-CAI uses frequency division multiplexing, which means that the terminal constantly occupies the frequency throughout each call transmitted or received. The terminal is incapable of searching for a better BS when the transmission quality deteriorates. The controller instructs the adjacent terminals to tune into the terminal in order to determine the best BS. Handover involves interrupting and reestablishing the radio link and, therefore, is noticeable to the subscriber. There is a type of transfer within a given cell to change from one frequency to another when a disturbance occurs on the current channel. CT2-CAI requires interrupting and reestablishing the radio link in this case. Dynamic channel allocation allows all cells to use all channels based on their need and the availability of channels.

10.4.1.1 Applications of CT2-CAI

Cordless telephone systems are suited to be used in residential and public applications. This is shown in Figure 10.7. Office application is limited to DECT and is covered in the next section. Of these three applications, residential use will provide the largest market outlet for CT2-CAI equipment in the long run. A substantial penetration of the cordless handset user base by CT2-CAI-based cordless handsets, however, requires quite a significant drop in the price of this technology. The price drop is expected as the global market is soaring from 12% in 1992 to about 40% in 1997. For the three cases, the coverage, type of call, and the user category are shown in Table 10.3.

10.4.1.2 Technical Characteristics of CT2-CAI

Coverage depends on the frequency used, the transmitting power, the sensitivity of the receivers, the shape and type of the buildings, and various factors related to equipment design. The equipment has a range of several hundred meters outdoors and is reduced to somewhere between 30m and 100m indoors [1]. CT2-CAI BSs must have one transceiver per radio channel and provide fewer channels. Therefore, several CT2-CAI channels are needed in cells to handle the traffic. The maximum traffic load is related to the size of the cells and the number of useful channels in each cell. Although there are 40 channels in the CT2-CAI system, crosstalk from adjacent channels cuts down the number of useful channels in a cell to a dozen. With 12 channels, 2% blocking probability, and 30-m cell size (office surrounding), the estimated traffic is about 2,000 Erlang/Km^2.

In CT2, a standard air interface allows mobile to interface with different BSs. CT2 allows digital speech transmission, which will in turn dynamically select the best available radio channel. The speech coder is the CCITT standard ADPCM that operates at 32 Kbps. Digital speech allows encryption techniques for privacy. CT2 is based on a digital technology using combined frequency division/time division transmission techniques. This time division duplexing operation is shown in Figure 10.8. Because power is limited to 10 mW, the operating range of phone is relatively small (nearly 100m) and interference to other users is also somewhat limited. Dynamic channel allocation ensures that best use is made of the available spectrum at any time. The system can operate in high traffic density surroundings, such as offices. When the subscriber wants to make a call, the radio searches around 40 channels and chooses the one that is most interference-free.

Carriers are spaced 100 kHz apart, and each conveys one conversation using time division duplex for a two-way conversation. A frame duration of 2 ms is shown in Figure 10.8. For 1 ms the mobile transmits to the BS, and the base transmits to mobile for the other 1 ms. Each slot contains 64 bits for user information and

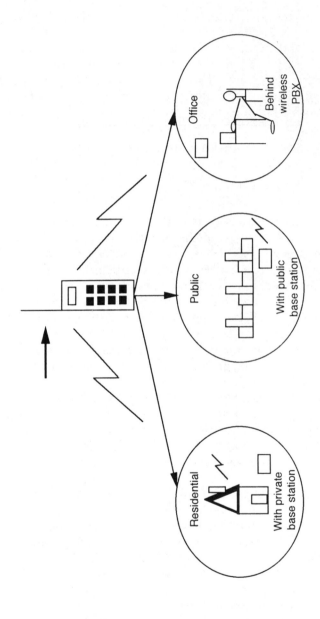

Figure 10.7 Three applications for the cordless handset.

Table 10.3
User Coverage for Cordless Telephones

Service Type	Coverage	Call Type	User
Cordless telephone (street wireless telephone)	Urban nonubiquitous cells from 100m to 200m	Incoming and outgoing calls	Business consumer
Cordless telephone (office wireless telephone)	Corporate ubiquitous cells from 10m to 50m	Incoming and outgoing call handover	Business
Cordless telephone (residential wireless telephone)	Ubiquitous cells from 50m to 300m	Incoming and outgoing calls	Consumer

Table 10.4
Technical Characteristics of CT2-CAI and DECT

Specification	CT2-CAI	DECT
Frequency band of operation (MHz)	864.1 to 868.1	1880 to 1900
Bandwidth (MHz)	4	20
Access method	TDMA/TDD/FDMA	TDMA/TDD/FDMA
Number of channels	40	10
Channel separation (kHz)	100	1,728
Channels/carrier	1	12 (full duplex)
Channel data rate (Kbps)	72	1,152
Voice coding rate (Kbps)	32	32
Voice coding type	ADPCM	ADPCM
Modulation	GMSK	GMSK
Hopping	No	No
C/I radio (dB)	20	12
Mobile power in Watts	0.01	0.25
Frame duration (ms)	2	10
Cell radius (Km)	0.2	0.15
Hand-off	No	Yes
Market	Residential, public telepoint	Business

4 bits of control information divided equally between two directions of transmissions. There are two guard bands that add an additional 8 bits. Thus, a frame of duration 2 ms contains a total of 144 bits, providing an effective data rate of 72 Kbps. With a channel spacing of 100 kHz, the bandwidth efficiency is 0.72 bits/s/Hz. The modulation scheme used is GMSK. The CT2 control channel is referred to as the D channel.

Time Division Duplexing (TDD) is implemented to generate a data flow to the radio module. The flow includes a B channel, containing the 32-Kbps encoded

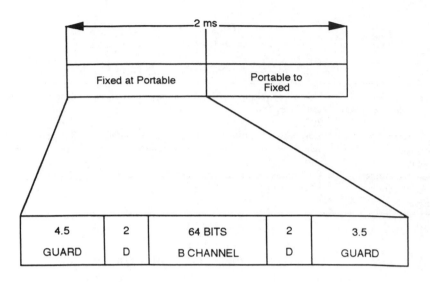

Figure 10.8 CT2 timeslot and frame [1].

speech signals, and a D channel, supporting signaling information used both in monitoring the performance of the radio link selecting service options and functions activated during the call and even in checking the identity of the mobile.

10.4.1.3 Implementation of CT2-CAI Handset

Radio terminal architecture, shown in Figure 10.9, is based on a transmission system composed of an antenna built into a radio transceiver unit, the coder-decoder converting the 32-Kbps digital signal into an analog signal, microphone, and earphone

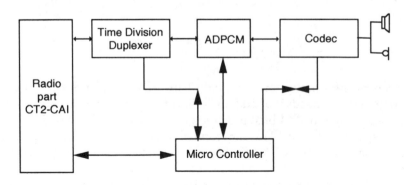

Figure 10.9 Cordless handset.

transducers. A microcontroller controls the entire transmission system, manages the CT2-CAI radio protocol, and processes keyboard and display unit signals. Data transmission is handled by a switch routing the 32-Kbps digital signal to the voice coder-decoder or data adapter. The battery life of the terminals depends on the power supply provided by a battery pack or, better still, a storage cell requiring a charger. It ranges from 3 to 10 hr of active calling and 30 to 100 hr of standby. The voice facilities generally allow for adjusting the earphone and ringing volume levels. The microphone features filters to reduce background noise. Some handsets have a telescoping antenna to eliminate hand capacitance effects perceptible on the fringes of an area of coverage. The new generation of handsets will benefit from a reduction in thickness of the display screen through the use of plastic components and from lighter plastic shells, which will significantly affect the cordless handset's weight.

10.4.2 DECT

In June 1992, the DECT standard was published by the ETSI, and the first commercial products came into market in 1993. Having been engineered for the most demanding cordless telephone applications, DECT is suitable for residential, public access, and local loop telephony. By interworking with different networks, DECT can cater a wide range of services to the user. For DECT, the need for open interface standards is particularly important. It is the only way to guarantee that terminal equipment may be brought to different locations and that the service can be provided by local networks.

DECT also interfaces with GSM, which leads to additional network capacity, especially in high-density surroundings such as business. The interworking between GSM and DECT will become a major milestone toward migration to the third-generation cellular systems. Thus, in this section, we will provide details of technical parameters of DECT followed by its applications and then the networking with the GSM system.

10.4.2.1 DECT Parameters and Salient Features

DECT works in the 1880-MHz to 1900-MHz frequency band. This band is divided into 10 carriers, and each of the carriers is subdivided into a cycle of 24 repeating timeslots (TDMA). Twelve of the timeslots carry traffic from BSs to portable devices, and the other twelve in the opposite direction (TDD). As in GSM the system also uses GMSK modulation. Speech is digitally encoded with 32-Kbps *ADPCM modulation* (ADPCM), with a channel data rate of 1.152 Kbps over a channel bandwidth of 1.728, which provides a bandwidth efficiency of 0. 67 bits/s/Hz. The BW efficiency is comparable to that of CT2-CAI.

The speech data rate of 32 Kbps is a much higher rate than that used in any of the world's digital cellular standards and, as a result, provides speech quality that is indistinguishable from a standard wired line telephone system. DECT double slot allocation is used to support ISDN service at 64 Kbps, and two or more slots may be combined in a single slot to provide 144 Kbps suitable for the 2B + D basic rate. Details of timeslot and frame structure are shown in Figure 10.10. As shown in the figure, the total frame duration of 10 ms is divided equally between two 5-ms segments. Each 5-ms segment is divided between 12 timeslots representing 12 channels each of 5/12 = 0.417-ms duration. Each channel data consists of 16 bits of preamble, 16 bits of synchronization, 64 control bits, 320 bits of information, and 64 guard bits. With a total of 480 × 24 bits of user data, over 10 ms, a total data rate of 1.152 Mbps is provided and is higher than the GSM data channel rate of 271 Kbps. Compared with CT2 phones, DECT doubles the transmission range (typical cell radius in buildings is 20m to 100m; outdoors it is up to 300m) and permits connection handover between BSs.

The mechanism for channel selection, known as *Continuous Dynamic Channel Selection* (CDCS), allows for systems to work in an uncoordinated environment. Every DECT portable device has, in principle, access to any channel. When a connection is needed, the portable device selects the channel that offers the highest quality connection. While the connection is maintained, the portable device continuously scans the radio environment and, if a better channel becomes available, switches the connection to that channel. The old and new connections overlap in time, therefore, making handover seamless [2,3].

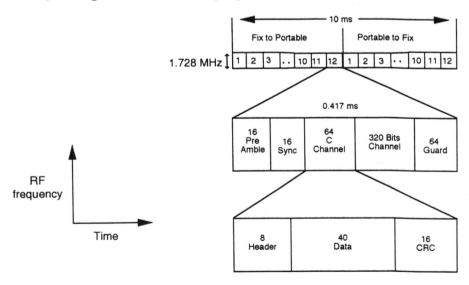

Figure 10.10 DECT slot and frame structure [1].

The DECT standard includes a number of security provisions including encryption of radio transmissions and authentication of portable devices. Within Europe, DECT is a mandatory standard. The frequencies used by DECT are reserved for DECT applications within all the CEPT countries. The standard also has widespread support within the industry. All the major PBX suppliers active in the European market have either launched DECT products or announced their intention to do so. The first commercial systems appeared on the market in 1993, and some manufacturers have already upgraded their product. DECT is also gaining significant interest outside Europe. In the United States a standard is being developed based on DECT for application in the PCS bands.

10.4.2.2 DECT Applications

The DECT standard was designed to support cordless telecommunications access in a number of existing and future telecommunications applications, such as *wireless local loop* (WLL), cordless telephony, and business cordless PBX. The WLL application system has a provision for high antenna gain to increase the range of the local loop.

The configuration of the WLL is shown in Figure 10.11(a). DECT-based WLL is optimized for urban and suburban areas where subscriber densities are high. Using directional antennas at the fixed locations (*Fixed Access Unit* [FAU]), the effective range of the BS can be extended up to 5 km. Here, at the *DECT Access Node* (DAN), a number of directional antennas are used to provide omnidirectional coverage. In medium- and high-density environments, DECT-based WLL systems are highly cost effective compared with cellular-based WLL systems. The crossover point occurs at a subscriber density of around 20 subscribers/Km^2. One reason for this is the TDMA radio format used in DECT, which allows a single radio transceiver to handle up to 12 simultaneous calls. This is also a very cost-effective service as it minimizes the up-front cost of the system [2,3].

For cordless telephone application, DLN acts as a traffic concentrator connected to various DECT BSs. This configuration is shown in Figure 10.11(b), where radial connections to several DECT BSs are used. Multiplexed data at 2 Mbps goes through the switch via a *Common Control Fixed Point* (CCFP).

Business cordless telephony (BCT) was the first application of DECT. The systems available today are either fully integrated cordless PBXs, including switching systems or an add-on system that can be interfaced to an existing PBX to form a hybrid system with both wired and cordless extensions. PBX has the highest traffic capacity requirements of any DECT application, with up to 10,000 Erlang/floor. The traffic densities are met using a multicell system on each floor with picocells, each having a range of about 30m to 70m. It is expected that the DECT PBX application will cover about 30% of all business applications by the year 2000. The

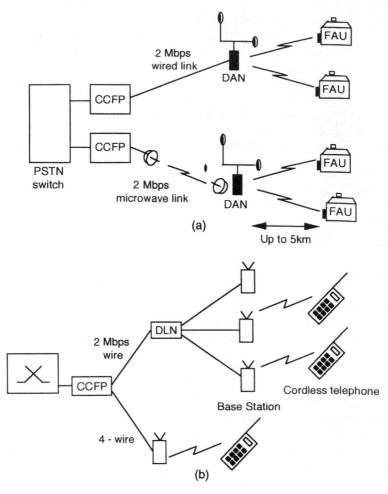

Figure 10.11 (a) Architecture of a DECT-based WLL system, (b) DECT-DLN acting as a concentrator, and (c) architecture of a cordless PBX system.

configuration for DECT PBX is shown in Figure 10.11(c). DECT equipment can be interfaced directly to the public network to provide a cordless center application. In this configuration, PSTN will directly interface with a DECT system; or in other words, the PBX will actually be located at the telephone central office.

10.4.2.3 DECT-GSM Interconnection

DECT-GSM interworking is the main means by which cellular network operators stand to gain from the technology. GSM network operators will benefit from DECT

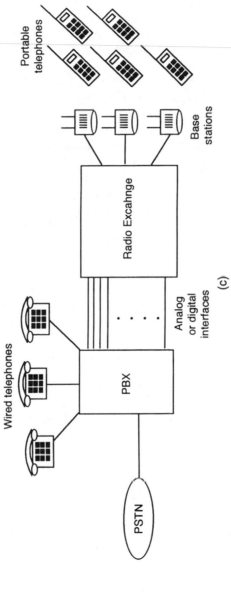

Figure 10.11 (continued).

in two ways. First of all, it can provide additional network capacity in high-density environments such as businesses. Second, it can support increased traffic during relatively quiet periods. Cellular networks experience their traffic peaks during the morning and evening rush hours, when people are on their way to and from work. During the working day, traffic levels fall; people are in their offices, and they make and receive calls using the fixed telephone network. Offering a cordless Centrex service, using the GSM network as the backbone and DECT as the access mechanism, is one way in which a GSM network could use spare capacity during the daytime. Ideally a mobile network that includes both DECT and GSM should provide users with dual-mode telephones capable of switching to DECT operation whenever a DECT system is in range. Such dual-mode telephones will become available in the very near future. The evolution toward this "enhanced GSM" is a step toward the third generation of systems. One scenario that involves combining DECT and GSM handset into one multimode handset is interesting. Integrating cordless and cellular systems with the possibility of roaming and even access-independent seamless services would be a major step toward the UMTS concept [1–3].

The interconnection of DECT with GSM through the A interface is shown in Figure 10.12(a). Since the coverage possibilities of DECT are limited, the wide-area coverage would be supported by normal GSM access. In "hot spots," that is, indoor applications and residential areas, the DECT access would be complementary to the GSM service area. The user would experience a seamless service and an overall improved system quality. From an interworking point of view the GSM PLMN will regard the DECT user as a normal GSM served user. The radio subsystem consists of *portable parts* (PP) and a *Radio Fixed Part* (RFP) attached to the MSC/VLR combination. Here the *fixed part* (FP) emulates the functionality of the BSC. The GSM system is attached to DECT through the standard A interface. The DECT system will require that the FP emulates the functionality of the BSC. This requires that the SS7 protocol stack with the GSM application on top be implemented in the IWU belonging to FP.

The ISDN+-based interconnection is shown in Figure 10.12(b). Here, the operator of a local DECT system could offer both local and public GSM services to the users. A DECT system that is attached to a PBX could offer private network services to its own private users and a provision for transparency of GSM services to visiting GSM subscribers with a DECT/GSM portable. Based upon an enhanced ISDN+ interface, the level of transparency could be more flexible. This configuration may allow for new service concepts where the DECT system is transparent for the actual GSM service but will only make use of the extended mobility provided by GSM. The ISDN+ interface is not yet completely defined, but the interesting concept is studied within the DECT and GSM communities. Figure 10.12(c) shows the general structure of GSM and the attachment of a DECT system. Here, the portable/mobile station has the possibility of either accessing the DECT system or the GSM network; that is, the interworking is made on a terminal basis. The terminal is a dual-mode

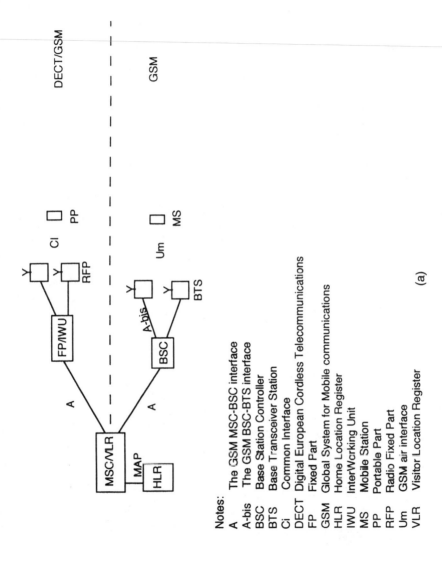

Notes:
A   The GSM MSC-BSC interface
A-bis   The GSM BSC-BTS interface
BSC   Base Station Controller
BTS   Base Transceiver Station
Ci   Common Interface
DECT   Digital European Cordless Telecommunications
FP   Fixed Part
GSM   Global System for Mobile communications
HLR   Home Location Register
IWU   InterWorking Unit
MS   Mobile Station
PP   Portable Part
RFP   Radio Fixed Part
Um   GSM air interface
VLR   Visitor Location Register

(a)

Figure 10.12 DECT/GSM internetworking (a) via A interface, (b) based on ISDN interface, and (c) upon a dual-mode interface.

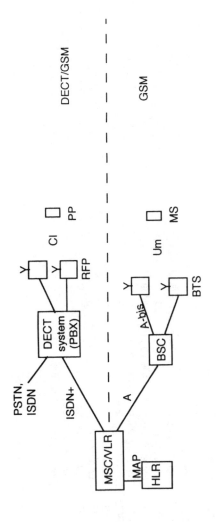

Notes:
A   The GSM MSC-BSC interface
A-bis   The GSM BSC-BTS interface
BSC   Base Station Controller
BTS   Base Transceiver Station
Ci   Common Interface
DECT   Digital European Cordless Telecommunications
GSM   Global System for Mobile communications
HLR   Home Location Register
MS   Mobile Station
PP   Portable Part
Um   GSM air interface
VLR   Visitor Location Register

(b)

Figure 10.12 (continued).

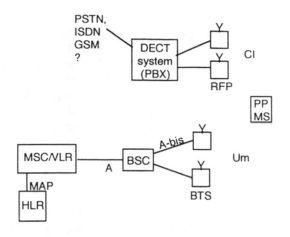

Notes:
A   The GSM MSC-BSC interface
A-bis   The GSM BSC-BTS interface
BSC   Base Station Controller
BTS   Base Transceiver Station
Ci   Common Interface
DECT   Digital European Cordless Telecommunications
HLR   Home Location Register
MS   Mobile Station
PP   Portable Part
RFP   Radio Fixed Part
Um   GSM air interface
VLR   Visitor Location Register

(c)

Figure 10.12 (continued).

unit that can operate both with DECT as well as with GSM systems. The DECT network may optionally be attached to GSM, but the DECT users may also be supported with local identities and services. The system can also be connected to another backbone network, for example, PSTN or ISDN for the provision of services.

10.5 EUROPEAN DCS 1800 SYSTEM

GSM redefined for 1800-MHz operation is the basis for PCN as it defines the complete digital communications system. The new standard was developed within ETSI's GSM technical committee, and phase 1 was approved in February 1991. In this section we define the salient requirements for DCS 1800 system

10.5.1 The Frequency Band of Operation and Channel Arrangement

The mobile transmit frequency range is from 1710 MHz to 1780 MHz, while the base transmit frequency range is from 1805 MHz to 1880 MHz. For each of these bands a guard band of 200 kHz is required at the bottom of both the bands in order to ensure that the radiation from this system will not fall into some other nearby operating system. Channel carrier frequencies are determined by the formulas

$$F_1(n) = 1710.2 + 0.2^*(n - 512) \text{ MHz} \qquad 512 \leq n \leq 885 \qquad (10.4)$$

$$F_u(n) = F_1(n) + 95 \text{ MHz}$$

The value of n is called the *Absolute Radio Frequency Channel Number* (ARFCN). In order to avoid interference from other operating systems, channels 512 and 885 are generally not used.

10.5.2 Mobile and BS Power Requirements

The MS maximum output power and power steps are shown in Table 10.5(a,b). There are two power classes with a normal tolerance level of ±2 dB, while the worst case tolerance is ±2.5 dB. Power step size is either ±2 dB, ±3 dB, ±4 dB, and ±5 dB starting at a peak level for the two classes of mobile.

The BTS transmitter maximum power per carrier measured at the combined input is shown in Table 10.6. The tolerance should be such that the power level can be decreased in nominal steps of 2 dB with an accuracy of ±1 dB. This will allow a fine adjustment of the coverage by the network.

10.5.3 RF Modulation Spectrum

The output RF modulation spectrum is shown in Table 10.7. The specification applies over the entire band and up to 2 MHz on either side of the band. The measurement bandwidth is 30 kHz up to 1800 MHz and 100 kHz beyond. The reference level in the table is the maximum level with respect to 30 kHz centered on the carrier frequency. For power levels between these specified levels, a linear interpolation can be used. At the nominal bandwidth of 200 kHz the spectrum must have decayed by 30 dB with respect to carrier. At the frequency offset of 400 kHz the specified value is at −60 dBc.

10.5.4 Power Ramping

The switching transients can be reduced by ramping the output power up and down when transmitting a burst. The transmitted information must not be affected by the

Table 10.5

(a) Mobile Power Class and Their Tolerance

Power Class	Maximum Peak Power (W)	Tolerance Normal (dB)	Tolerance Extreme (dB)
1	1	±2	±2.5
2	0.25	±2	±2.5

(b) Power Steps and Their Tolerance

Power Control Level	Peak Power (dBm)	Tolerance Normal (dB)	Tolerance Extreme (dB)
0	30	±2	±2.5
1	28	±3	±4
2	26	±3	±4
3	24	±3	±4
4	22	±3	±4
5	20	±3	±4
6	18	±3	±4
7	16	±3	±4
8	14	±3	±4
9	12	±4	±5
10	10	±4	±5
11	8	±4	±5
12	6	±4	±5
13	4	±4	±5

Table 10.6

CS Transmit Level and Its Tolerance

Transmit Power Class	Peak Power (W)	Tolerance (dB)
1	20	−0, 3
2	10	−0, 3
3	5	−0, 3
4	2.5	−0, 3

process of power ramping, which is performed at the beginning and end of the timeslot using the mask shown in Figure 10.13(a,b). The timeslot in the figure corresponds to a duration of 156.25 μsec for normal burst, frequency correction burst, synchronization, and dummy burst. A GP of 8.25 bursts are inserted between bursts. The remaining 148 bits form the active part of the bursts. In the access burst the GP after the burst is 68.25-bits long, leaving an active part of 88 bits. The useful part of the burst in all cases is one bit period shorter than the active part, and it

Table 10.7
Output RF Spectrum Due to Modulation

Transmit Power Level (dBm)	Measurement Bandwidth at Specified Frequency Offset (kHz)						
		30 kHz				100 kHz	
	100	200	250	400	600 to 1800	>1800 to 6000	>6000
≥43	+0.5	−30	−33	−60	−70	−70	−80
41	+0.5	−30	−33	−60	−70	−70	−80
39	+0.5	−30	−33	−60	−70	−70	−80
37	+0.5	−30	−33	−60	−70	−68	−80 BTS
35	+0.5	−30	−33	−60	−70	−66	−80
≤33	+0.5	−30	−33	−60	−70	−65	−80
30	+0.5	−30	−33	−60	−60	−65	−75
28	+0.5	−30	−33	−60	−60	−63	−73
26	+0.5	−30	−33	−60	−60	−61	−71 MS
≤24	+0.5	−30	−33	−60	−60	−59	−69

begins half way through the first bit period. During the useful part of the burst, the amplitude of the modulated RF spectrum must remain approximately constant. As seen from Figure 10.13(a,b), approximately 74 dB of power ramp occurs during a 28-μs interval, which corresponds to 7.6 bit intervals. After ramping up its RF power, a MS has 542.8 μs to transmit information for normal, frequency, and time synchronization and dummy bursts and 321.2 μs for access burst.

10.6 CONCLUSIONS

Microcellular-based CT2, DECT, and DCS 1800 systems are the present generation systems on which the future PCS systems will be based. For ordinary and fixed telephone systems, DECT-based WLL is an economic solution for medium- and high-density applications, with the added attraction of enabling subscriber mobility. For business subscribers, cordless PBX will become the industry standard in the next few years. Finally, for cellular operators, DECT offers a chance to expand capacity beyond limits of current digital cellular standards and to continue keeping pace with the booming demand for mobile communications.

Problems

10.1 Narrate the main objectives of second-generation cellular systems. How do they differ from the first-generation systems?

10.2 Tabulate the reasons for the expected explosive growth of cellular system.

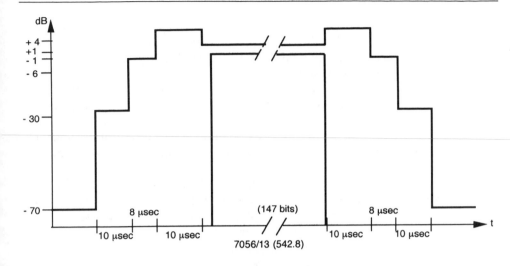

(a)

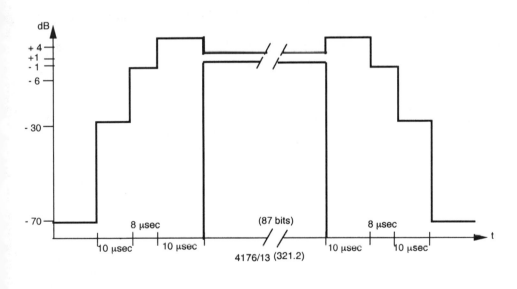

(b)

Figure 10.13 Time mask for (a) normal duration bursts (NB, FB, DB, SB) and (b) access bursts (AB).

10.3 Why is cellular usage higher in newer areas where the cellular has recently been introduced versus the old established area? Why shouldn't the rate of usage be the other way round?

10.4 Give reasons to justify that the predicted fivefold increase in subscriber growth will only double the revenue by the turn of the century. What conclusions can you draw from this?

10.5 What does one mean by convergence of cellular with transportation, financial, and entertainment industries? Name some other industries where the convergence will also apply.

10.6 Narrate different areas of cordless telephone 2 (CT2) and DECT applications. Justify that they are in actuality microcellular systems. Describe in a tabular form some advantages and disadvantages of CT2 and DECT systems.

10.7 Narrate a set of operational requirements for the complete mobility of CT2.

10.8 What is the basic concept of seamless handover?

10.9 From the frame format for CT2 given in Figure 10.8, compute the channel data rate. What is the frame efficiency of the system?

10.10 From the frame format of the DECT system given in Figure 10.10, compute the channel and the control channel data rates.

10.11 From the user's point of view, what is the difference between a CT2 service and DECT service?

10.12 Compare the speech coding rate, transmission power, access method, and the frequency of operation for CT2 and DECT systems.

References

[1] Mehrotra, A., *Cellular Radio Performance Engineering,* Norwood, MA: Artech House, 1994.

[2] Kandiyoor, Suresh, "DECT: Meeting Needs and Creating Opportunities for Public Network Operations," *International Telecommunication Conf.,* Session 1.3, 1995.

[3] Hjern, Magnus, "DECT/GSM Interworking," *International Telecommunication Conf.,* Session 6.2, 1995.

CHAPTER 11
▼▼▼

FUTURE OF PCS SYSTEMS

11.1 INTRODUCTION

Recent developments in the domains of service engineering, network management, and intelligent networks and the rapid advancement of component technology have prompted an increase in the pressure to integrate fixed and mobile networks. There is also a great desire to have one multiapplication, handheld, terminal. Increasing scope and sophistication of the multimedia services is now expected by the customer. For all these reasons, demand performance has advanced beyond the capability of second-generation cellular technology. The very success of second-generation equipment in becoming more cost effective and increasingly cost attractive raises the significant prospect that they will reach an early capacity and service saturation. These pressures will force the emergence of third-generation systems.

European industries, research laboratories, and universities have joined hands to fulfill the ultimate goal of today's communication engineers—that is, to provide communication services from any person to any person at any place and at any time without any form of delay through any medium using one pocket-sized unit at minimum cost with acceptable quality and security via a PIN. With the current rapid advances in communication technology, it can be stated without qualification that the objective of today's communications engineers is to achieve a *Future Wireless Personal Communications System* (FWPCS), which was yesterday's myth (before

1970) and will be tomorrow's reality (beyond 2000). Recognizing the strategic importance of R&D telecommunications as a major driving force of socioeconomic progress, the European Community launched the RACE program in 1988. Its follow-up, launched in 1995 and referred to as ACTS, was conceived as a program that reflects a change of emphasis toward a market-driven R&D program. In the context of the ACTS program, R&D in mobile and personal communications services and networks is called upon to play an essential role. The specific issues that are now addressed relate to the service and capacity evolution of the second-generation mobile communication systems and networks and aim at the development of two third-generation technological platforms, namely, *Universal Mobile Telecommunications System* (UMTS) and *Mobile Broadband Systems/Wireless Customer Premises Networks* (MBS/WCPN), which will respond to the needs of providing seamless services across various radio environments and operational conditions as well as the extension toward providing broadband multimedia services. The full richness of mobile communications depends on multimedia presentation, which is at present beyond the performance of current wireless communications systems. Hence the objective is to progressively extend mobile communications to include multimedia and high-performance capabilities and permit their natural integration and interworking with future wired networks.

The aim is to specify a third-generation mobile communications that features compact, light-weight terminals, a quality and range of services compatible with those of today's fixed networks, high infrastructure capacity, and coverage of all environments system by the end of the decade for both consumer and business users, that is, at home, in the office, in the car, and in the street.

In the United States, PCS will be implemented with a wide variety of technologies around 1900 MHz and 800 MHz. Wide-area mobility will be served both by TDMA as well as CDMA. Low-mobility systems will be served by many proprietary radio access technologies.

In view of this, the objective of this chapter is to describe the role of RACE, ACTS, and UMTS, which are discussed in Section 11.2. Section 11.2.3 provides the present solution approaches to various PCS problems. The role of the different key players in the PCS system and their interactivity is discussed in Section 11.2.4. The U.S. effort for PCN system and the recent spectrum option by FCC is discussed in Section 11.3. Lastly, main elements of various mobile satellites in design today are discussed in Section 11.4. Section 11.5 provides the conclusions of the chapter.

11.2 PCS SYSTEMS OF THE FUTURE

The UMTS and FPLMTS are third-generation mobile telecommunications systems that aim to unify the present diverse systems into a seamless radio infrastructure capable of offering a wide range of services around the year 2000. UMTS and

FPLMTS will operate in many different operating environments including the integration of fixed and mobile networks. They will deliver a much wider range of services than we see today, with the quality we have come to expect from the fixed telecommunications networks. FPLMTS will offer highly cost-effective radio access networks, an inherently modular nature allowing networks to be configured with the selected capabilities required to provide a fully personalized mobile network [1–6].

A small, low-cost, light-weight, and convenient pocket-sized communicator will provide mass market access with a range of terminal types supporting more advanced services. FPLMTS will support a number of different operating environments, ranging from very high capacity indoor picocells to large outdoor Macrocells and satellite global coverage as shown in Figure 11.1. Close integration between the satellite and terrestrial components of FPLMTS will facilitate the initial deployment of service via satellite where there is little or no existing fixed infrastructure. The technological capabilities of the third-generation system will be measured in terms of terminal mobility and data rate. Deployment of the terrestrial infrastructure will then follow later in some areas depending on population density.

The vision of FPLMTS and UMTS dates back to the mid-1980s when the telecommunications systems' increasingly diverse range of mobile and service offerings led to parallel studies within the ITU and RACE communities in an attempt to integrate this variety into a single unit. Thus, the separate services of paging, cordless, wireless PBX, cellular, and mobile data would be seamlessly combined into a single user-friendly communications capability that would also offer advanced high data

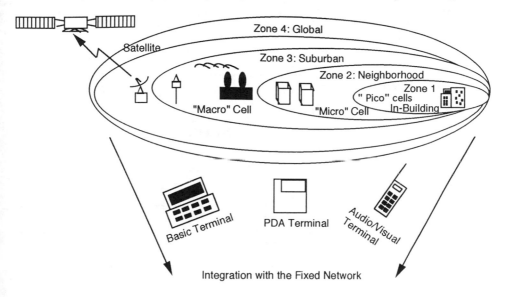

Figure 11.1 System environments.

rate services including multimedia. In effect, FPLMTS would also offer the mobile element of the *broadband integrated services digital network* (B-ISDN).

Given this, a likely outcome toward the year 2000 is as follows. In Europe, GSM/DCS 1800 will be the dominant cellular network standards in which operators and manufacturers will have invested billions of *European Common Currency* (ECUs). In the United Kingdom alone, four companies using 900-MHz and 1800-MHz spectrum will support in excess of 12 million customers. In the United States, PCS will be implemented with a variety of technologies around 1900 MHz and 900 MHz. Wide-area mobility systems will be served by both CDMA and DCS 1900.

European as well as U.S. operators will no longer be constrained to a particular technology. This will allow U.S. and European operators to work vigorously toward FPLMTS/UMTS. The very success of second-generation equipment in becoming more cost effective and increasingly cost attractive raises the significant prospect that they will reach an early capacity and service saturation throughout the world. These pressures will lead to the emergence of third-generation systems. Figure 11.2 provides different systems of first and second generation whose combination will

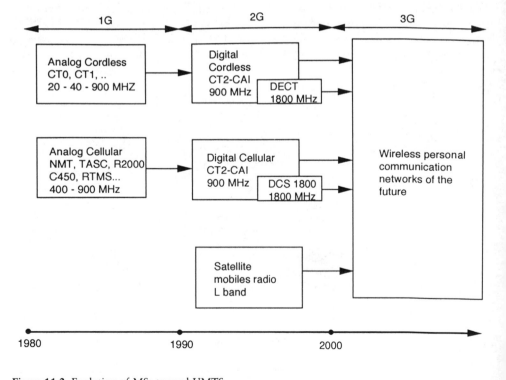

Figure 11.2 Evolution of MSs toward UMTS.

provide the seed for wireless personal communication networks of the future. Evolutionary systems include developments at both 900 MHz and 1800 MHz.

At the global level, the work carried out within the ITU has been instrumental in the definition of FPLMTS/IMT2000. In Europe, R&D on third-generation technology, commonly referred to as the UMTS and *Mobile Broadband Systems* (MBS), falls under the European community RACE and ACTS Programs. The responsibility for coordinating European technical specifications for UMTS belongs, however, to the ETSI, a Special Mobile Group.

All of these technologies will, of course, continue to develop and strengthen their individual markets for personal communications and allow operators, service providers, and manufacturers to achieve important economies-of-scale through their wider application and further refinement. However, there are several factors spurring rapid demand for performance advances beyond the capability of second-generation technology.

- Rapid advances of component technology;
- The pressure to integrate fixed and mobile networks;
- The developments in the fields of service engineering, network management, and intelligent networks;
- The desire to have one multiapplication, handheld terminal.

Third-generation mobile communications systems aim to integrate all the different services of second-generation systems, provide competitive service provision of over 50% of the world population, and cover a much wider range of broadband services (such as voice, data, video, and multimedia) consistent and compatible with the technology developments within the fixed telecommunications networks. The progressive migration from second generation toward third will start at the turn of the century, which will encourage new customers while ensuring services to the existing users.

Figure 11.3 illustrates the diversity of radio coverage areas that will be required from local (private or business) to continental or even planetwide. The coverage will also be tailored as a part of the subscriber agreement. For example, personal mobility could be supported by giving each user his own mobile terminal or by using a data storage device, such as a SIM card that can be inserted into a variety of terminals to provide a "personal" number service. Figure 11.4 portrays the technological capabilities of different generations of cellular systems, including third-generation UMTS/FPLMTS. In defining the technologies for mass communications in the year 2000 or beyond, the market must account for personal communications from a mobile and wireless point of view. Though the evolution of second-generation systems into "2+" will open up some new value-added services, the third-generation system is required to provide a "wireline quality with wireless freedom" globally to

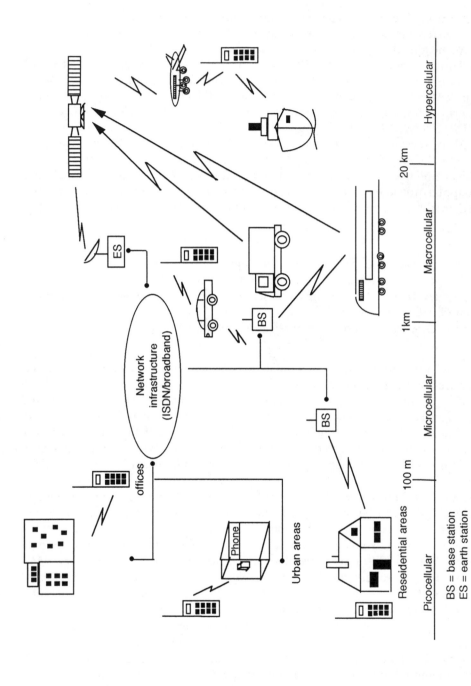

Figure 11.3 Coverage of the UMTS.

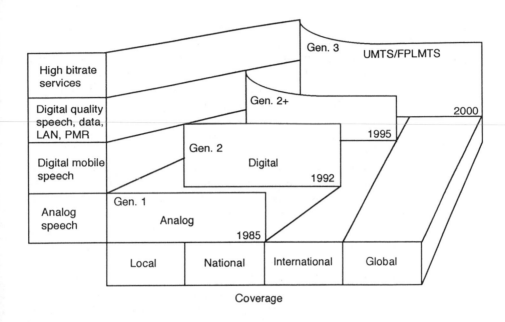

Figure 11.4 Service and coverage capabilities of mobile telecommunication system generations.

a large population, a large number of global roamers, and a wide range of novel services up to high bit rates.

11.2.1 European Role

In Europe, R&D on UMTS falls under the European community RACE program. Responsibility for coordinating European technical specifications for UMTS belongs, however, to the ETSI Special Mobile Group. Its main aim is the introduction throughout Europe of an integrated broadband telecommunication network that is capable of meeting the traffic requirements of the early 21st century. The RACE radio mobile project concerns the mobile radio part of and is paving the way for standardization of the UMTS system. The RACE program has two phases. RACE I concentrated on system engineering, outline specifications, and key technologies; RACE II is concentrating on system integration and the prototyping of new services and applications. Both phases involve a set of related but autonomously organized and managed projects, and each project is conducted by a consortium of independent organizations, including telecommunications operators and equipment manufacturers, research institutes, and universities. Clusters of projects are organized into project lines, which provide a focus for collaboration around specific objectives. The first phase of the RACE program, from 1988 to 1992, involved many European operators, manufacturers, and universities. The many research projects of this first phase contributed to a first definition of the UMTS system [5–12].

Just as the ETSI was beginning its work on standardizing UMTS, the European Commission launched RACE II, a second phase of the program to run until the end 1994 or possibly the end of 1995. This encompasses several radio mobile projects. The objective of these various projects was to turn scientific knowledge into empirically proven concepts and to contribute effectively to the process of standadization. The various projects covering different areas of technology are listed in Table 11.1 and described in the following paragraphs.

The MONET project is concerned with the network aspects of UMTS and is reviewing the various access methods against a common benchmark. The ATDMA and CODIT projects are probing the possibilities of the TDMA and CDMA access methods for UMTS so that they can be compared and tradeoffs can be made. The objective of project MAVT is to identify new video and audio coding algorithms for the transmission of high-quality audio, moving, and still video in a mobile environment. The MBS project is looking at the more futuristic use of the 60-GHz frequency band for broadband transmissions. Project TSUNAMI is concerned with developments of adaptive "smart" antenna components, such as phased array, beamforming network, and signal processing algorithms. The project GIRAFE investigates microelectronics for the front end of the mobile transceiver. The PLATON project is concerned with the fixed design and development of software tools for UMTS network engineering.

ACTS' aim is to consolidate European strength in digital broadband communications across a diverse and competitive communications environment. The areas of activities of ACTS are related to the following.

- Interactive digital multimedia systems and services;
- Photonic technologies;
- High-speed networking;
- Intelligence in networks and service engineering;
- Quality, security, and safety of communication service and systems.

Table 11.1
Different Areas of Technology Pursued in GSM Project

Study Topic	Project
Network principles (Fixed network issues)	MONET
Radio access study (TDMA and CDMA issues)	ATDMA and CODIT
Audio and video encoding	MAVT
Broadband mobile	MBS
Smart antennas	TSUNAMI
Receiver front end	GIRAFE
Network planning tools	PLATON

11.2.2 Key Objectives of UMTS/FPLMTS

The objective of both UMTS in Europe and FPLMTS is to develop a multifunction, multiservice, multiapplication, digital mobile system that will provide personal communications up to 2 Mbps, support universal roaming, and offer broadband multimedia services. UMTS/FPLMTS will be designed such that they have a terrestrial and a satellite component with a suitable degree of commonality between them, including the radio interfaces. The system will have following features and objectives.

- Use 2-GHz band allocated by WARC 92;
- Make efficient and economical use of the radio spectrum consistent with providing service at acceptable cost;
- High traffic capacity;
- Offer universal mobility, that is, uninterrupted coverage irrespective of location, journey, and service required, which will require a gradual changeover from multiplicity of systems to unification of radio interface and mobility management functions;
- Support a wide range of terminals, including the cellular telephone, the personal communicator, and the portable microcomputer;
- Provide high-quality and integrity, which is comparable to the fixed network;
- Use of a small pocket terminal worldwide;
- The provision of service by more than one network in any coverage area;
- Provide an open architecture that will facilitate the easy introduction of advances in technology and different applications;
- Provide a modular structure that will allow the system to start from as small and simple a configuration as possible and grow as required with time.

The architecture of such a UMTS/FPLMTS is shown in Figure 11.5. In order to support a multifunction, multiservice, multiapplication digital system, a broadband multimedia service with capacity up to 2 Mbps will be required. At the time UMTS/FPLMTS service is available, ATM will be an established transmission technique. Hence, the UMTS/FPLMTS environment must support ATM-cell transmission through to the user's terminal. This compatibility will enable service providers to offer a homogeneous network where users can receive variable bit rate services regardless of their access media. As shown in the figure, the UMTS network would be a mixture of cellular public networks, satellite networks, business networks (wireless PABX), domestic "networks," and mobile vehicle-borne networks. These various subnetworks could be interconnected using an ISDN infrastructure that will be initially narrowband and subsequently broadband. As shown in the figure, the subscriber will be able to access both narrowband ISDN services as well as public UMTS/FPLMTS networks [8–16].

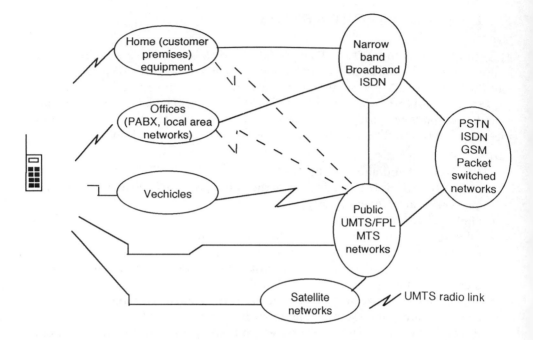

Figure 11.5 Architecture of the UMTS.

11.2.3 Solution Approach for Future Systems

In this section we will discuss some solution approach to the future PCS systems. The areas discussed are as follows.

- SIM roaming;
- Intersystem roaming;
- Future mobile service and database requirements;
- Database requirements;
- Interface standardization for UMTS;
- Future pricing of services;
- Future mobile terminals.

11.2.3.1 SIM Roaming

One solution of roaming can be solved by having a SIM card that works on multiple standards. Such a card will be highly desirable for future use and will allow the subscriber not to carry the handset from place to place. Wherever he goes, he can rent a cellular telephone and it will work with his SIM card [8,9].

In this case the SIM card will have data according to several standards. For example, such a dual-mode SIM card can operate on a GSM system when inserted in a GSM terminal; that is, a GSM terminal would access the GSM directory on the SIM. Similarly, a PDC terminal would access the PDC directory on the SIM when the SIM is inserted in a PDC terminal. In the future this concept can be expanded and one SIM card can work on many standards.

This solution has many advantages. First, it allows for usage of already existing terminals, thus not requiring new terminals to be developed for end users to be able to roam between different standards. Second, the usage of such a SIM card can enable the mobile subscriber to have the necessary data for all cellular standards on one SIM card, enabling a mobile subscriber to choose different systems based on his choice and the availability of the system. This multisystem SIM card has many advantages. First, it allows for usage of already existing terminals, thus not requiring new terminals to be developed in order for the end users to be able to roam between different standards. Second, using such a SIM card can enable the mobile subscriber to have the necessary data for all cellular standards on one and the same SIM card, which will enable a mobile subscriber to choose different systems based on their choice and the availability of the system. Figure 11.6 shows such a SIM card working on systems A and B.

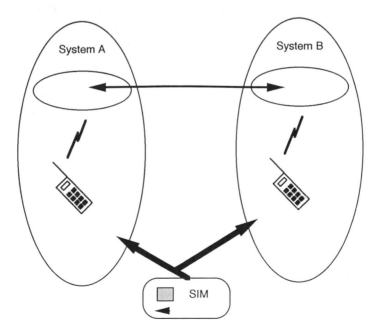

Figure 11.6 SIM roaming.

11.2.3.2 *Intersystem Roaming*

Currently a number of cellular standards exist, each of them supporting seamless roaming and call delivery within each standard. This includes air interface procedures, whereby a MS can access a visited network in the same way as it would access its home network, and network interfaces, allowing networks run by different operators to be interconnected using commonly agreed signaling procedures. One example of this is GSM, which supports international roaming. Other examples are the AMPS standard, which today supports roaming within North America, and PDC, which supports roaming within Japan. In most of the aforementioned cellular standards, the MAP is the signaling protocol that handles the roaming signaling in the respective network. In order for a standard to provide roaming, it is equally important that the standard includes a fully specified air interface as well as the necessary network interface(s). However, all the standards mentioned here contain their own specific air interface specifications as well as their specific MAP protocols, which does not allow roaming between two different systems. Thus it is not possible to roam between GSM and AMPS, nor it is possible to roam between GSM and PDC. The scheme presented here will allow, at a future date, roaming between two different systems, each signaling within its own system and in its own MAP protocol.

The scheme shown in Figure 11.7 introduces new modes that have the purpose of providing interworking between two different systems. The nodes that perform interworking between two different cellular standards are the *Interworking Location Register* (ILR) and the *Interworking Mobile Switching Center* (IMSC). ILR works

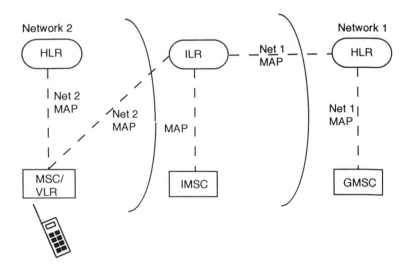

Figure 11.7 General network architecture.

as follows. In the view of the HLR in the home network (Network 1 in the figure), the ILR is seen as a VLR. In view of the MSC/VLR in the visited network, the ILR is seen as an HLR. This implies that the ILR can ensure that the two interconnected standards can cooperate. The following example provides the basic scheme for international roaming.

Example

Let us consider an example of two independent systems having ILR/IMSC interface between them. Let us also assume that a subscriber belonging to the GSM system switches on his power. The location updating procedure from the GSM terminal will be handled by the MSC/VLR in the GSM network. The GSM MSC/VLR will communicate with the ILR using the normal GSM MAP protocol, assuming the ILR to be the GSM HLR described. The ILR will, when receiving the GSM MAP location update procedure from the GSM MSC/VLR, perform a PDC MAP location registration procedure toward the PDC HLR to update the location data for the subscriber in the PDC HLR. When the GSM MAP procedure is completed between the GSM MSC/VLR and the ILR, the GSM location is stored in the GSM MSC/VLR and the ILR in accordance with the GSM standard. Now a call from a fixed subscriber to a PDC subscriber, roaming in GSM, will work as follows. The GMSC in the PDC network will, when receiving the call from the fixed network, ask the PDC HLR to terminate call-routing information. The PDC HLR will thus provide the PDC GMSC with routing information, which enables the PDC GMSC to route the call to the IMSC. The ILR will, when asked for GSM routing information, provide a GSM roaming number to the IMSC. The IMSC then forwards the call to the GSM MSC/VLR using the received GSM roaming number and the call is completed in the GSM MSC/VLR as a normal GSM mobile-terminated call. The main reason for having IMSC is that the functional interface between the GMSC and the MSC/VLR differs in the GSM and PDC systems. To enable routing a call originally set up to a PDC subscriber when the subscriber is roaming in a GSM network, the IMSC has the role of interworking between PDC and GSM, specifically, handling call routing to roaming subscribers [10–14]. This scheme is shown in Figure 11.8.

The approach described will allow subscribers to roam internationally by making use of the multimode terminals (multimode terminals will allow a subscriber to access different systems) as is shown in Figure 11.9. This will allow a mobile to have a worldwide service just by having a single multimode terminal.

11.2.3.3 *Future Mobile Service and Database Requirements*

As discussed, future UMTS will consist of a wide variety of interconnected networks and facilities. For mobile-originated transactions, let us assume that a terminal can

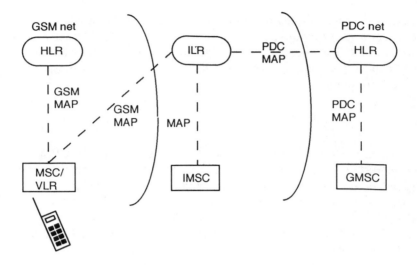

Figure 11.8 Network architecture for roaming from PDC to GSM.

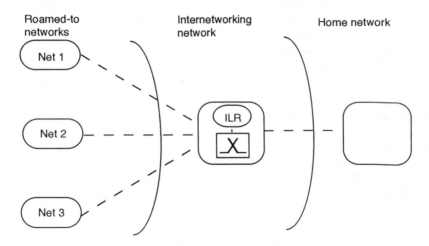

Figure 11.9 Multimode terminal.

access the base station, which responds to a request for service, and can then proceed to negotiate conditions and price. Since payment can be by electronic cash over the air, a user does not have to subscribe to, or belong to, the network they use. The BS, or the network to which it is attached, will have sufficient intelligence to carry out an electronic cash or credit transaction. This is likely to be a commonplace function in the near future for many types of remote shopping. Since cash is transferred electronically, the subscriber does not have to belong to the specific network from which he wants the service. We then have the problem of locating the mobile terminal and establishing the connection via one or more competing networks. Let us provide an example of how the mobile can potentially be located in UMTS/FPLMTS of the future.

Suppose I urgently wish to contact you. I call your personal number. It is routed to your home because after 6 p.m. you are most often at home. The bell in your home rings. After five seconds with no response, I am asked if I wish to pay for a search or be connected to an answering machine or voice-mail service. Because I need to speak to you urgently, I ask for a "Grade I search" (assume Grade I is a worldwide search, no expense spared, cost up to my credit limit; lesser grades possibly can be activated also with a limitation of an area search such as one country or few countries). The telecom operator to whom I am connected may now contact a Location Agency. The Location Agency will consult a data base to see what it knows about you. Based on knowledge of your habits and any recent calls to or from you, the search may then proceed in one or more of the following ways.

You might be working late: try your office. Ask all PCN and cellular operators within the country "can you connect to X, and if so, what will be the charge." Ask local pager operators to page you, but suppose the Location Agency database knows its last contact with you was an hour before from XYZ hotel, in New York. A call is made to your personal number via the hotel switchboard. No response is obtained from either your room extension or the hotel's cordless telephone system. Therefore, a New York radio-paging operator is asked to page you. A response comes back from your pocket phone.

The Location Agency computer will then calculate the cost of a call, using its knowledge of networks and organizations to determine the lowest cost options, and then call me back and offer to connect me to you at a cost of say $2 per minute for voice or $5 per megabyte of data transmission. In the next section we will discuss the associated database requirements based on this complex problem.

11.2.3.4 Database Requirements

In order to locate an individual, a comprehensive database must contain all relevant information for the user. The database might also keep success statistics for some or all of the "usual places" for an individual. For instance, it could generate and keep hour-by-hour probabilities. Other information in the database might include to which services the user subscribes or authorizes use, for example. The usual information about the subscriber that has to be stored in the database is shown in Table 11.2.

It should be noted that there will possibly be many Location Agencies maintaining the data base. Each individual agency will only have to maintain a fraction

Table 11.2
Location Agency Data Base Entries

Possible Entries for the Subscriber in the Data Base

Subscriber name
Home number
Office number
Personal number
Car phone number
Usual location: Home except:

Mon to Fri	08.00 to 17.00	Office
Sat	10.00 to 12.00	Gaithersburg Shopping Center, MD
Sun	09.00 to 13.00	Golf Club at Montgomery Village, Gaithersburg

Locations put in by user as a form of "follow me." Last known location (e.g., 04/30/95 from 12.00 to 16.00 hrs (Sanfrancisco))

of the world capacity. Apparently the activity (load) factor has to be estimated. Data base information sources can be from telephone companies, mobile network operators, national numbering authorities, and directly the users for whom the data base has been created. We now provide an example of a rough estimate for the total database storage requirement and the number of location requests/sec.

Example

Assume the world population to be about 6 billion and only 20% will use the service. Also, assume that on the average there are four transactions/day/person, and 10% of those will require locating the destination terminal. Then, the number of location processes/sec = $6 \times 10^9 \times 0.2 \times 4 \times 0.1 = 480 \times 10^6$ per day = 5,500 queries/sec. Since this will vary according to the time of the day, we can assume a safe factor of two, which will bring us to about 11,000 queries/sec. If each user requires about 300 bytes on the average, then the total storage requirement for 6×10^9 people is $300 \times 0.2 \times 6 \times 10^9 = 3,600$ Gbytes. Assuming once again that only 20% of the world population has the service and each entry is updated on the average four times a day, the total number of transactions/day = $0.2 \times 6 \times 10^9 \times 4 = 2.4 \times 10^9$. Assuming that the peak load is double, the average total peak load becomes 4.8×10^9.

11.2.3.5 Interface Sandardization for UMTS

The third-generation mobile communications system will require standardization at various interfaces of the mobile network. This will facilitate interconnections between networks and operability over the network/terminal interfaces and enable free movement of the mobile. All user interfaces (IF_1), radio interfaces (IF_3), network interfaces (IF_4), and service provision interface (IF_5) are to be standardized. Service providers, users, regulators, operators, and equipment manufacturers can, for their advantage, agree to standardize voluntary functions or interfaces to promote cost-efficient implementation of networks or services. Thus, various internal interfaces designated as IF_v will also be standardized. Figure 11.10 shows the standardized interfaces for future systems.

11.2.3.6 Future Pricing of the Services

Tradeoffs will be made on future cost of the services between the cost of the handset, monthly rental, and call charges. The general practice is to heavily discount the handset to a minimum entry cost for the subscriber. However, the call charges and the monthly rental has to be raised in order to protect high dealer commission and operator's subsidy. Some operators of PCS service will have lower monthly and call charges and also include free off-peak local calls to attract customers. But those users have to pay higher handset cost initially.

11.2.3.7 Future Mobile Terminals

Over a single day a mobile radio user can face different situations based on his environment and the services requested. At home or in other confined private areas, the customer will use a private indoor access point with a low-power transmitter. On the way to his office, he will use the public cellular system and be able to maintain communication where he might be using a medium- to high-power transmitter. At the office, they communicate over a wireless PBX with a low-power terminal.

In order to satisfy the stated requirement, the basic interoperability between the system and the associated terminals must be assured. This in turn demands terminals with multimode characteristics. The multimode terminal will be completely transparent to the user. The modes of operation of such a terminal have to be completely independent of the access technology. For the cases where multiple types of services are available, the terminal should also be able to select the best system. The user's procedure to access the system should remain the same for each system. The major technological areas of design are shown in Table 11.3.

Based on Table 11.3 there are three main elements of multimode terminals: antenna, RF platform, and DSP and microcontrollers. For antenna a compromise has to be made between its length and efficiency. From a marketing point of view, it is better to have a smaller size or even an integrated antenna in the casing; for efficiency, it is better to have a longer antenna. In operational mode, the antenna efficiency suffers also from a nonoptimized orientation with respect to the direction

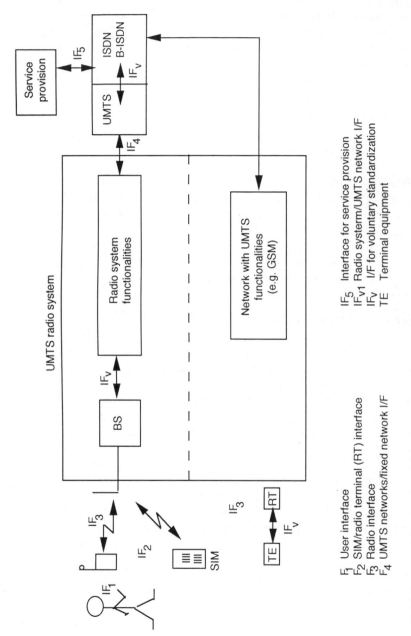

Figure 11.10 Interoperability interfaces of third-generation mobile communications system.

IF_1 User interface
IF_2 SIM/radio terminal (RT) interface
IF_3 Radio interface
IF_4 UMTS networks/fixed network I/F

IF_5 Interface for service provision
IF_{v1} Radio systerm/UMTS network I/F
IF_v I/F for voluntary standardization
TE   Terminal equipment

Table 11.3
Technological Design Issues of the Multimode Terminals

Technological Areas	Issues for Multimode Terminal
Antenna	Multiband integration required
Radio	Multiband and multibandwidth
ASICs, DSP, and microcontroller	Reconfigurable functions
Human Machine Interface	Keypad and display should be user-friendly
Algorithms	Should save battery, include power control, best access technology selection
Software and control	Multistandard management and control of multiservice providers

of the BS antenna. In multimode terminals, the antenna should adapt to different frequency bands. It is necessary to achieve tradeoffs between all working frequencies. Whip antennas are used for cellular and cordless. For satellite systems, the antenna could be a quadrifillar or a patch antenna. The major challenge is to make a dual-mode cellular/satellite antenna with a small size and high efficiency.

The complexity of the generic architecture shown in Figure 11.11(a) is dependent upon the number of IF stages. Thus, the level of integration is somewhat limited. However, this is a well-established, proven technology. In this configuration, gain control, noise figure, selectivity, and spurious can be well controlled. There are also skilled engineers with a well-established knowledge base [12–14].

The alternate receiver implementation is by direct conversion to baseband. Here the received signal is amplified and directly down-converted to baseband, as shown in Figure 11.11(b). The advantages are as follows.

- Limited number of RF stages and no IF stages;
- Only one oscillator;
- Optimum power consumption;
- No image filtering required;
- Overall less-complex, reduced component counts.

The major part of signal processing is done at a low frequency and gives direct access to I and Q vectors to ease modulation and demodulation. The drawbacks of this architecture are the possible leakage of the local oscillator and DC offset rejection. The first problem can be solved by proper shielding techniques and the design of the LNA and mixer with good reverse isolation. The second problem can be solved by digital compensation techniques.

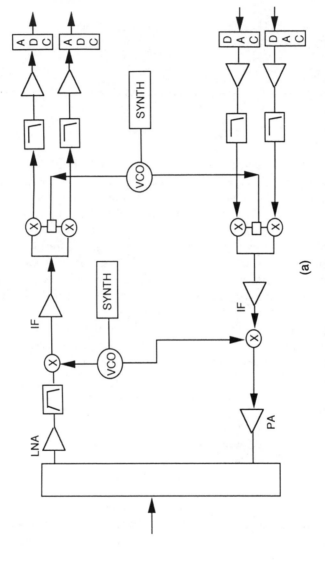

(a)

Figure 11.11 (a) IF and (b) zero IF receiver architectures and (c) dual-mode terminal architecture.

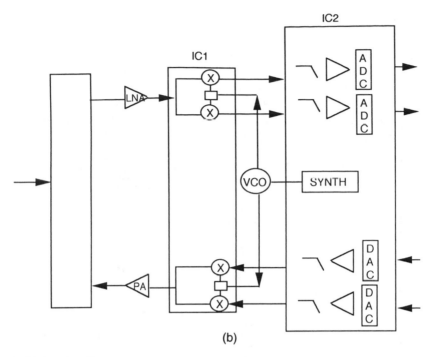

(b)

Figure 11.11 (continued).

ASICs, DSPs, and Microcontrollers

DSPs are software programmable. They are used for speech coding, modulation, and encryption. Today, for PCS applications, ASICs are developed around a DSP core for its raw computing power. This improves drastically the performance of the ASICs. ASICs are used for various purposes. They are used for dedicated tasks such as digital channel filtering, synchronization, equalization, and channel coding/ decoding. For multimode terminals these elementary dedicated machines have to be installed to cope with the different standards, and they are activated according to the selected mode.

Microprocessors are used to implement protocol stacks and system software. They are adapted to perform control-oriented input/output operations as well as the *Human Machine Interface* (HMI) management, display control, and keypad control.

Combining various components provides the layout for a multimode (dual-mode for this discussion) subscriber terminal shown in Figure 11.11(c).

The optimum architecture is based on three major subsystems: primary front end, radio front end, and low or baseband frequency processor. In the primary front

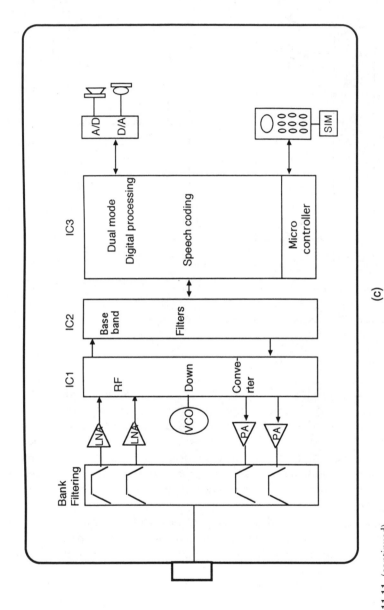

Figure 11.11 (continued).

end, the *Low Noise Amplifier* (LNA), the *Low Power Amplifier* (PA), and a set of duplexers or switches appear to be dependent on the specific system. It should be noted that a unique LNA capable of supporting a worst case noise figure and large dynamic range requirements is needed to adapt as a common element for all systems and is more expensive than duplicating circuitry. The same observation can be made for the PA. Different systems require different solutions; that is, depending of the modulation, the PA could be class A, B, or C. Again the duplication appears to be the cost-effective solution. It should be noted that unique LNA or PA must also adapt to different frequency bands and different frequency bandwidths.

The transmit and receive RF signals are up and down converted by a I and Q converter (IC1). The baseband signal at the output of IC1 is filtered by analogue filters (IC2). These filters perform the channel filtering and anti-aliasing filtering before the analog-to-digital and after digital-to-analog conversion. IC3 could be a VLSI made around a DSP core. For the receive functions, the DSP will perform channel estimation, equalization, demodulation, decoding deinterleaving, and speech decoding. For the transmit functions, DSP will perform speech coding, channel coding, interleaving, and modulation.

A microcontroller is added to IC3 to control the aforementioned process and the HMI interface (for example, keypad, display, and SIM). IC3 will support the software, which implements the protocol stacks, for different modes of operation. The software will control management functions for multimode operation. This architecture is designed to be totally independent of the access mode and can support FDMA, TDMA, and CDMA.

11.2.4 Role of Different Players

The deployment of future mobile communications systems will require a substantial increase in the number of users and a much broader range of services. Also, there will be unrestricted choice of equipment suppliers due to various standardization. While earlier equipment were built to manufacturer's specifications, the new equipment will conform to international standards. Thus, there will be multiple identities whose roles and relationships have to be understood. The key players are as follows.

- Users;
- Network operators;
- Service providers;
- Equipment suppliers;
- Standardizing bodies;
- Regulators.

Figure 11.12 shows the links between the key players for future UMTS. The users' participation is essential as they are the source for dictating requirements. This will be through standardizing bodies and representative associations.

By definition, network operators are the holders of a license reflecting the national regulatory body's specification, which for most countries are the local Post and Telecommunications. Their specifications provide standards and frequencies to which the network equipment must adhere and the necessary interfacing conditions to fixed networks. Where appropriate, they will negotiate bilateral agreements with other operators, particularly in other countries if applicable (Europe is a good example), to allow their subscribers access to the same service outside their country of origin without applying for other subscriptions.

In today's market there is a clear distinction between a service provider and a network operator. The service provider is a direct link to the end user from initial contact to after-sales service while enjoying total freedom in product promotion and pricing. A service provider essentially provides sale of the service, sale of the terminal, and where appropriate, installation (for example, installation of a mobile system in the car).

Relations between equipment suppliers and other players depend on the type of equipment: terminals, infrastructure equipment switching, radio, OMC, subscriber management, and billing systems. As barriers come down and manufacturers' specifications disappear, suppliers will find themselves facing greater competition. The need to increase market share will necessitate international alliances and partnership agreements among equipment providers.

While international cooperation on the use of frequencies is a necessary way of controlling reciprocal interference, harmonization of frequency use within identical

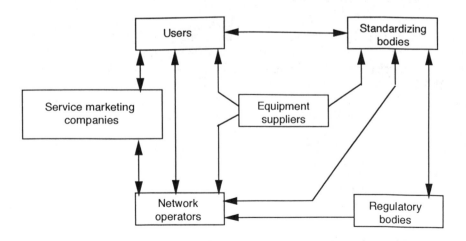

Figure 11.12 Relations between different players.

provides continuous global coverage with dual satellite visibility in some major regions. The MEO provides several advantages beyond low cost: the propagation time delay is reduced to about 68 ms to 83 ms, which is substantially less than GEO delay and is imperceptible in human conversations [16]. However, the combination of Odyssey ground station processing, the vocoder delay, and PSTN terrestrial networks can bring the voice delay up to 200 ms, which is still less than the 270-ms delay encountered in GEO satellite. Also the satellite's orbital position places it between the Inner and Outer Van Allen Radiation Belts, reducing the radiation shielding required on the satellite. In this environment the satellite is designed for 15 years of on-orbit lifetime, while LEO satellites have a typical lifetime of five to seven years. Only nine satellites are required to ensure that one satellite is visible at all times. By adding three satellites, two or three satellites are visible at times, and service can be provided to most of the world's land mass continuously.

11.4.3.1 Technical Requirements

The Odyssey system design is driven by several key requirements of personal telephone users. Other driving requirements are manufacturing ability and reliability of the key system components from the user's point of view. The portable handheld personal telephone, cost effectiveness, and ability to generate revenue are also drivers. These key requirements are as follows.

- Full duplex voice communication with 24-hr availability;
- High-quality voice encoding;
- Communication capability covering all of the global land masses;
- Low cost to end user/low cost handiest;
- Dual-mode compatibility with terrestrial cellular systems;
- Global roaming capability with one phone number;
- Handset size and battery capacity consistent with current technology;
- Alpha numeric paging and data transmission capability;
- Low-power handset that will meet health and safety standards;
- Minimum life cycle capital cost, low number of satellites;
- Low voice time delay compared to GEO satellite;
- No satellite on-board processing or satellite crosslinks;
- Frequency sharing by CDMA;
- Reliable continuity of service, low risk of call dropout;
- Flexible service coverage.

High quality of voice is guaranteed by having a Mean Opinion Score of 3.5 or more. Economical design is important so that the subscriber service charge can be priced in line with terrestrial service charges. Economy is achieved through low

the satellite would employ antennas with 27-dB gain and transmitter peak output powers to 1.0W. Receiver LNAs would have 3.0-dB noise figures.

11.4.3 Odyssey

TRW, Inc. and Teleglobe of Canada, in limited partnership, are developing Odyssey with a constellation size of 12 satellites orbiting in three planes at a higher, *medium Earth orbit* (MEO) of 10,354 km. There are four satellites in each of three orbit planes inclined at 55 degrees as shown in Figure 11.15. The indicated constellation

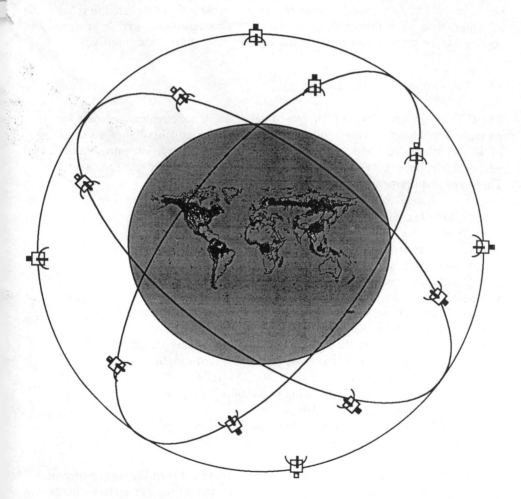

Figure 11.15 Odyssey constellation.

standards is somewhat politically driven. ETSI was created in Europe to fulfill this role. Their members include users, operators, manufacturers and regulatory bodies.

Regulators define the rules governing competition between the various operators by granting national operating licenses, and they are responsible for ensuring that the licensing conditions are observed throughout the life of the system.

11.3 U.S. EFFORT FOR UNIVERSAL PERSONAL COMMUNICATION SERVICES

Over the last few years, the *Federal Communications Commission* (FCC) has created tremendous opportunities for wireless equipment manufacturers with its allocation of frequencies below 3 GHz for future PCS applications. In addition, Congress has radically changed the way the FCC will give out licenses for wireless services in the future. A 1993 law assures that auctions, rather than lotteries or comparative hearings, will be used in the future. That same law will force the federal government to give up 200 MHz of its spectrum to commercial users. Originally, the FCC gave out licenses on a first-come first-serve basis. However, when radio and television broadcasting became successful, many applications were filed for the same channel. In response, the FCC developed a process called *comparative hearing* to decide which applicant is "best qualified" to be the licensee. This process involves weeks of formal trial-type hearings before a judge and is both slow and expensive. The tremendous demand for cellular licenses a decade ago forced the Congress to allow the FCC to use lotteries toward licenses. But that process produced widespread fraud as some promoters took money from the unsuspecting public and never bothered to file license applications, while others stuffed the ballot box with applications using different names but really controlled by the same entity. For the last fifteen years, maybe longer, economists have argued with lawyers about the virtues of auctioning the spectrum. The federal government auctions mineral rights and grazing rights; why not wireless transmission rights? It took the current budget deficit crisis to convince politicians that auctioning the spectrum would be a good policy. According to studies by the Congressional Budget Office, billions of dollars in revenues would flow into the U.S. Treasury if the FCC were to auction off spectrum instead of giving it away in lotteries. So the 1993 Budget Reconciliation Act gives the FCC authority to use competitive bidding for new radio services [14,15].

PCS is defined by the FCC as a family of emerging wireless communications services that will be provided in the 900-MHz and 2-GHz bands. The commission has allocated 3 MHz in the 900-MHz band for narrowband PCS, 120 MHz in the 2-GHz band for broadband PCS, and 20 MHz in the 2-GHz band for unlicensed PCS. The FCC has adopted the Rand McNally *Major Trading Areas* (MTAs) and *Basic Trading Areas* (BTAs) to define PCS service areas. Each of these allocations will likely host the deployment of many new wireless services. Since 1989, more than 100 companies have contributed to the PCS proceedings, opened by the U.S. FCC, and more than 220 companies have received experimental licenses for testing PCS-type services. From those companies, the FCC selected four companies for pioneer's preferences: one for narrowband PCS at 900 MHz (granted June 24, 1993) and three for broadband PCS at 2 GHz (granted December 23, 1993).

Thus, the objective of this section is to elaborate on recommended spectrum for PCS services, major and basic trading areas within US, different types of services (narrowband, wideband, and unlicensed services), and the selected pioneers for PCS services.

11.3.1 Spectrum for PCS Services

The spectrum allocated to broadband licensed PCS is divided into seven blocks, designated A through F. Each block consists of a pair of frequency ranges, one for the transmissions from the handsets to the BS and the other for the reverse journey. Blocks A, B, and C are 30 MHz each (pairs of 15-MHz bands); and D, E, F are 10 MHz each (pairs of 5-MHz bands). The total allocated bandwidth is 120 MHz and is shown in Figure 11.13. The 30-MHz blocks A and B are used in large regions called MTAs, while the other blocks will be used in smaller BTAs [14]. These terms are geographical areas in the United States defined by Rand McNally's *Commercial Atlas and Marketing Guide*. The PCS channel plan is shown in Table 11.4. Broadband PCS is intended for licensed voice, data, or other transmissions, eventually including video and multimedia.

The FCC has also allocated 3 MHz of radio spectrum around 900 MHz to narrowband PCS. This band is primarily used for advanced paging services, which have three subbands of 1 MHz each (901 to 902, 930 to 931, and 940 to 941).

Additionally, a frequency band from 1910 MHz to 1930 MHz (20 MHz) is allocated for unlicensed usage. This is used by low-power, limited-range devices owned and operated by users on their own premises [15].

11.3.2 Trading Areas

The FCC has established 543 PCS service areas in the United States. For their borders, the FCC has adopted the divisions defined in Rand McNally's *1992 Commercial Atlas and Marketing Guide* (123rd edition) within the 48 contiguous states and Hawaii: 47 MTAs and 427 BTAs. To these areas, FCC has added four more areas similar to the major trading areas, and five more areas similar to the smaller BTAs to cover Alaska and the other U.S. territories. According to Rand McNally, a MTA consists of more than one BTA.

Assuming that there are two PCS operators per MTA and five per BTA, this suggests that the United States could have altogether 102 and 2,460 licensees,

Table 11.5 lists the service areas and the available licenses. Licenses are divided on nationwide, regional (into 47 MTAs), and local (into 487 BTAs) bases.

11.3.4 Wideband PCS

In contrast to narrowband PCS, the FCC has allocated 120 MHz of radio spectrum in the 2-GHz band to licensed broadband PCSs capable of transmitting almost any form of communication. This spectrum is allocated in 10-MHz, 20-MHz, and 30-MHz blocks as shown in Table 11.4. In this context, broadband does not necessarily imply video; rather, it refers to the fact that channels at 2 GHz occupy more bandwidth than do those at 900 MHz, and this wider bandwidth will allow transmission at a higher data rate. Compared to narrowband PCS, broadband PCS is distinguishable both in frequency band in which it is offered as well as in the capacity available to the licensee.

A PCS operator will have to provide service to 33% of the population within five years; 66% in seven years; and 90% within 10 years. This may be a problem, as the 2-GHz spectrum is currently used for point-to-point microwave and PCS licensees will have to "buy out" the existing users and move them to higher frequency bands.

11.3.5 Unlicensed PCS Services

The unlicensed PCS allocation in the 2-GHz band is evenly divided between isochronous (principally voice) and asynchronous (data) uses as shown in Figure 11.13. A total of 20 MHz from 1910 MHz to 1930 MHz has been allocated for unlicensed use. Unlicensed PCS is not considered as a mobile service by the FCC. Devices operating in this band will be low-powered and owned and operated by end users within their premises, such as wireless local area networks, wireless private branch

Table 11.5

Channel Allocation for Narrowband Licenses

Licensed Service Area	Available Channels
Nationwide	3-kHz to 50-kHz paired with 12.5 kHz
	5-kHz to 50-kHz paired with 50 kHz
	3 kHz to 50-kHz unpaired
Regional (47 MTAs)	7-kHz to 50-kHz paired with 12.5 kHz
	4-kHz to 50-kHz paired with 50 kHz
	4-kHz to 12.5-kHz unpaired for use by existing paging licenses
Local (487 BTAs)	2-kHz to 50-kHz paired with 12.5 kHz
	4-kHz to 12.5-kHz unpaired for use by existing paging licenses

exchanges (PBXs), and personal data assistants. Because of concerns regarding interference to incumbent microwave users in the 2-GHz band, the deployment of these devices where the device is fully portable and not tied to any identifiable location may be delayed until the successful completion of band-clearing efforts to relocate the microwave users from the unlicensed PCS allocation. The FCC's rules governing unlicensed PCS are now subject to numerous petitions for reconsideration.

11.3.6 Pioneer's Preference

Since 1989, more than 100 companies have contributed to the PCS proceedings opened by the U.S. FCC, and more than 220 companies have received experimental licenses for testing PCS-type services. From those companies, the FCC selected four companies for pioneer's preferences: one for narrowband PCS at 900 MHz (granted June 24,1993) and three for broadband PCS at 2 GHz (granted December 23, 1993).

A pioneer's preference eliminates the need for the company to compete with others in the licensing process. In effect, it guarantees the company a license, which confers exclusive rights to spectrum-based assets of tremendous potential value.

Mobile Telecommunications Technologies Corp. (Mtel), of Jackson, MS, was granted the narrowband PCS pioneer's preference. Mtel had developed a multicarrier modulation technology capable of transmitting a 24-kbps simulcast signal in a single 50-kHz channel, which is 10 times the rates of the current radio paging transmission. This method will allow Mtel to market a new two-way-paging-with-acknowledgment service along with several advanced messaging services.

The broadband PCS proceeding recognized the efforts of *American Personal Communications* (APC), Washington, DC; Cox Enterprises Inc., Atlanta, GA; and Omnipoint Communications Inc., Colorado Springs, CO. Each recipient has been granted a license to cover a specific MTA using the 30 MHz of spectrum block A (1850 MHz to 1865 MHz and 1930 MHz to 1945 MHz).

APC developed a *Frequency Agile Sharing Technology* (FAST) for efficient sharing of the 2-GHz PCS radio spectrum among existing microwave users; its license will cover the area encompassing Washington, DC, and Baltimore, MD.

Cox showed how to use cable-television facilities for backbone communications linking PCS microcells. Its cable microcell integrator equipment receives radio voice channels and then modulates and multiplexes the channels onto the CATV plant. Cox has demonstrated cell-to-cell hand-off with equipment operating at 2 GHz and interacting with a cable TV system. The company's license will cover the Los Angeles and San Diego areas.

Omnipoint produced the first 1850-MHz to 2200-MHz handheld telephone. Their equipment uses the spread spectrum technology of *code division multiple access* (CDMA) with TDD to separate users in time. The Omnipoint devices spread their transmissions across either 5 MHz or 10 MHz of bandwidth, using frequency offsets and codes to separate cells. Omnipoint's license will cover New York City, most of New York State, and northern New Jersey.

11.4 SATELLITE MOBILE COMMUNICATIONS

On January 31, 1995, the U.S. FCC issued licenses for the construction, testing, and U.S. operations of three satellite-based systems, namely, Iridium, Globalstar, and Odyssey, which intend to offer mobile voice and data services on a worldwide basis. The spectrum for these systems are 1.6-GHz uplink and 2.5-GHz downlink and was allocated at the World Administrative Radio Conference in 1992. Current projections call for the first such system to begin service by the middle of 1999. A fourth proposal, Inmarsat-Pnet, is in the advanced stage of planning [16–19].

Yet another mobile satellite system, Orbcomm, will provide data-only services on a global basis, again using spectrum allocated in 1992. Orbcomm received FCC approval in October 1994. The first two satellites were launched in early 1995, service is scheduled to begin in the final quarter of 1996, and the full constellation of satellites is to be completed within a few years. Thus, this section will briefly discuss the technical requirements and architectures of Iridium, Odyssey, Globalstar, and Inmarsat-Pnet satellites.

11.4.1 Potential Markets

The potential for MSS market is enormous and divergent. The most promising targets for MSS will be: (1) emerging markets (developing countries), where they can provide a permanent telecommunications network while infrastructure is being developed; and (2) developed countries like the United States, Japan, and western European countries where they will complement existing cellular service and new PCS coverage. Consequently, we expect to see unprecedented growth in wireless communications, with *Personal Communications Satellite Services* (PCSS) capturing a significant portion of this growth.

Forecasts call for the wireless market to grow to over 200 million subscribers by the year 2000, from today's 40 million plus. Based on past experience, these numbers are probably conservative. One should note that when cellular phones first hit the market back in 1982, forecasters predicted that it would take until the year 2000 to get the first one million sales. However, we can see where we are today in terms of cellular growth. Almost all predictions have turned out to be untrue in this area.

Other potentially significant user groups include international business travelers (so-called global roamers), the military, police, customs, and other government agencies. But in assessing the market potential of MSS, we tend to discount the significance of the international business traveler segment. They may become the main group of users for MSSI. Overall, we see a very significant market for MSS. The Odyssey program alone is forecasting that it will attract two million voice subscribers by the year 2002 and nearly five million by 2008. Odyssey also hopes to have about one million data subscribers.

One should note that communications is essentially a service business. Service rates are an important factor that determines the success of any mobile satellite business. We get some insight into the economic elasticity of mobile communications service by looking at two segments: cellular and INMARSAT. After 16 years INMARSAT has over 40,000 subscribers paying an average rate of $7.50 per minute. Cellular, after 10 years, has over 40 million subscribers paying an average service rate of $0.83 per minute (considering roaming charges). This suggests that decreasing prices by a factor of 10 increases demand by a factor of 1000. TRW has performed market surveys with 6100 participants. The data from this survey confirms that the shape of this elasticity is in the vicinity of the cellular prices. The survey shows strong demand for a universal satellite-based service. The Odyssey service cost is supposed to be less than one dollar per minute with a subscriber base of 2.3 million subscribers. At higher service rates, a sharp drop in the number of subscribers is expected.

11.4.2 Iridium

Motorola's Iridium is regarded as the most ambitious in terms of size and complexity. It comprises 66 satellites orbiting in six planes, supported by a cellular-type network architecture. The crosslink capability permits point-to-point direct subscriber access from anywhere in the world in a seamless environment. Thus, Iridium could conceivably operate independently of the PSTN and will use TDMA for subscriber-link modulation.

After considerable tradeoffs and optimizations, Motorola has chosen an orbital altitude of nominally 780 km, with 66 satellites configured in six polar orbital planes of 11 satellites each. The satellites are networked together via crosslinks and with a system control facility, gateways, and subscriber units [19]. The overall system is shown in Figure 11.14.

The system is being designed to use the principles of cellular telephony to provide continuous line-of-site coverage from and to any spot on Earth. Three antennas on a satellite would form a honeycomb pattern of 48 beams on the ground. As the satellites orbit the Earth, the user would be handed-off from one beam to the next. Some handoffs would be intrasatellite, that is, within the 48-beam footprint of a single satellite. Once a user is at the edge of a beam of one satellite, the handoff would be intersatellite. Instead of stationary cells and mobile users as in ground cellular systems, the Iridium is designed to have relatively stationary users and moving cells. Even the Concorde airliner, which cruises at 2,253 Km/h, moves relatively slowly compared to a satellite that orbits the Earth in 100 min, an effective ground-

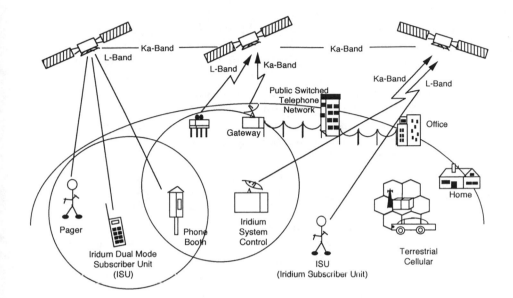

Figure 11.14 Iridium system.

speed of about 24,000 Km/h. The satellites in the six orbits would converge as they approach the poles, and their beams would overlap. Some of the outer beams would be "turned off" to eliminate the overlap and to save power.

The satellites would be networked together via 23-GHz crosslinks to allow the relay of calls. Each satellite would have four crosslinks, allowing each to communicate with the satellite immediately ahead and behind in its own plane as well as to the nearest satellites in each of the two adjacent planes.

11.4.2.1 Ground Architecture

A gateway, shown in Figure 11.14, is designed to interconnect the network with the PSTN. The gateway is intended to handle call set-up, caller location, and collection of the necessary data to support billing. When the power is on, the mobile subscriber would communicate to the nearest satellite, which then will notify the access control processor, which in turn updates the location register of the home gateway [19].

Since gateways interface with their respective PSTN, calls could not only be placed between two system subscribers but also between any subscriber and any PSTN telephone. In the current design, the gateways typically would employ a minimum of three 3.3-m tracking dish antennas that are separated by about 30 Km. At least two dishes would be needed because, as a satellite disappears over the horizon, another will have appeared at a different location above the horizon; and the handover from one to another is to be instantaneous.

The dual-mode handheld subscriber unit is designed to be compatible with both terrestrial cellular and Iridium systems. Where the terrestrial cellular will be available, the subscriber unit will access that system. Where a terrestrial cellular system is not available, barring regulatory restrictions, the dial tone of Iridium will be available. The portable/handheld unit is being designed to operate for 24 hours on a single recharge in a combination of standby (able to receive a "ring" indicating an incoming call) and active modes.

The Communications Network

The system would employ communication links in two portions of the spectrum. The up/downlinks between the satellites and the subscriber units would operate in the 1610.0-MHz to 1626.5-MHz region of L-Band. This band was only recently allocated for satellite-based personal communications at the *1992 World Administrative Radio Conference* (WARC) in Torremolinos, Spain.

The satellite crosslinks as well as the up/down feederlinks to the gateways are designed for operation in the 20-GHz to 30-GHz region of K-band. Licenses are required for operation in each country; some are pending now.

L-Band and K-Band Links

L-Band phased arrays form the 48 spot beams have gains between 17 dBi and 25 dBi depending upon the beam location from nadir. The handheld portables would employ a short, simple antenna with a near omnidirectional pattern with a gain of +1 dB and *right circular polarization* (RCP). The maximum satellite L-band transmitter *high-power amplifier* (HPA) output power per channel would be 3.5W *Peak Effective Power* (PEP), while the handheld portable would produce a maximum HPA output of 3.7W PEP.

Receiver LNA with 0.8-dB noise figures will complete the major link to budget allocations that would net a link margin of 15 dB in an open-terrain, LOS environment at full transmitter power. Transmitter power would be automatically adjusted to maintain the desired level of $E_b/(N_o + I_o)$ received at the opposite end of each link. The 48 beams would provide *Spatial Division Multiple Access* (SDMA), which when combined with *Time Division Multiple Access* (TDMA) and *Frequency Division Multiple Access* (FDMA) would allow a great degree of dynamic reallocation of capacity by geographical location in response to demands for service. The combination of SDMA, TDMA, and FDMA provides very efficient spectral use.

The intersatellite links (K-band link) would employ antennas with 37-dB gain and transmitter peak output powers to 3.4W. For the space-to-gateway feeder links, investment cost of the satellite, ground segment, and the networks, which are major considerations for all satellite programs [16]. Cost to the subscriber is supposed to be competitive with ground-based cellular system cost. The Odyssey handset will be compatible with terrestrial cellular signal formats. This will be achieved by the addition of microelectronics chips to existing handset designs to produce interoperability with both cellular and Odyssey. The chip sets will be matched to the standards of various regions of the world. In Europe, the handsets will be interoperable with GSM. In the United States, the handsets will work with the *American Digital Standard* (ADS), *Advanced Mobile Phone Service* (AWS), or Odyssey. The Odyssey handset will meet all communication system design driver requirements listed in the previous section.

Odyssey would provide high-priority service for premium customers who require reliable mobile communications via a small handset. Subscribers can have a single telephone that does not require any special access codes. Priority is given to the use of the terrestrial-based cellular services. When a call is placed, the handset first senses the presence or absence of cellular frequencies and attempts to place a call through the local cellular network. If the call is blocked, then it will be routed through Odyssey.

Another feature that decreases investment costs and works to decrease the overall RF propagation delays is the elimination of on-board processing and intersatellite RF links. The onboard RF processing in the satellite increases the complexity, hence the satellite cost, design, and manufacturing. On-board processing also adds time delays in the overall propagation of the RF signal to and from the user. Utilizing a CDMA signaling structure between the satellite and users allows for the more efficient use of the allocated spectrum, especially in view of competing systems where the assigned frequency spectrum for this service is shared by more than one system.

The handset will transmit approximately 0.5W of average power. This transmit power level will be adequate for both voice and digital data transmission. Although the Odyssey orbital altitude is greater than LEO, satellite antenna gain compensates for greater path loss. The transmit power level provides an appropriate margin against loss due to rain or vegetation, for example. It is important to point out that since the Odyssey system operates with high elevation angles of greater than 20 deg, less margin is required for path loss parameters than with very low orbiting systems, which must operate at shallow elevation angles. Thus, link reliability is improved.

11.4.3.2 Ground Architecture

The Odyssey ground segment provides the voice and data services to the user through the Odyssey satellites. It interconnects the Odyssey mobile subscribers to the PSTN through the *Odyssey Earth Stations* (OESs) and local gateways. The ground segment will provide the interconnections and database operations necessary to offer the subscribers global roaming that is transparent to the user. The main components of the ground segment are OES, *Satellite Control Center* (SCC), *Service Operations Center* (SOC), local gateways, and the *Ground Network* (OGN). The architecture of Odyssey is shown in Figure 11.16.

Frequencies for satellite-based personal mobile communications were designated at the 1992 WARC. Uplink transmissions from user to satellite are conducted in the L-band (1610 MHz to 1626.5 MHz), while downlink transmissions are done in the S-band (2483.5 MHz to 2500 MHz). This assigned frequency spectrum is currently undergoing negotiation between the FCC and potential users of these frequency bands. The Odyssey frequency plan will reflect any results from these negotiations.

11.4.4 Globalstar

The Globalstar system is based on a constellation of 48 LEO satellites, orbiting in eight planes at 52 deg inclination, at an altitude of 1,406 Km, with an orbit period of 113 min. A consequence of this low orbit is that transmission delay is minimal compared to that resulting from *Geostationary Earth Orbit* (GEO) since radio path distances are greatly reduced. Each satellite is also significantly smaller than a typical

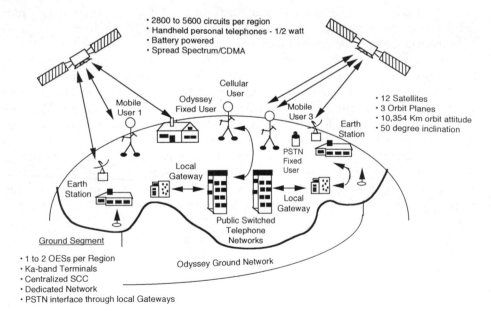

Figure 11.16 Odyssey system.

telecommunications satellite, weighing under 500 kg and with a volume under $2m^3$. Coverage will be provided between approximately 70 deg latitude N and S so that essentially only the polar regions are not served. Although its architecture is similar in some aspects to Iridium, Globalstar will be interconnected with existing, earth-based telecom infrastructure through 100 to 200 Gateway stations. The large number of Earth stations are made necessary when the satellite does not have crosslinks [18]. The service will be available to fixed subscribers, primarily for remote areas where alternative means of communication may not exist, and is planned to begin in 1998 and be in full commercial operation in 1999.

The system employs advanced digital radio transmission technology, using CDMA techniques with a transmission bandwidth of 1.25 MHz and operates on 1610 MHz to 1626.5 MHz (uplink) and 2483.5 MHz to 2500 MHz (downlink).

The user terminal transmit power level is a maximum of +26 dB (EIRP), which is similar to that of the modern handportable cellular phone. The subscriber sets are available in fixed, mobile, and handportable configurations.

11.4.4.1 Ground Architecture

Calls are routed via satellite between each user terminal and a network of gateway, ground stations, any one of which may be capable of serving users within a radius of 1,000 Km. The system incorporates a position location capability that can be used to allocate service between overlapping gateway coverage areas. In order to maximize circuit quality, Globalstar is designed to operate via two satellites simultaneously for the majority of time, using coherent diversity combining, soft hand-off techniques, and automatic power control [18]. The system architecture is shown in Figure 11.17(a).

Gateways will comprise up to four steerable antenna arrays, each of 5.5-m diameter, and will include provisions for call monitoring and routing, with interconnection either directly to the local PSTN or indirectly via the terrestrial cellular network. In both cases, subscriber location registers are required for both home and visiting subscribers. Since the overall network capacity is reduced when there are too many operational gateways in close proximity, service to smaller countries may be provided via shared gateways in nearby countries, using international circuits to reroute traffic back to the appropriate operating center.

Compatibility with GSM

By 1999, GSM is widely predicted to remain the world's dominant digital cellular standard and Globalstar is supposed to provide service via dual-mode handsets to GSM subscribers when outside the coverage areas of GSM networks. In addition, the use of common administration, billing and customer care systems, and other

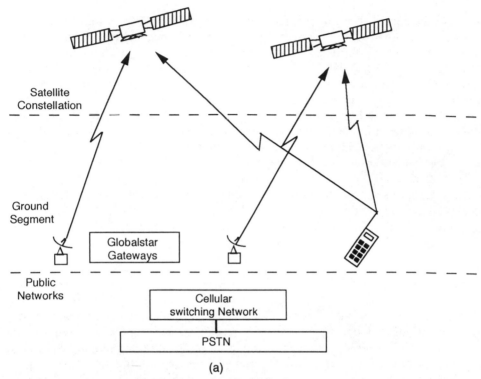

(a)

Figure 11.17 (a) Basic system architecture and (b) Globalstar integration with GSM.

telecommunications infrastructure, for both GSM networks and the Globalstar networks, should provide significant economies of scale for the operators of combined GSW Globalstar networks. Figure 11.17(b) shows the integrated GSW Globalstar architecture.

Thus, it is important that Globalstar will be able to support or interface easily with those uniquely powerful and well-proven features of GSM that have been specified and developed primarily for the international roaming market. These features include the following.

- Mobility management (HLR and VLR);
- International roaming;
- Subscriber administration;
- SIM card technology;
- Security and authentication (EIR and AUC);
- Accounting and billing;
- Short message service.

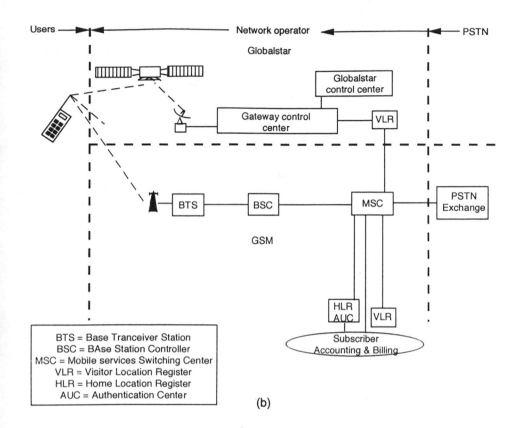

Figure 11.17 (continued).

Clearly it is desirable that celestial service be seen by the subscriber as a logical and seamless extension of total service, as far as is practicable. This puts a number of BS requirements on the integration of the Globalstar and GSM technologies.

The experience of total networks in migrating users from analog to digital networks has shown subscribers to be very reluctant to changing numbers. It is important that existing terminal subscribers can maintain their GSM numbers when transferring to dual-mode GSM Globalstar service. In addition, it is important that they can be reached on the same number, regardless of whether they are registered on the GSM network or the Globalstar network.

Existing terrestrial subscribers will view Globalstar's service as an extension to the total network capability. The most obvious manifestation of this will be combined billing and customer care services [18].

Due to greater system capacity, better in-building, coverage, and probable lower cost, the GSM network will generally be used as a first preference wherever it is available. Automatic network selection will be provided between C-Globalstar and GSM, with appropriate service indication and facilities for manual override were required being available to the user. GSM users and network operators will expect the levels of security provided in any new network.

11.4.5 Inmarsat-P

Inmarsat-P from the *International Maritime Satellite Organization* (INMARSAT) plans to design a constellation of ten satellites deployed in two orbital planes inclined at a 45 deg circular orbit at a *medium Earth orbit* (MEO) of 10,355 km. The new company, *Intermediate Circular Orbit* (ICO) is legally and physically distinct from Inmarsat; with a separate board and management it is privately owned. But there are some strong linkages with Inmarsat, which is the largest investor in the company. Inmarsat will also acts as wholesaler of aeronautical and maritime services [17].

The system will support dual-mode cellular/satellite handheld terminals capable of delivering voice, fax, and narrowband data to both domestic and international travelers who roam outside compatible cellular coverage area. Satellite users, general aviation aircraft, small vessels, and semi-fixed installations in rural and remote areas of the countries will also be served. The ground system will comprise of 12 Earth stations known as *satellite access nodes* (SANs) that will be interconnected by terrestrial links. The P-Net (ground network) will be linked to the PSTNs switches, known as gateways, and operated by national service providers. User terminals will be supplied by mobile telephone manufacturers.

Inmarsat-P system is supposed to provide service globally. The only constraint on service availability is the need for the user to have LOS path to a satellite. Calls cannot be made or received when an obstruction such as a mountain, a building, or dense foliage lies between the terminal and the satellite. Calls from within the building are only possible if the user is standing near a clear glass window facing a direct LOS to a satellite [17].

The system will support basic services available to all users. Supplementary services and optional offerings may not be available in all countries. Supplementary services include call barring, call waiting, and call forwarding.

Basic services are voice telephony, data services, and SMS. Voice service will be comparable in quality with that delivered by present digital cellular systems, except that call setup time may be more than cellular due to this being a satellite system. Users will be able to elect to receive calls into a voice mailbox that can be retrieved later. The system will support a wide range of data services to handheld, other mobile terminals, and semifixed installations: one-way and two-way, real-time and store-and-forward, at rates from 2.4 Kbps to 64 Kbps. The data source will be from the following.

- Handheld with peripheral such as a PCMCIA card at 2.4 Kbps to 4.8 Kbps;
- Handhelds data terminals at 9.6 Kbps;
- Vehicular cradle mounted at 19.2 Kbps;
- Other sources of data include semifixed and vehicular at 64 Kbps.

As stated, data applications will include fax, messaging (including electronic mail), file transfer, one-way broadcasting of messages, and information services.

Short Messaging Service: The SMS is designed to deliver paging and call notification messages when the terminal is unfavorably located (high loss-obstructed area). The arrival of a message into a data or voice mailbox will be called to the attention of the user by either a normal signal or the higher penetration signal.

11.4.5.1 Technical Characteristics .

The chosen frequency bands for the user link service links are 1980 MHz to 2010 MHz and 2170 MHz to 2200 MHz, known as the 2-GHz bands. For feeder links the system needs about 100 MHz in each direction. The most likely bands are 5150 MHz to 5250 MHz (uplinks) and 6975 MHz to 7075 MHz (downlinks). This orbital configuration is expected to provide the advantages of an average elevation angle of more than 40 deg from mobile terminal to satellite and path diversity, with a high probability that an average user will see more than one satellite at 10 deg or more elevation angle.

The high elevation angle reduces the chances of signal blockage. The second results in a high probability that the user will have another satellite to switch to when the one satellite goes below the horizon or behind an obstruction [17].

Due to the intermediate altitude of the satellite, the propagation delay will be less than 200 ms, which is well below 270 ms for GEO orbits and will meet the user acceptability.

Each satellite will communicate with user handsets via a service link antenna with a minimum aperture of 2m. Due to limitations on the handsets, the transmit power satellite antenna is designed with 163 spot beams. This in turn will provide a basic link power margin in excess of 8 dB.

Each satellite has 750 carriers and with six channels per carrier (TDMA access) provides a total of 4,500 telephone channels. The complete constellation is claimed to support a maximum of 2,400 million voice minutes per year.

The satellites will be based on Hughes HS601 design, with attitude and orbital control systems modified to suit the special requirements of intermediate circular orbit. Communications payload mass is approximately 550 kg, with a total power requirement of 5 kW. The total satellite mass is around 2,000 kg, permitting direct-injection launches of two spacecraft at a time. The expected lifetime for the satellite is about 10 years.

The basic user terminal will be handheld and similar in size and weight to cellular terminals. The volume and weight are expected to be less than 300 cc and 300 gm, respectively. It is expected that a range of terminals will be developed in due course to meet the requirements for other applications and environments.

The handheld will have an antenna no more than 10 cm in length and 0.8 cm in diameter and an integrated battery giving at least one hour of talk time. Most handhelds will be dual-mode and capable of working with both satellite and cellular/ PCS networks. It is expected that the terminals will be developed to be compatible with all major terrestrial analogue and digital standards. Additional features will include:

- Satellite/cellular service availability indicator;
- Screen and keys to support viewing of SMS and other text messages;
- Mailbox status indicators;
- Data/fax port.

11.4.5.2 Ground Architecture

The major components of the ground segment are shown in Figure 11.18. The main elements are *Satellite Access Node* (SAN), HLR, NMC, and gateways. The satellites will communicate with ground networks via a dozen SANs located approximately two per continent. Each SAN will comprise four antennas and associated satellite communications equipment, a switch (MSSC) to route traffic within the P-Net and to the gateways, and user mobility management databases. Each SAN will track the satellites visible to it and direct traffic to the one currently offering the most robust link and longest call duration before hand-off.

The P-Net, the SANs, and the terrestrial backbone network connecting them will allow each call to be routed to the SAN best able to handle it [17]. Managed by a NCC, the P-Net will also carry system operational communications.

The gateways will act as switches between the P-Net and the PSTN. In most cases the gateways will use existing switches, There is no technical limit to the number of gateways that can access the ICO network.

The user mobility data base in each SAN will hold details of every user terminal currently logged on to that SAN. The data base will perform the same function as a GSM VLR. At least one SAN and possibly more will also maintain a database of all registered users in the network, which is equivalent to a GSM HLR.

Satellite tracking, telemetry, and control will be carried out by about five TT&C stations. They will

- Track the satellites and adjust their orbits to maintain the correct constellation geometry;

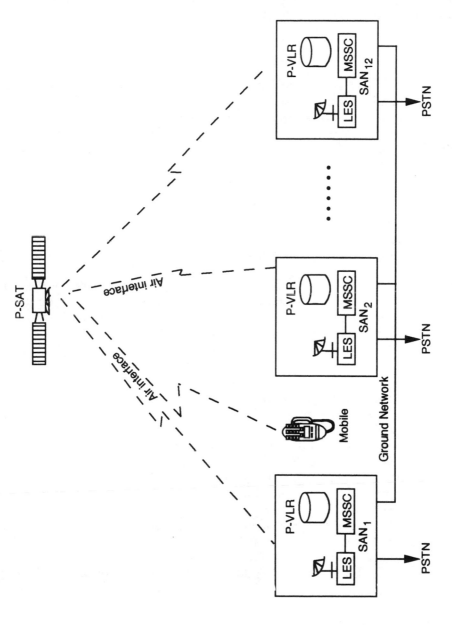

Figure 11.18 Ground segment layers of P-net.

- Monitor and adjust the satellites' power levels, temperature, stability and other operating characteristics;
- Restore the constellation in the event of spacecraft failure.

11.5 CONCLUSIONS

True mobility for the user can only be achieved by multiple entries on a SIM card, the design of multimode terminals, and the availability of fast and large databases. Both LEO and MEO satellites will be required to provide global coverage for the mobile, especially for those remote areas of the world where a cost-effective cellular systems can not be provided.

Problems

11.1 Name the different European agencies engaged in the development of future PCS system.

11.2 Name different programs active in Europe for the development of PCS systems.

11.3 Name different projects in Europe and their importance in the context of PCS systems. Justify that those are critical areas of importance. Can you think of some other important area of importance?

11.4 What are the main objectives of UMTS/FPLMTS?

11.5 What information do you think is essential for SIM to work on different systems?

11.6 Based on Figure 11.8 explain every necessary steps the system has to take to complete an incoming call from GSM system when the GSM subscriber is in the area of PDC system as a roamer.

11.7 For the following scenarios find the transmission rate:
Per user data stored in the database: 500 Bytes;
Total number of subscriber to the system 200,000;
Average number of transactions per day per user is 5.

11.8 Why do you think it is essential to standardize all relevant interfaces in the future PCS systems?

11.9 What are the potential problem areas in the design of multimode terminals? What are their potential solutions?

11.10 Name various advantages and disadvantages of zero IF receiver architecture?

11.11 Justify functions and the role of different players in PCS systems

11.12 Explain the concept of wideband and narrowband services for PCS systems. What type of data transmission speed you think can be achieved in 1.25 MHz channel bandwidth?

11.13 What is the basic distinction between MTS and BTA in USA?

11.14 Explain the concept of paired channel in narrowband applications in USA. What do you mean by 50 kHz paired with 12.5 kHz channel? In what application you think 50 kHz unpaired channel used?

11.15 Where is unlicensed PCS use?

11.16 Draw a table of advantages and disadvantages for LEO and MEO satellites for PCS services. Why do you think GEO may not become popular in this application?

11.17 Why is the life of MEO satellite more than LEO satellite?

11.18 Provide simple reasoning to arrive to a conclusion that about ten MEO satellites are enough for world wide coverage.

11.19 Is mobile satellite a direct competitor to ground based cellular system? Since mobile satellite can provide coverage throughout the world why do you need ground based cellular systems?

References

[1] Kandiyoor, Surech, "DECT: Meeting Needs and Creating Opportunities for Public Network Operators," Telecom 1995, Session 1.3, pp. 79–82.

[2] Hjern, Magnus, "DECT/GSM Interworking," Telecom 1995, session 6.2, pp. 337–340. Spaniol, Otto, et. al., "Impact of Mobility on Telecommunication and Data Communication Networks," *IEEE Personal Communications,* October 1995, pp. 20–33.

[3] Gabelgaard, Bent, "Converging GSM and IN—Deploying PCS Services in GSM," Telecom 1995, Session 5.1, pp. 21–25.

[4] Schwarz da Silva, Joao, "Mobile and Personal Communications—The European R&D Perspective," Telecom 1995, Session 1.3, pp. 51–55.

[5] Prasad, Ramjee, "European Radio Propagation and Subsystems Research for the Evolution of Mobile Communications," *IEEE Communications Magazine,* February 1996.

[6] Grant, Michael, "Personal Mobility vs. Terminal Mobility: Development of Personal Numbering Services in Europe's Personal Communications Industry," Telecom 1995, Session 1.3, pp. 41–43.

[7] Cayla, Guy, "Wireless Local Loop: At Last, the Last Mile," Telecom 1995, Session 6.2 pp. 31–335.

[8] Samalam, Ashoke Kumer, " Will—Convergence of Transmission and Switching Technologies," Telecom 1995, session 6.2, pp. 363–365.

[9] Short, Michale J., "Globalisation of Mobile—Trends in Digital Cellular," Telecom 1995, session 1.3, pp. 91–98.

[10] Damosso, Eraldo, and Georges de Brito, "Cost 231 Achievements as a Support to the Development of UMTS: A Look into the Future," *IEEE Communications Magazine,* February 1996, pp. 90–96.

[11] Mouly, Michel, and Marie-Bernadette Pauted, "Current Evolution of the GSM Systems," *IEEE Personal Communications,* October 1995, pp. 9–19.

[12] Gibson, Rodney W., "Location Agencies for Universal Mobile Telecommunications Services," Telecom 1995, session 1.3, pp. 45–49.

[13] Ramsdale, Peter, "The Path to Third Generation Personal Communications Systems," Telecom 1995, session 5.4, pp. 174–180.

[14] MacFarlane, David, "Third Generation Mobile—UMTS and FPLMTS," Telecom 1995, session 5.4, pp. 185–188.

[15] Sherry, Howard, "A United States Perspective PCS Standards," Telecom 1995, session 6.2, pp. 351–355.

[16] Spitzer, Christopher J., et. al., " Odyssey Personal Communications Satellite System," *Third International Mobile Satellite Conf.,* 1993, pp. 297–302.

[17] Hart, Nick, et al., "Inmarsat's Personal Communication System," *Third International Mobile Satellite Conf.,* 1993, pp. 303–304.
[18] Wiedeman, Robert A., "The Globalstar Mobile Satellite System for Worldwide Personal Communications," *Third International Mobile Satellite Conf.,* 1993, pp. 291–295.
[19] Hatlelid, John E., "The Iridium System Personal Communications Anytime, Anyplace," *Third International Mobile Satellite Conf.,* 1993, pp. 285–289.

GLOSSARY

AB	Access Burst
ACCH	Associated Control Channel
Active MS	The state of a MS when processing a call
Adaptive Frame Alignment	Means of ensuring that timeslots received at the BS from active MSs at different distances from the BS are in time alignment. Transmit timeslot advance period in the MS is necessary to ensure this. The value can be assessed by measuring the "round-trip delay," which depends on the distance of the MS from the serving BS.
AGCH	Access Grant Channel
Algorithm A3	Cryptographic algorithm that produces SRES, using RAND and K_i
Algorithm A5	Cryptographic algorithm that produces cipher out of clear using K_c
Algorithm A8	Cryptographic algorithm that produces K_c using RAND and K_i
ARFCN	Absolute RF Channel Number. An integer that defines the absolute RF channel number.
ARO	Automatic Request for Retransmission

AUC	Authentication Center
Authentication	The corroboration that a peer entity is the one claimed
Authentication center (AUC)	The component of the fixed part of the PLMN that contains subscriber authentication keys (K_i) and generates security related parameters (RAND, SRES, K_c) depending on implementation
BA	BCCH Allocation
BAIC	Barring of All Incoming Calls
BAOC	Barring of All Outgoing Calls
Basic services	The telecommunication services excluding the supplementary services
Base station area	The part of the GSM area to which a base station gives service
Base Station Identity Code (BSIC)	A block of code consisting of the PLMN color code and a base station color code. One Base Station can have several color codes.
BCCH	Broadcast Control Channel
BCCH allocation (BA)	The radio frequency channels allocated in a cell for BCCH transmission
Bearer service ————	A type of telecommunication service that provides the capability for the transmission of signals between user-network interfaces
Bit number	Number that identifies the position of a particular bit period within a timeslot
BS	A land station in the land mobile service
BSC	A network component in the PLMN with the functions for control of one or more BTSs
BSIC	Base Station Identity Code
Base Station System (BSS)	The system of base station equipment that is viewed by the MSC through a single interface as being the entity responsible for communicating with mobile stations in a main area. The equipment of a BSS may cover one or more cells. A BSS may consists of one or more base stations. The BSS shall consist of one Base Station Controller and several Base Transceiver Stations.
BTS	A network component that serves one cell and is controlled by a Base Station Controller. The BTS contains one or more transceivers.

Burst	The physical content of a timeslot
CA	The radio frequency channels allocated to a particular call
CBCH	Cell Broadcast Channel
CC	Country Code
CCH	Channels that carry system management messages
CELL_RESELECT_ HYSTERESIS RXLEV	Hysteresis required for Call Reselection
CFU	Call Forwarding Unconditional
Cell	The area of radio coverage locally defined as seen by the mobile station with a base station identity code and uniquely defined by the network with a call global identification
Cell Coverage Area	Area within which a defined quality of reception is provided, that is, a planned radio coverage of a cell
Cell Global Identification	A block of code that uniquely identifies a cell within all GSM PLMNS. It consists of the LAI and cell identity
Cell Identity (CI)	A block of code that identifies a cell within a location area
Channel	A means of one-way transmission. A defined sequence of periods (e.g., timeslots) in a TDMA system: a defined frequency band in an FDMA system; a defined sequence of periods and frequency bands in a frequency hopped system.
Cipher key	A sequence of symbols that controls the operation of encipherment and decipherment
Cipherkey setting	Mutual agreement between the Mobile Station and the fixed part of the system upon a common cipherkey (K_c) to be used in a subsequent encipherment/ decipherment process
Ciphertext	Unintelligible data produced through the use of encipherment
Class 1a, 1b, 2	The classification of speech encoder bits depending on the degree of protection needed. Class 1a and class 1b bits have protection, while Class 2 bits have no protection. Error detection is performed on class 1a bits.
Collection charge	The collection charge is the charge in its national currency collected by a GSM PLMN operator or an authorized agent from its customers for the use of the service. The establishment of the collection charge is a national matter.
Current Serving	BS on one of those channels (TCH, DCCH, or CCCH) the MS is currently operating

Customer	The customer is the individual or entity who obtains a service from a GSM PLMN operator or an authorized agent and is responsible for payment of all charges and rentals due.
Dummy Burst (DB)	Means of filling a timeslot with an RF signal when no information is to be delivered to a channel
DCCH	Dedicated Control Channel
Decipherment	The transformation by cryptographic techniques to produce Plaintext from Ciphertext
DET	Detach
Discontinuous Reception (DRx)	Means of saving battery power by periodically and automatically switching the MS receiver on and off
Discontinuous Transmission (DTx)	Means of saving battery power and reducing interference by automatically switching the transmitter off when no speech or data are to be sent
Dm	Control Channel (ISDN terminology applied to mobile service)
Downlink	Physical link from the BS toward the MS (BS transmits, MS receives)
DTAP	Direct Transfer Application Part
DTE	Data Terminal Equipment
DTMF	Dual Tones Multi-Frequency (signaling)
Eb/No	Ratio of energy per modulating bit to the noise spectral density
EIR	Equipment Identity Register
Encipherment	Transformation by cryptographic techniques to produce Ciphertext from Plaintext
FACCH	Fast ACCH
FACCH/F	Full Rate Fast Associated Control Channel
FACCH/H	Half Rate Fast Associated Control Channel
FCCH	Frequency Correction Channel
FEC	Forward Error Correction
FN	Frame number identifies the position of a particular timeframe within a hyperframe.
Frequency Correction Burst (FCB)	Period of RF carrier less than one timeslot whose modulation bit stream allows frequency correction to be performed easily within a mobile station

Gateway MSC (GMSC)	An MSC that provides an entry point into the PLMN from another network or service. A gateway MSC is also an interrogating node for incoming PLMN calls.
GSA	GSM System Area
GSM	Group Special Mobile or Global System for Mobile Communication
GSM MS	GSM Mobile Station
GSM PLMN	GSM Public Land Mobile Network
GSM PLMN operator	An administration or its licensed operator(s) that provides a GSM PLMN and its telecommunication services
GSM service area	The area in which a mobile station can be reached by a fixed subscriber without the subscriber's knowledge of the actual location of the mobile station within the area. A service area may include the areas served by several PLMNS.
GSM system area (GSA)	The group of GSM PLMN areas accessible by GSM mobile stations
Guard Period	Period at beginning and end of timeslot during which MS transmission is attenuated
HANDO	Handover
Handover	The action of switching a call in progress from one radio channel to another radio channel. Handover is used to allow established calls to continue by switching them to another radio resource whenever mobile stations move from one base station area to another. Handover may take place between the following GSM entities; timeslot RF carrier, cell, base station, BSS, MSC. The prefixes "inter" and "Intra" are used to describe the type of handover according to context (e.g., intercell handover). These entities are listed in nested order such that "inter-BSS handover" is implicitly inter-BSC, for example. Additionally, the words "internal" or "external" may prefix the term intra-BSS handover with the following meanings "Internal intra-BSS handover" is an intra-BSS handover that takes place without reference to the MSC (although the MSC will be informed on completion). "External intra-BSS handover" is an intra-BSS handover which is controlled by an MSC.
Handover Access Burst	Access burst used during handover
Handover Execution	Signaling message sequence that causes the MS to continue the call in another (predetermined) cell

Handover Margin	Hysteresis factor that minimizes repetitive handovers between adjacent cells
Handover Strategy	Procedure defined by an algorithm that prescribes how handover shall be carried out (e.g., handover to a cell which offers better signal strength or to a cell that allows communication with less power)
HLR	Home Location Register
HOLD	Call Hold (Supplementary Service)
Home location register (HLR)	The location register where the current location and all subscriber parameters of a mobile station are permanently stored
HPLMN	The PLMN where a subscription is held and therefore which contains the HLR of the subscriber
HPU	Hand Portable Unit
HSN	Hopping Sequence Number
Hyperframe	A hyperframe consists of 2,048 superframes. It is the longest recurrent time period of the frame structure (12,533.76s).
IAM	Initial Address Message
IDN	Integrated Digital Network
Idle Mode	State of an MS not actively processing a call
IMEI	International Mobile Station Equipment Identity. Uniquely identifies the mobile station as a piece or assembly of equipment.
IMSI	International Mobile Subscriber Identity. It can serve as a key to derive subscriber information such as directory number(s) from the HLR.
Inter-Cell Handover	Handover between cells. Change of serving call.
ISDN	Integrated Services Digital Network. An integrated services network that provides digital connections between user-network interfaces.
IWF	Interworking Function. The general term used to describe the interoperation of networks services and supplementary services, for example.
K	Constraint Length of the Convolutional Code
K_c	Cipher Key
LAC	Location Area Code
LAN	Local Area Network

Land Mobile Service	A mobile service between base stations and land mobile stations or between land mobile stations
Land Station	A station in the mobile service not intended to be used while in motion
LAPDm	Link Access Protocol on the Dm channel
Local PLMN (LPLMN)	The LPLMN is the HPLMN or VPLMN depending on the location of the MS at the time.
Local Mobile Station Identity (LMSI)	Unique identity temporarily allocated to visiting mobile subscribers in order to speed up the search for subscriber data in the VLR, when the MSRN allocation is done on a per call basis
Location area	An area in which a mobile station may move freely without up-dating the location register. A location area may comprise one or several base station areas.
Location area Identification	The information indicating the location area in which a cell is located
Location Information	The information indicating where a mobile station is located in the system area
LR	The function whereby PLMNs keep track of the location information of Mobile Stations located in the system area
Location updating procedure	The procedure by which location register updating takes place
Logical channels	Two classes of logical channels are defined. Control channels (CCH), which carry system management messages and traffic channels (TCH), which carry users' speech or data.
MA	Mobile Allocation. The radio frequency channels allocated to a MS for use in its hopping sequence.
MACN	Mobile Allocation Channel Number
MAI	Mobile Allocation index
MAIO	Mobile Allocation Index Offset
MAP	Mobile Application Part. The internetworking signaling between MSCs and LRs and EIRs; a part of CCITT Signaling System No. 7.
MCC	Mobile Country Code
MCI	Malicious Call Identification
ME	Mobile Equipment
MM	Man Machine

MMI	Man Machine Interface
MNC	Mobile Network Code
MO	Call or short message originated from the Mobile Station
Mobile Equipment (ME)	The ME is the Mobile Station (MS) without the SIM.
Mobile International ISDN number (MSISDN)	Uniquely defines the mobile station as an ISDN terminal. It consists of three parts: the country code (CC), the national destination code (NDC), and the subscriber number (SN).
Mobile Station feature	A MS feature is a function or a piece of equipment that directly relates to man machine operation of the MS. Three categories of MS features are distinguished: basic, supplementary, and additional features.
Mobile Service	A radio communication service between mobile and land stations or between mobile stations
MS	A station in the mobile service intended to be used while in motion or during halts at unspecified points
MS-PWR-CLASS	MS Power Class. Parameter defining the power class of an MS.
MS_RANGE_MAX	Mobile Station Range Maximum. Handover criterion to determine serving cell.
MSC	Mobile-services Switching Center. The MSC performs the functions of switching, routing and control of the call and charging and accounting. The MSC also controls the interworking with fixed networks.
MSC area	The part of the PLMN covered by an MSC. An MSC area may consist of one or several location area.
MSCM	Mobile Station Class Mark
MSISDN	Mobile Station ISDN Number
MSRN	A code that is allocated to a mobile station when registered with a VLR for the purpose of routing calls to the MSC in the area where the Mobile Station is located. The MSRN is used by the home location register for rerouting calls to the mobile station.
MT	Call or short message intended to be delivered to the Mobile Station
MTP	Message Transfer Part

Multiframe (Multiple time-frame)	Two types of multiframe are defined in the system: a 26-frame multiframe with a period of 120 ms and a 51-frame multiframe with a period of 3060/13ms.
NB	Normal Burst
Network access charge	Part of the collection charge, intended to cover cost of service which is not dependent on the actual use of networks telecommunication services. It may consist of an initial fee and a subscription fee.
Network Management (NM)	NM is all activities which control, monitor and record the use and the performance of resources of a telecommunication network in order to provide telecommunication services to customers/users at a certain level of quality.
Network Management Control (NMC)	The NMC node of the GSM TMN provides global and centralized PLMN monitoring and control, by being at the top of the TMN hierarchy and linked to subordinate OMC nodes.
Network utilization charge	Part of the collection charge that is intended to cover the network/telecommunications services. The charge is registered on a per case basis.
NM	Network Management
NMC	Network Management Center
NSAP	Network Service Access Point
NT	Network Termination
O&M	Operations and Maintenance
OACSU	Off-Air-Call-Set-Up
OCB	Outgoing Calls Barred
OMC	Operations and Maintenance Center
OSI RM	OSI Reference Model
Off-Air-Call-Set-Up (OACSU)	The procedure by which a telecommunication connection is established whilst the RF link between the MS and the BS is not occupied.
Operations and Maintenance Center (OMC)	The OMC node of the GSM TMN provides dynamic O&M monitoring and control of the PLMN nodes operating in a geographical area controlled by a specific OMC.
PAD	Packet Assembly/Disassembly facility
Paging, Paging procedure	The procedure by which a GSM PLMN fixed infrastructure attempts to reach a Mobile Station within its location area before any other network-initiated procedure can take place
PCH	Paging Channel

PDN	A network established and operated by an administration for the specific purpose of providing data transmission services to the public. Circuit-switched (CSPDN), packet-switched (PSPDN), and leased circuit data transmission services are feasible depending on national regulations Public Data Networks.
Peak power	Measure of the maximum RF power when averaged over one radio frequency cycle, during the useful part of the burst.
PIN	Confidential information which may be used in the authentication between subscriber and SIM to corroborate that the subscriber is the one claimed
Plaintext	Unciphered data
PLMN	A network, established and operated by an Administration or its licensed operator(s), for the specific purpose of providing land mobile communication services to the public. It provides communication possibilities for mobile users. For communication between mobile and fixed users interworking with a fixed network is necessary.
Power Class of MS	Class of maximum output power to which a MS belongs; for example, handheld, portable, and transportable mobile, defined by the maximum peak power that can be output by a MS. Five power classes are defined.
Power Control Message	Message that controls the MS transmitted RF power level
PSPDN	Packet Switched Public Data Network
PSTN	Public Switched Telephone Network
Quarter bit Number	Timing of quarter bit periods (12/13 us) within a timeslot
Queuing	The procedure in which calls, originating from a MS or terminating in are kept pending for reasons of congestion or when the called party is occupied
RA	Random mode request Information field
RAB	Random Access Burst
RACH	Random Access Channel
Radio frequency channel (RFCH)	A partition of the RF spectrum allocation with a defined bandwidth and center frequency
RAND	Random number to be used as challenge in a challenge response protocol
RFC	Radio Frequency Channel

RFCH	Radio Frequency Channel
RFN	Reduced Frame Number
Round-trip delay	Delay period between transmit and receive instant of a time-slot in the BS, determined by the response behavior of the MS and MS-BS distance timeslot in the BS
RXLEV	Received Signal Level
RXQUAL	Received Signal Quality
SABM	Set Asynchronous Balanced Mode
SACCH	Slow Associated Control Channel
SACCH/C4	Slow SDCCH/4 Associated Control Channel
SACCH/C8	Slow SDCCH/8 Associated Control Channel
SAPI	Service Access Point Indicator
SB	Synchronization Burst
SCCP	Signaling Connection Control Part
SCH	Synchronization Channel
SDCCH	Stand-alone Dedicated Control Channel
SDCCH/4	Stand-alone Dedicated Control Channel/4
SDCCH/8	Stand-alone Dedicated Control Channel/8
Security feature	Security features protect: (a) the access to the mobile service; (b) Any relevant item from being disclosed on the radio path, mainly in order to ensure the privacy of user related information.
Service Provider	The organization through which the subscriber obtains GSM telecommunication services. This may be a network operator or possibly a separate body.
Serving BS	BS with which the MS is currently communicating
Serving Cell	Cell containing the serving BS
SID-frame	Frame that carries the averaged background noise information
Signaling Interworking	Signaling interworking is the function required to connect the signaling protocols used in a GSM PLMN to those used in the appropriate fixed network.
Signed Response (SRES)	Response to a challenge in the challenge response protocol
SIM	Subscriber Identity Module
SMSCB	Short Message Service Call Broadcast
SP	Signaling Point

SRES	Signed Response (used for authentication)
SS	Supplementary Service
STP	Signaling Transfer Point
Stream cipher	Bit-by-bit binary addition of Plaintext bit stream and cipher key bit stream.
Subchannel number (SCN)	One of the parameters defining a particular physical channel in a BS
Subscriber	The definition of this term is identical with that of the term "customer."
Subscriber authentication key	Subscriber individual confidential information used in authentication and in the cipher key generation process
Subscriber Identity authentication	The corroboration by the PLMN that the subscriber identity (IMSI, TMSI), transferred by the mobile subscriber within the authentication procedure at the radiopath is the one claimed
Subscriber Identity confidentiality	The property that the subscriber identity (IMSI) is not made available or disclosed on the radio interface
Subscriber Identity Module (SIM)	Removable module that is inserted into a mobile equipment; it is considered part of the Mobile Station and contains security-related information (IMSI, K_i, PIN), other subscriber related information, and the Algorithms A3 to A8.
Supplementary service	A modification of or a supplement to a basic telecommunication service
Synchronization burst (SB)	Period of RF carrier less than one timeslot whose modulation bit stream carries information for the MS to synchronize its time-frame start to that of the received signal
TA	Terminal Adaptor
Tail bits	Known bits that are added at the end of the information stream to allow use of a convolutional code to build a (non-systematic) block code with the minimum distance properties
TC	Transaction Capabilities
TCAP	Transaction Capabilities Application Part
TCH	Tariff Channel
TCH/F	A Full Rate TCH
TCH/F2.4	A Full Rate Data TCH (<2.4 Kbps)
TCH/F4.8	A Full Rate Data TCH (4.8 Kbps)

TCH/F9.6	A Full Rate Data TCH (9.6 Kbps)
TCH/F	A Full Rate Speech TCH
TCH/H	A Half Rate TCH
TCH/H4.8	A Half Rate Data TCH (4.8 Kbps)
TCH/H	A Half Rate Speech TCH
TDMA Frame Number	The number of a particular TDMA frame in the cyclic MMA frame numbering range (2715647)
TE	Terminal Equipment
Teleservice	A type of telecommunication service that provides the complete capability, including terminal equipment functions
Temporary Mobile Subscriber Identity (TMSI)	A unique identity temporarily allocated to visiting mobile subscribers in order to support the subscriber identity confidentiality service
Terminal adaptation function (TAF)	The terminal adaptation function is a functional entity associated with an MS. The TAF provides the functionality necessary to permit interworking between an MT and terminal equipment (TE). The function of the TAF depends on the service and the type of TE. The TAF is required to convert the protocols provided by the MT to those used by the TE. The terminal adaptation functions are described in GSM 07.01, 07.02, and 07.03.
Timeslot Number (TN)	Identifies a particular timeslot within a TDMA frame
Timing Advance	A signal sent by BTS to the MS. It enables the MS to advance the timing of its transmissions to the BS so as to compensate for propagation delay.
TMN	Telecommunications Management Network
TMSI	Temporary Mobile Subscriber Identity
TN	Timeslot Number
Traffic channels	Channels that carry users' speech or data
Training sequence	Sequence of modulating bits employed to facilitate timing recovery and channel equalization in the receiver
Training sequence code	Parameter used to select one of a number of training sequences

Transceiver (Trx)	A network component that can serve full duplex communication on full-rate traffic channels according to recommendation GSM 05.02. If Slow Frequency Hopping (SFH) is not used, then the TRx serves the communication on one RF carrier.
TS	Timeslot
TXPWR	Tx power level request MS-TXPWR-REQUEST; and Tx power confirmation by mobile MS-TXPWR-CONF
Ul	Unnumbered Information (Frame)
Uplink	Physical link from MS towards BS (MS transmits, BS receives)
User	The user is the individual or entity designated by the customer, individually or by class, as having access to the service and having such authorization, individually or by class, as may be required by the GSM PLMN operator or an authorized agent concerned.
VAD	Voice Activity Detector
VLR	The location register where all relevant parameters concerning a mobile station are stored as long as the mobile station is in a location area controlled by this register
VMSC	Visited MSC
VPLMN	Visited PLMN
Voice activity detection (VAD)	A process used to identify presence or absence of speech data bits. This is used along with DTX.
Work Station (WS)	The remote device via which O&M personnel executes input and output transactions for NM purposes

▼▼▼

ABOUT THE AUTHOR

Asha K. Mehrotra received his B.S., M.S., and Ph.D. degrees in electrical engineering from Bengal Engineering College, Sibpur India; Nova Scotia Technical College, Halifax, Canada; and Polytechnic Institute of New York (formerly Brooklyn Polytechnic) in 1961, 1967, and 1981, respectively. He has recently joined Hughes Network Systems as an advisory engineer where he is working on a mobile satellite project (Inclined Circular Orbit, ICO). Prior to this Dr. Mehrotra was a senior principal MTS at The Analytic Sciences Corporation (TASC), in Reston and Rosslyn, Virginia. At TASC he dealt with various classified and commercial projects and played lead engineering role in the IRIDIUM project with Motorola. In the past he has worked on the MILSTAR satellite project at MITRE as a member of the technical staff and with TDRSS satellite at Space Communication Company as a manager of the systems analysis group. His past experience includes areas of HF communication, switching systems, and telephone transmission systems. His hardware experience spans the design of central office equipment, PBXS, and facsimile systems. In the past, Dr. Mehrotra has taught graduate courses in computer science and electrical engineering at Virginia Polytechnic Institute and State University and George Mason University. He has been an adjunct professorial lecturer at George Washington University for the past 17 years, where he teaches graduate courses in communication engineering. He has delivered many seminars including presentations in GSM at Bell South, Lockeed Martin, Plexsys International, and U S West. He has also presented seminars overseas. He is a member of the IEEE communication and computer society.

▼▼▼

INDEX

The Artech House Telecommunications Library

Vinton G. Cerf, Series Editor

Wireless LAN Systems, A. Santamaría and F. J. López-Hernández

Wireless: The Revolution in Personal Telecommunications, Ira Brodsky

Writing Disaster Recovery Plans for Telecommunications Networks and LANs, Leo A. Wrobel

X Window System User's Guide, Uday O. Pabrai

For further information on these and other Artech House titles, contact:

Artech House
685 Canton Street
Norwood, MA 02062
617-769-9750
Fax: 617-769-6334
Telex: 951-659
email: artech@artech-house.com

Artech House
Portland House, Stag Place
London SW1E 5XA England
+44 (0) 171-973-8077
Fax: +44 (0) 171-630-0166
Telex: 951-659
email: artech-uk@artech-house.com

WWW: http://www.artech-house.com